Nitric Oxide in the Nervous System

NEUROSCIENCE PERSPECTIVES

Titles in this series

Roger Horton and Cornelius Katona (eds), Biological Aspects of Affective Disorders
Trevor Stone (ed), Adenosine in the Nervous System
Judith Pratt (ed), The Biological Bases of Drug Tolerance and Dependence
Michel Hamon (ed), Central and Peripheral 5-HT_3 Receptors
John Waddington (ed), $D_1 : D_2$ Dopamine Receptor Interactions
Thomas Barnes (ed), Antipsychotic Drugs and Their Side-Effects
Eva Giesen-Crouse (ed), Peripheral Benzodiazepine Receptors
David Nicholson (ed), Anti-Dementia Agents
Alan Harvey (ed), Natural and Synthetic Neurotoxins
John Wood (ed), Capsaicin in the Study of Pain
Yossef Itzhak (ed), Sigma Receptors
John Mayer/Ian Brown (eds), Heat Shock Proteins in the Nervous System
Chris Carter (ed), Neuropharmacology of Polyamines
Roger Pertwee (ed), Cannabinoid Receptors

Nitric Oxide in the Nervous System

edited by

Steven R. Vincent
Department of Psychiatry,
The University of British Columbia,
Vancouver, British Columbia, Canada

ACADEMIC PRESS
Harcourt Brace & Company, Publishers
London San Diego New York
Boston Sydney Tokyo Toronto

ACADEMIC PRESS LIMITED
24/28 Oval Road,
London NW1 7DX

United States Edition published by
ACADEMIC PRESS INC.
San Diego, CA 92101

This book is printed on acid free paper

A catalogue record for this book
is available from the British Library

ISBN 0–12–721985–4

Typeset by Servis Filmsetting Ltd, Manchester, UK
Printed and bound in Great Britain
by TJ Press, Padstow, Cornwall

Contents

Contributors ix

Series Preface xi

Preface xiii

1 Molecular characterization of nitric oxide synthase 1
D. S. Bredt

1.1 Introduction 1
1.2 Isolation of NOS 2
1.3 Molecular cloning 3
1.4 NOS cofactors 6
1.5 Physiologic functions for brain NOS 10

2 Biochemistry and molecular pharmacology of nitric oxide synthases 21
B. Mayer

2.1 Introduction 21
2.2 Biochemical properties of NO synthases 22
2.3 Identification of cofactors 25
2.4 Mechanisms of NO synthesis 26
2.5 Regulation of NOS activity 31
2.6 Molecular pharmacology of NOS inhibition 34

3 The NO receptor: characterization and regulation of soluble guanylyl cyclase 43
D. Koesling, P. Humbert and G. Schultz

3.1 Introduction 43
3.2 Primary structure of soluble guanylyl cyclase 44
3.3 Catalytic activity 46
3.4 Isoforms 47
3.5 Regulation 48

4 Cyclic GMP receptor proteins: role in nervous system and other tissues 51
T. M. Lincoln

4.1 Cyclic GMP-dependent protein kinases 51

4.2 Cyclic GMP-regulated ion channels 64
4.3 Cyclic GMP-regulated phosphodiesterases 69
4.4 Role of cyclic GMP kinase in the nervous system 74

5 Localization of nitric oxide neurons in the central nervous system 83
S. R. Vincent

5.1 Introduction 83
5.2 Caveats and cautions 84
5.3 Localization of NOS in the mammalian central nervous system 86
5.4 Non-mammalian species 93
5.5 NOS expression during development and following injury 94
5.6 Localization of soluble guanylyl cyclase, the NO receptor 95
5.7 Conclusions 96

6 Nitric oxide and excitatory amino acid-coupled signal transduction in the cerebellum and hippocampus 103
P. L. Wood

6.1 Historical background 104
6.2 Nitric oxide synthase (1.14.23) 105
6.3 Cerebellum 107
6.4 Hippocampus 115
6.5 Conclusions 118

7 Nitric oxide signalling, long-term potentiation and long-term depression 125
E. M. Schuman

7.1 Introduction 125
7.2 Long-term potentiation, the hippocampus and memory 126
7.3 Nitric oxide and LTP 130
7.4 Long-term depression in the hippocampus 143
7.5 Long-term depression in the cerebellum 144
7.6 Conclusions 147

8 Nitric oxide signalling in the hypothalamus 151
S. Amir

8.1 Introduction 151
8.2 NOS in the hypothalamus 152
8.3 Corticotropin-releasing hormone neurosecretory system 152
8.4 Luteinizing hormone-releasing hormone neurosecretory system 155
8.5 Vasopressin neurosecretory system 155
8.6 Somatostatin release 156

	8.7	Autonomic regulation	156
	8.8	Functional considerations	158
	8.9	Conclusions	160
9	**Glial cells as nitric oxide sources and targets** *S. Murphy, D. M. Grzybicki and M. L. Simmons*		163
	9.1	Introduction	164
	9.2	Evidence for a constitutive, calcium-activated nitric oxide synthase in astrocytes	165
	9.3	Induction of NOS in glia *in vitro*	167
	9.4	Regulation of the inducible NOS in glia	172
	9.5	Induction of NOS *in vivo*	176
	9.6	Glia as targets for NO	183
	9.7	Summary	185
10	**Nitric oxide toxicity in neuronal injury and degeneration** *P. D. Varner and J. S. Beckman*		191
	10.1	Introduction	191
	10.2	The short half-life of nitric oxide	193
	10.3	Nitric oxide is not highly reactive or toxic	194
	10.4	Nitric oxide and transition metals	195
	10.5	Diffusion of nitric oxide	196
	10.6	Haemoglobin is a major trap for nitric oxide *in vivo*	197
	10.7	Superoxide in cerebral ischaemic injury	197
	10.8	Peroxynitrite as an oxidant	198
	10.9	The interaction of superoxide and nitric oxide in cerebral ischaemia	199
	10.10	Inhibition of nitric oxide synthesis in cerebral ischaemia	200
	10.11	Superoxide dismutase and cerebral ischaemia	200
	10.12	Superoxide dismutase-catalysed nitration	201
	10.13	Motor neuron disease, superoxide dismutase and peroxynitrite	202
11	**Nitric oxide and the regulation of cerebral arterial tone** *N. Toda*		207
	11.1	Introduction	207
	11.2	NO derived from vasodilator nerve	208
	11.3	NO derived from the endothelium	217
	11.4	Conclusion	222
12	**Nitric oxide in the autonomic and enteric nervous systems** *M. J. Rand and C. G. Li*		227
	12.1	Discovery of NO-mediated neuroeffector transmission	228

12.2 Criteria for establishing nitrergic transmission 230
12.3 Tissues innervated by autonomic nitrergic nerves 245
12.4 Nitrergic transmission in the gastrointestinal tract 250
12.5 Mechanisms of nitrergic transmission 260
12.6 Interactions between NO and other autonomic transmitters 267
12.7 Concluding remarks 268

13 Nitric oxide and the neural regulation of the penis 281
N.O. Sjöstrand and E. Klinge

13.1 Introduction 282
13.2 Basic mechanisms of penile erection, smooth muscle effectors and haemodynamics 282
13.3 Efferent nerves 284
13.4 The debate on the neurotransmission of penile erection and the search for a transmitter 286
13.5 Nitric oxide as the principal transmitter candidate 291
13.6 Nitric oxide as likely mediator of cholinergic effects 294
13.7 Final considerations 299

Index 307

Contributors

Shimon Amir Center for Studies in Behavioral Neurobiology, Concordia University, 1455 de Maisonneuve Boulevard West, Montreal, Québec, H3G 1M8, Canada

Joseph S. Beckman Departments of Anesthesiology and Biochemistry, The University of Alabama at Birmingham, Birmingham, AL 35233, USA

David S. Bredt Departments of Physiology and Neuroscience, The University of California at San Francisco, School of Medicine, 513 Parnassus Ave., San Francisco, CA 94143-0444, USA

Dana M. Grzybicki Department of Pathology, University of Iowa College of Medicine, Iowa City, IA 52242, USA

Peter Humbert Institut für Pharmakologie, Freie Universität Berlin, Thielallee 67-73, D-14195 Berlin, Germany

Erik Klinge Department of Pharmacy, Division of Pharmacology and Toxicology, The University of Helsinki, P.O. Box 15, Kirkkokatu 20, SF-00014 Helsinki, Finland

Doris Koesling Institut für Pharmakologie, Freie Universität Berlin, Thielallee 67-73, D-14195 Berlin, Germany

Chun Guang Li Pharmacology Research Laboratory, Department of Medical Laboratory Science, Royal Melbourne Institute of Technology, Victoria 3001, Australia

Thomas M. Lincoln Department of Pathology, Division of Molecular and Cellular Pathology, The University of Alabama at Birmingham, Birmingham, AL 35294, USA

Bernd Mayer Institut für Pharmakologie und Toxikologie, Universität Graz, Universitätsplatz 2, A-8010 Graz, Austria

Sean Murphy Department of Pharmacology, University of Iowa College of Medicine, Iowa City, IA 52242, USA

Michael J. Rand Pharmacology Research Laboratory, Department of Medical Laboratory Science, Royal Melbourne Institute of Technology, Victoria 3001, Australia

Günter Schultz Institut für Pharmakologie, Freie Universität Berlin, Thielallee 67-73, D-14195 Berlin, Germany

Erin M. Schuman California Institute of Technology, Division of Biology, Pasadena, CA 91125, USA

Martha L. Simmons Department of Pathology, The University of California at San Francisco, San Francisco, CA 94143, USA

Nils O. Sjöstrand Departments of Physiology and Pharmacology, Division of Physiology I, Karolinska Institutet, S-171 77 Stockholm, Sweden

Noboru Toda Department of Pharmacology, Shiga University of Medical Sciences, Seta, Ohtsu 520-21, Japan

Pamela D. Varner Department of Anesthesiology, The University of Alabama at Birmingham, Birmingham, AL 35233, USA

Steven R. Vincent Department of Psychiatry, The University of British Columbia, Vancouver, B.C., V6T 1Z3, Canada

Paul L. Wood Department of Pharmacology, Cocensys, Inc., 213 Technology Drive, Irvine, CA 92718, USA

Series Preface

The neurosciences are one of the most diverse and rapidly changing areas in the biological sphere. The need to understand the workings of the nervous system pervades a vast array of different research areas. By definition research in the neurosciences encompasses anatomy, pathology, biochemistry, physiology, pharmacology, molecular biology, genetics and therapeutics. Ultimately, we are striving to determine how the human brain functions under normal circumstances and perhaps more importantly how function changes in organic disease and in altered states of mind. The key to many of these illnesses will unlock one of the major therapeutic challenges remaining in this era.

The difficulty lies in the vastness of the subject matter. However I try, I find it almost impossible to keep abreast of the changes occurring in my immediate sphere of interest, let alone those exciting advances being made in other areas. The array of journals dealing with neurosciences is enormous and the flow of reprints needed to keep me updated is daunting. Inevitably piles of papers accumulate on my desk and in my briefcase. Many lie there unread until sufficient time has passed for their content to be overtaken by yet more of the ever rising tide of publications.

There are various approaches that one can take to deal with this problem. There is the blinkered approach in which you pretend that literature outside your area does not exist. There is the ignore it totally option. Indeed, one colleague of mine has ceased to read the literature in the belief that, if there is a publication of critical importance to his research, someone will tell him about it. I am not that brave and instead I arrived at what I thought was the ideal solution. I started to read critical reviews of areas of current interest. But I soon came unstuck as I realized that, for many subjects of importance to the neurosciences, such authoritative works did not exist.

Instead of simply moaning that the literature was incomplete, I conceived the idea of *Neuroscience Perspectives*. For purely selfish reasons I wanted to have available a series of individually edited monographs dealing in depth with issues of current interest to those working in the neuroscience area. Already a number of volumes have been published which have been well received and the series is thriving with books on a range of topics in preparation or in production. Each volume is designed to bring a multidisciplinary approach to the subject matter by pursuing the topic from the laboratory to the clinic. The editors of the individual volumes are producing balanced critiques of each topic to provide the reader with an up-to-date, clear and comprehensive view of the state of the art.

As with all ventures of this kind, I am simply the individual who initiates a chain of events leading to the production of the series. In reality, it is key individuals at Academic Press who are really responsible for the success of *Neuroscience Perspectives*. In particular, Dr Carey Chapman and Leona Daw have the uneviable task of recruiting editors and authors and keeping the ship on an even keel.

Finally, I hope that *Neuroscience Perspectives* will continue to be enjoyed by my colleagues in the neurosciences. Already the series is being read, understood and enjoyed by a wide audience and it is fast becoming a reference series in the field.

Peter Jenner

Preface

The gas nitric oxide (NO) has burst upon neuroscience only recently, and yet it has permeated into almost every avenue of current research. The unique properties of this novel messenger have revolutionized our way of thinking about neurotransmission. These special properties have also led neuroscientists to invoke NO to explain many previously unexplained phenomena in neurobiology. Fortunately, the development of numerous pharmacological agents is now allowing these hypotheses to be tested.

It is hoped that this volume will provide a synopsis of what we now know about NO. How and where NO is produced, how it acts at the molecular level to activate the synthesis of cGMP, and the possible targets of cGMP in the nervous system are reviewed. The roles of the NO/cGMP signal transduction pathway in the central and peripheral nervous systems, in glial cells and in neuropathology are then explored. Together, these reviews should lead to further work elucidating the varied functions of that most fugacious of neurotransmitters, NO.

CHAPTER 1

MOLECULAR CHARACTERIZATION OF NITRIC OXIDE SYNTHASE

David S. Bredt

Departments of Physiology and Neuroscience, University of California at San Francisco, School of Medicine, 513 Parnassus Ave., San Francisco, CA 94143-0444, USA

Table of Contents

1.1 Introduction 1
1.2 Isolation of NOS 2
1.3 Molecular cloning 3
1.3.1 Brain NOS 3
1.3.2 Inducible NOS 5
1.4 NOS cofactors 6
1.5 Physiologic functions for brain NOS 10

References 16

1.1 Introduction

The free radical nitric oxide (NO) is now recognized as a major, albeit highly atypical, neuronal messenger. Traditional neurotransmitters are hydrophilic small molecules which are packaged and released from synaptic vesicles. Their spatial signalling is restricted to synaptic clefts where they bind to membrane receptors and are inactivated by uptake pumps or degratory enzymes. NO, however, cannot be stored and released from lipid lined packets. Instead, NO is generated when needed by a complex family of nitric oxide synthase enzymes. NO signalling is not restricted to defined synapses; rather, NO readily diffuses through cell membranes where it binds and influences numerous protein targets.

Because NO cannot be stored, released or inactivated after synaptic release by conventional mechanisms, it must be regulated at the level of biosynthesis. Indeed, NO synthase (NOS) is one of the most regulated enzymes in biology. NOS activity is regulated at multiple levels. All NOS isoforms display complex transcriptional regulation, while post-transcriptional alterations of NOS by protein phosphorylation, myristylation and other modifications alter both the enzyme's activity and subcellular distribution.

Nitric Oxide in the Nervous System
ISBN 0-12-721985-4

1.2 Isolation of NOS

Biological functions for NO were first recognized in the immune and cardiovascular systems. Furchgott and Zawadzki (1980) found that blood vessel relaxation in response to acetylcholine requires the endothelium which releases a labile substance that diffuses to the adjacent smooth muscle. This endogenous relaxing factor appeared to be similar to NO, the active metabolite of nitroglycerin and other organic nitrates, as they all dilate blood vessels by stimulating cGMP formation through activation of guanylyl cyclase (GC) (Arnold *et al.*, 1977; Ignarro *et al.*, 1981). Indeed, endothelial cells were found to generate sufficient NO from L-arginine to fully account for endothelial-derived relaxing activity (Palmer *et al.*, 1987; Ignarro *et al.*, 1987).

In unrelated work, studies of nitrosamines as carcinogens demonstrated the existence of endogenous nitrates. Both germ-free rats and humans excrete large amounts of nitrates in excess of dietary intake; this excretion increases dramatically during bacterial infections (Green *et al.*, 1981a,b). These nitrates, measured in urine, were found to arise from macrophages through oxidation of the amidine nitrogen of L-arginine with the stoichiometric production of L-citrulline and a reactive substance subsequently shown to be NO. The capacity of macrophages to kill tumour cells in culture depended upon external arginine, and this effect was blocked by certain methylated arginine derivatives which also blocked the formation of nitrite (Hibbs *et al.*, 1987). NO was identified as the active substance (Marletta *et al.*, 1988; Stuehr *et al.*, 1989).

NO derived from neurons was first appreciated when cerebellar cultures stimulated by excitatory amino acids were found to release a substance with properties similar to the endothelial-derived relaxing factor (Garthwaite *et al.*, 1988). Synaptosomal preparations from brain were found to generate NO from arginine (Knowles *et al.*, 1989). A definitive involvement of NO in neuronal signalling was demonstrated by the ability of NOS inhibitors, such as L-^{w}*N*-nitroarginine (L-NNA) and L-^{w}*N*-methyl-arginine (L-NMA), to block the pronounced stimulation of cGMP levels in brain slices that is associated with activation of *N*-methyl-D-aspartate (NMDA) subtype glutamate receptors (Bredt and Snyder, 1989; Garthwaite *et al.*, 1989).

Because NO is highly reactive and unstable, much of our understanding of the disposition and functions of NO derive from the characterization of NOS. Though NO was first discovered in endothelial cells and macrophages, biochemical and molecular characterization of NOS from these sources was difficult due to the minute quantities in these tissues. Initial enzymatic characterization of NOS in macrophages and endothelium pointed to important similarities and differences. Both enzymes use NADPH and arginine as substrates and generate citrulline as the co-product with NO (Marletta *et al.*, 1988; Palmer and Moncada, 1989). NOS activity in endothelial cells is potently activated by calcium and is referred to as constitutive because the enzyme protein is present in resting endothelium. Macrophage NOS is not regulated by calcium and it is often referred to as the inducible form because it is only present in cells following activation by certain combinations of cytokines.

Initial characterization of NOS in brain homogenates suggested a pharmacologic similarity to the endothelial activity. NOS was readily detected in untreated brain tissue and absolutely requires calcium for activity (Knowles *et al.*, 1989). Initial efforts to purify the enzyme were unsuccessful because of a rapid loss of enzyme activity upon purification. Enzyme activity could be reconstituted from anion exchange columns by combining certain fractions, suggesting the existence of enzyme subunits. These data were explained by the observation that calmodulin is absolutely required for NOS activity (Bredt and Synder, 1990). Purification of brain NOS (bNOS) was achieved using a 2′5′-ADP-linked sepharose column eluted with NADPH. The purified NOS migrated as a single 160 kD protein on SDS PAGE indicating that a single polypeptide in association with calmodulin is the synthetic machinery for NO production. Using this approach, other groups purified bNOS (Mayer *et al.*, 1990; Schmidt *et al.*, 1991), macrophage NOS (macNOS) (Hevel *et al.*, 1991; Stuehr *et al.*, 1991; Yui *et al.*, 1991; Evans *et al.*, 1992) and endothelial NOS (eNOS) (Pollock *et al.*, 1991).

1.3 Molecular cloning

Molecular cloning of cDNA for brain (Bredt *et al.*, 1991a; Nakane *et al.*, 1993; Ogura *et al.*, 1993), endothelial (Janssens *et al.*, 1992; Lamas *et al.*, 1992; Sessa *et al.*, 1992), and macrophage (Lowenstein *et al.*, 1992; Lyons *et al.*, 1992; Xie *et al.*, 1992) forms of NOS has revealed elaborate means for regulation of NOS function. The predicted amino acid sequences for the NOSs contain numerous consensus binding sites for regulatory co-factors, while the genomic structure reveals complex gene promoters and alternate splicing forms (Figure 1).

1.3.1 Brain NOS

Molecular cloning of the brain isozyme was possible following its isolation. A two-step PCR cloning strategy was used with oligonucleotide primers based on tryptic peptides sequenced from purified bNOS (Bredt *et al.*, 1991a). The resulting PCR product was used to obtain NOS clones from a rat brain cDNA library. The open reading frame predicts a polypeptide of 160 kDa which shows 36% identity to cytochrome P450 reductase in its C-terminal half while the N-terminal half of NOS is not homologous to any other known genes (Figure 1). The region of homology to cytochrome P450 reductase includes highly conserved consensus sequences for binding to NADPH, flavin adenine dinucleotide (FAD) and flavin adenine mononucleotide (FMN).

Midway through the predicted amino acid sequence of NOS there is an amphipathic alpha helix domain which fits the consensus sequence for calmodulin

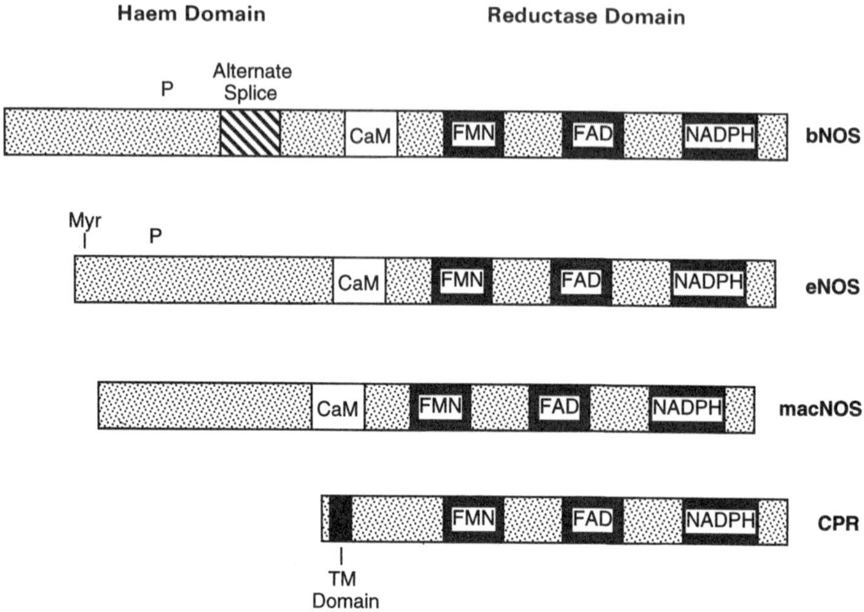

Figure 1 Linear model of co-factor recognition sequences within the cloned NOS isozymes and cytochrome P450 (CPR). Predicted sites for alternative splicing, myristylation (Myr), calmodulin binding (CaM) and protein kinase A phosphorylation (P) within the NOS sequences and the transmembrane domain in the CPR sequence are noted.

binding. Carafoli and co-workers confirmed that this region is responsible for interaction with calmodulin. Synthetic peptides corresponding to this region potently bind calmodulin in a calcium-dependent manner (Vorherr *et al.*, 1993). A consensus sequence for protein kinase A phosphorylation is present at amino acid number 372. Whether this serine is actually phosphorylated by protein kinase A or other enzymes is not yet known.

bNOS has also been cloned from human (Nakane *et al.*, 1993) and mouse (Ogura *et al.*, 1993) cerebella using low stringency screening with the rat bNOS cDNA. The rat bNOS shares 94% and 98% amino acid identity with the human and mouse, respectively. Northern blot analysis reveals a greater abundance of bNOS mRNA in human skeletal muscle than in human brain. Recent studies (discussed below) suggest a role for NO in modulating skeletal muscle contraction. The human gene was mapped to chromosome 12 and is not closely associated with any known disease marker. Cloning of the mouse bNOS reveals alternative splicing of the mRNA in brain (Ogura *et al.*, 1993). In the mouse cerebellum 10% of NOS mRNA has a 415 nucleotide deletion corresponding to nucleotides 1510–1824. Interestingly, this is in a region of NOS which is highly conserved between the various isoforms. Functional expression of this shorter alternately spliced bNOS isoform reveals that it does not generate NO from L-arginine. This short form, however, does display reductase activity and generates reactive oxygen species in a calcium-dependent manner. Regulation of this alternative splicing of bNOS remains an important unanswered issue.

1.3.2 Inducible NOS

NOS was cloned from cultured murine macrophage independently by three labs. One group conducted expression cloning using a macNOS-specific antibody (Xie *et al.*, 1992) while the other labs used bNOS as a homologous probe (Lowenstein *et al.*, 1992; Lyons *et al.*, 1992). The macNOS cDNA encodes a polypeptide of 133 kDa which contains consensus sequences for binding calmodulin, NADPH, FMN and FAD. The amino acid sequence of murine macNOS is approximately 50% identical to both bNOS and eNOS. The protein kinase A phosphorylation site conserved between bNOS and eNOS is absent.

NOS has not yet been cloned from human macrophages. Interestingly, NOS catalytic activity and protein have rarely been identified in human macrophages (Denis, 1993; Kobzik *et al.*, 1993) despite extensive searches in many laboratories. It appears that dissimilar mechanisms regulate macNOS in human and rodent macrophages. However, an inducible calcium-independent NOS activity has been well characterized in human hepatocytes following treatment with lipopolysaccharide (LPS), interferon-gamma (IFN-γ), tumour necrosis factor-alpha (TNF-α) and interleukin-1B (IL-1B). Cloning of this human cDNA reveals 82% identity with murine macNOS suggesting that it represents a similar if not identical isoform (Geller *et al.*, 1993). An inducible NOS gene was independently cloned from articular chondrocytes activated with IL-1B (Charles *et al.*, 1993). These human inducible NOS cDNAs are essentially identical and their genes map to chromosome 17.

The complex regulation of murine macNOS expression is largely encoded in a promoter immediately 5′ to the first exon (Lowenstein *et al.*, 1993; Xie *et al.*, 1993). There are two interdependent regulatory regions upstream of the TATA box, which itself is 30 base pairs upstream of the major transcriptional start site. The first of these regions lies about 50–200 base pairs 5′ to the start site. It contains LPS-related elements such as a binding site for NF–IL-6 and the kB binding site for NFkB, indicating that this region regulates the LPS-induced expression of macNOS. A more distant regulatory site about 1000 bases upstream of the start site does not itself directly regulate macNOS expression, but provides a 10-fold synergistic activation when coupled with the first regulatory region. This second region contains motifs for IFN-γ-related transcription factors and is likely responsible for IFN-γ-mediated gene activation.

Endothelial NOS (eNOS) was cloned independently by three labs, all of which used low stringency screening strategies based on the DNA sequence of bNOS (Janssens *et al.*, 1992; Lamas *et al.*, 1992; Sessa *et al.*, 1992). The open reading frame for eNOS predicts a polypeptide of 140 kD which shares 60% identity with bNOS. Consensus binding sites for FAD, FMN, NADPH, calmodulin and protein kinase

A phosphorylation are conserved between the brain and endothelial isoforms. A unique motif in the eNOS gene is a consensus sequence for N-terminal myristoylation. This post-translational modification accounts for the particulate localization of the eNOS protein, which does not have a hydrophobic membrane-spanning domain. When incubated with cultured endothelial cells, [^{3}H]myristate is directly incorporated into eNOS, and mutation of the myristoylation sequence renders eNOS soluble (Buscoini and Michel, 1993). Presumably the myristoyl group is inserted into the plasma membrane which accounts for the enzyme's particulate location.

Cloning of the genomic human eNOS reveals a large gene which contains 25 exons spaning 21 kilobases on the 7q35–36 region of chromosome number 7 (Marsden *et al.*, 1993). The 5′ promoter region of the human gene contains AP–1, AP–2, NF–1, heavy metal, acute-phase response shear stress, and sterol-regulatory elements (Robinson *et al.*, 1994). These 5′ sequence motifs fit with a recent study showing induction of eNOS in cerebral blood vessels following ischaemia (Abu-Soud and Stuehr, 1993).

1.4 NOS cofactors

NOS catalyses a five electron oxidation of one of the two equivalent amidine groups of L-arginine to form NO and L-citrulline. The odd electron stoichiometry of the reaction is highly unusual and allows for the generation of the free radical, NO. The cloning of NOSs (discussed above) reveals consensus sequences for NADPH, FAD and FMN binding. The potent interaction of NADPH with NOS (K_m=300 nM) was exploited in the purification of NOS, which heavily relies on affinity chromatography with an NADPH analogue, 2′5′-ADP.

While NADPH is a stoichiometric substrate, the two flavins co-purify with NOS in a ratio of 1 equal measure each of FAD and FMN per NOS monomer (Hevel *et al.*, 1991; Mayer *et al.*, 1991; Bredt *et al.*, 1992). FAD slowly dissociates from macNOS and must be exogenously supplied for maximal activity. The presence of flavins in NOS explains the capacity of the enzyme to participate in 'odd electron' biochemistry. Flavins function as excellent one electron donors and acceptors. Unpaired electrons become highly delocalized through the conjugated isoalloxazine structure of the flavin allowing for a stable semiquinone radical. By this mechanism, electrons can be passed one at a time between the flavins and from the flavins to chemical substrates. The close homology of NOS with cytochrome P450 reductase suggests that electrons follow the same path through NOS as they do through cytochrome P450 reductase, that is NADPH initially reduces FAD, which in turn reduces FMN. In fact, NOS and cytochrome P450 reductase share a domain thought to be involved in this electron transfer (Bredt *et al.*, 1992).

NOS shares many functional properties with cytochrome P450 reductase. bNOS catalyses a rapid NADPH-dependent reduction of cytochrome *c*. In the absence of arginine NOS can transfer electrons from NADPH to O_2 and form O_2^- and H_2O_2 (Pou *et al.*, 1992; Klatt *et al.*, 1993). The formation of these reactive oxygen intermediates may contribute to glutamate neurotoxicity and neurodegeneration. Interestingly, L-NNA but not L-NMA blocks the formation of O_2^- and H_2O_2. Deletional mutagenesis indicates that amino acids 787–1429 fully account for NOS reductase activity (Bredt and Synder, 1994). This region can therefore properly be referred to as the reductase domain.

Cytochrome P450 reductase ultimately transfers its reducing equivalents to the haem-containing cytochrome P450 enzymes. Similarly, all known NOS isoforms contain 1 equal measure of iron-protoporphyrin IX per NOS monomer (McMillan *et al.*, 1992; Stuehr and Ikeda-Saito, 1992; White and Marletta, 1992). NOS displays a typical cytochrome P450-reduced CO difference spectra with wavelength absorbance maximum at 445 nm indicative of a haem-binding cysteinyl ligand. Purified NOS is inhibited by CO, which is also consistent with the participation of a cytochrome P450-type haem in the reaction. NO also inhibits purified NOS, perhaps by NO interacting with the enzyme's haem prosthetic group. This has been suggested to represent a mechanism for NO to feedback inhibit its own synthesis (Rogers and Ignarro, 1992; Assreuy *et al.*, 1993; Rengasamy and Johns, 1993).

All other cytochrome P450 enzymes, from bacteria to human, share a cysteinyl peptide consensus for coordination to the haem. In NOS, the classic P450 haem-binding cysteinyl peptide sequence is absent. Close inspection reveals that the amino acids surrounding cysteine 414 show some of the expected homology. Comparison with P450BM3, the bacterial haem and flavin-containing P450, shows close alignment of cysteine 675 with the putative haem binding site in P450BM3. Site-directed mutagenesis of these cysteines may clarify the site of haem binding.

Haem–substrate binding is also the initial event in catalysis with the various cytochrome P450 hydroxylase enzymes. This mechanism appears to be shared by NOS as optical difference spectroscopy indicates that haem must bind to the substrate arginine prior to participating in the oxidative reactions (McMillan and Masters, 1993). Unlike other mammalian cytochrome P450 enzymes, NOSs are unique because they are localized to the cytosol and their flavin- and haem-containing domains are fused in a single polypeptide. Presumably, early in evolution, cytochrome P450 reductase donated electrons to NOS and later a combination event took place joining the two. Recently, a bacterial fatty acid mono-oxygenase, P450BM3, also has been identified as a soluble, self-contained P450 system (Narhi and Fulco, 1986). This enzyme, however, is not calmodulin regulated and shows no homology to NOS in the haem domain suggesting that it may have arisen from an independent gene fusion.

NOS activity is also controlled by tetrahydrobiopterin (H4B). macNOS is absolutely dependent upon H4B (Kwon *et al.*, 1989; Tayeh and Marletta, 1989), while purified bNOS retains substantial activity in the absence of added H4B (Bredt and Snyder, 1990; Schmidt *et al.*, 1991). This difference in behaviour is explained by the tight binding of H4B to bNOS such that H4B co-purifies with the enzyme (Schmidt *et al.*, 1992). In aromatic amino acid hydroxylase enzymes, H4B functions directly in the substrate hydroxylation; therefore, it was initially assumed to subserve a similar role for NOS. Experiments by Kaufman and coworkers (Giovanelli *et al.*, 1991) pointed to a different role for H4B in NOS; they proposed that H4B stabilizes NOS. Their conclusions were based on experiments with purified bNOS which showed that H4B functions catalytically, is not recycled, and does not affect the initial rate of NOS. Based on experiments with pterin analogues used to probe the macNOS reaction, Marletta and colleagues (Hevel and Marletta, 1992) also suggest that H4B functions to stabilize NOS.

NOS enzymes can be discriminated by their dependence upon calcium. In the brain, calcium influx through glutamate receptor channels or voltage-activated calcium channels stimulates calmodulin thereby activating NOS. This mode of activation explains the ability of glutamate or neuronal depolarization to stimulate NO formation in a matter of seconds. In endothelial cells, neurotransmitters acting at G-protein coupled receptors activate the phosphoinositide cycle to mobilize intracellular Ca^{2+} which stimulates NOS. Thus, calcium-regulated NOS accounts for the capacity of NO to mediate rapid events such as neurotransmission and blood vessel dilatation. The calcium requirement for NOS activity reflects the affinity of calcium for calmodulin (EC_{50}=200–400 nM). Following activation by calcium, calmodulin binds tightly to NOS (K_d=5–10 nM). Calmodulin regulates the electron transfer and oxygen activation activities of NOS (Abu-Soud and Stuehr, 1993). Arginine binding, however, is unaffected by calcium or calmodulin (White and Marletta, 1992). The brain and endothelial enzymes are inhibited by calmodulin antagonists such as trifluoperazine (IC_{50}=10 mM) (Bredt and Synder, 1990).

Initial studies showed that inducible NOS of macrophage is neither stimulated by Ca^{2+} nor blocked by calmodulin antagonists. Cloning of inducible NOS, however, revealed calmodulin recognition sites (Figure 2). Calmodulin is in fact very tightly bound to inducible NOS with the binding unaffected by chelation of Ca^{2+} (Cho *et al.*, 1992). Physiologically, activated macrophages generate NO at resting intracellular Ca^{2+} levels. Calmodulin binds so tightly to inducible NOS that it can be considered an enzyme subunit rather than a mechanism for regulation. This calcium-independent regulation by calmodulin may reflect that macNOS is an evolutionary later addition to the NOS gene family. macNOS may have acquired calcium independence via specific mutations in its calmodulin binding domain. Alternatively, it is possible that calmodulin regulates macNOS by mechanisms which have not yet been elucidated.

NOS activity is also regulated by phosphorylation. Consensus sequences for phosphorylation by cAMP-dependent protein kinase are evident in bNOS and eNOS (Figure 2). These are not as obvious in macNOS. Consensus sites for phosphorylation by other kinases have not been characterized in detail. However, biochemical studies indicate that neuronal NOS can be phosphorylated by cAMP-dependent protein kinase, protein kinase C (PKC), cGMP-dependent protein kinase, and Ca^{2+}/calmodulin-dependent protein kinase (Brüne and Lapetina, 1991; Nakane *et al.*, 1991; Bredt *et al.*, 1992). Stimulation of PKC by phorbol esters causes a rapid inactivation of bNOS in tissue culture cells stably transfected with NOS,

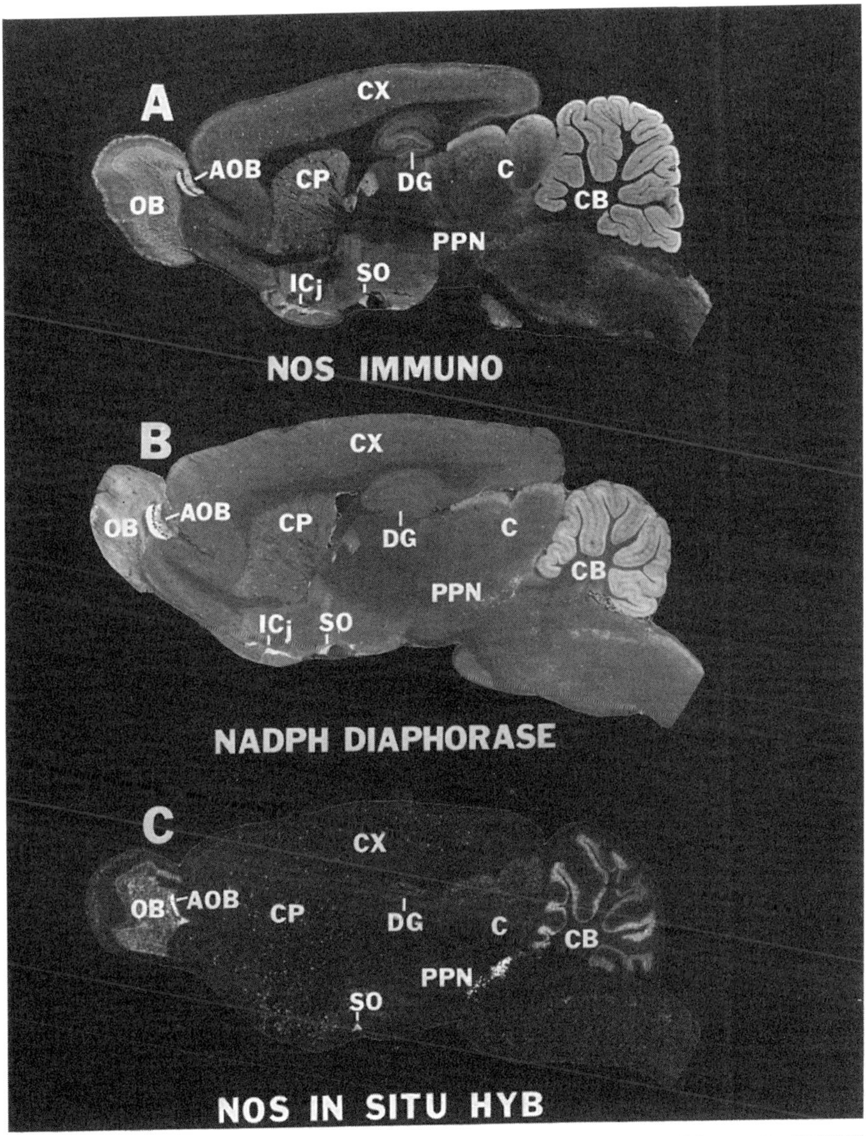

Figure 2 Localization of NOS in brain. Histologic localizations of NOS protein, NADPH diaphorase activity and mRNA in brain. Adjacent sagittal brain sections were processed for (A) NOS immunohistochemistry, (B) NADPH diaphorase histochemistry and (C) NOS *in situ* hybridization. All three methods show densest staining in the accessory olfactory bulb (AOB), pedunculopontine tegmental nucleus (PPN) and cerebellum (CB), with lesser staining in the dentate gyrus of the hippocampus (DG), main olfactory bulb (OB), superior and inferior colliculus (C) and supraoptic nucleus (SO). Intensely staining isolated cells are apparent scattered throughout the cerebral cortex (CX), caudate putamen and basal forebrain. Some regions enriched in NOS protein and diaphorase staining are devoid of NOS mRNA suggesting that in these regions the NOS protein has been transported in nerve fibres distant from its site of synthesis. These regions include the molecular layer of the cerebellum, the islands of Callejae (ICj) and the neuropil of the caudate putamen (CP) and cerebral cortex (CX).

suggesting that PKC down-regulates NOS activity. PKC-mediated down-regulation of NOS is potentiated by the immunosuppressant drug, FK-506, which blocks the protein phosphatase calcineurin (T.M. Dawson *et al.*, 1993).

Phosphorylation of the endothelial enzyme regulates both its enzymatic activity and its subcellular distribution. Unlike bNOS and macNOS which are largely cytosolic, eNOS is predominately localized to the plasma membrane (Pollock *et al.*, 1991). Michel *et al.* (1993) found that eNOS is rapidly phosphorylated in intact endothelial cells in response to bradykinin. Phosphorylated eNOS is exclusively localized to soluble fractions. Since eNOS activity is entirely membrane associated, phosphorylated eNOS should be catalytically inactive. Instead, catalytically active, non-phosphorylated NOS is localized to the plasma membrane where it presumably generates NO that is released into the extracellular environment. While neuronal NOS has been thought to be predominantly soluble, about 50% of NOS activity in brain homogenates is particulate and cannot be solubilized even with strong salt treatment (Hiki *et al.*, 1992). Thus, in neurons as well as blood vessels the active form of NOS may be the unphosphorylated enzyme localized to the plasma membrane. Because NO itself cannot be stored and released from synaptic vesicles, subcellular translocation of NOS may represent the mechanism to ensure that NO is released to the cell's exterior.

1.5 Physiologic functions for brain NOS

Following purification of neuronal NOS, antibodies were developed for immunohistochemical staining (Bredt *et al.*, 1990). In the peripheral nervous system, NOS is present in discrete populations of neurons which are closely associated with smooth muscle cells (see Chapter 12). In the myenteric plexus of the gastrointestinal (GI) tract NOS is selectively expressed in inhibitory motor neurons which have short, anally projecting axons which run in parallel with the inner circular smooth muscle layer (Costa *et al.*, 1992). NO released from these cells mediates the non-adrenergic non-cholinergic relaxation of the gut. NO also appears to regulate relaxation of the stomach as well as numerous sphincters throughout the GI tract (Desai *et al.*, 1991).

NOS fibres are prominently found surrounding the adventitial layer of certain blood vessels (Nozaki *et al.*, 1993). Large cerebral blood vessels contain far more bNOS than the smaller, more distal vessels. This connection between cerebral blood vessel diameter and NOS levels fits with physiologic studies showing a prominent non-adrenergic non-cholinergic relaxation mechanism restricted to large cerebral vessels (Toda, 1981; see Chapter 11).

NOS is abundantly expressed in the adventitia surrounding arteries of the corpora cavernosa. Electrical stimulation of the cavernous nerve in intact rats produces prominent penile erection which is blocked by low doses of intravenously administered NOS inhibitors (Burnett *et al.*, 1992; Rajfer *et al.*, 1992). Accordingly,

NO is presumably the transmitter of these nerves regulating penile erection in mammals (see Chapter 13).

bNOS is present in a few non-neuronal cell types. The macula densa of the kidney, which regulates fluid reabsorption by the kidney, contains abundant bNOS. Release of NO from macula densa cells is stimulated by tubular fluid reabsorption and mediates a vasodilating feedback response which regulates body fluid volume and blood pressure homeostasis (Wilcox *et al.*, 1992). In skeletal muscle bNOS is restricted to the cell membrane of type II fibres which are fast-contracting and low in mitochondrial content (Kobzik *et al.*, 1994). Actively contracting skeletal muscle releases NO as measured by reduction of cytochrome *c*. Inhibition of NOS potentiates skeletal muscle contraction while addition of NO depresses contractile force. Whether NO influences smooth and skeletal muscle by similar mechanisms is not yet clear.

In the central nervous system, bNOS is present exclusively in neurons (Figure 2; see Chapter 9). NOS is distributed in discrete neuronal populations with divergent patterns such that in the cerebral cortex only 1% of neurons are stained in morphology of medium aspiny neurons, while in the cerebellum essentially granule and basket cells contain NOS (Bredt *et al.*, 1991b). These patterns do not resemble a vascular distribution, and NOS in brain is not predominantly associated with smooth muscle cells, suggesting that NO subserves other functions in CNS. If NO transmission occurred exclusively through guanylyl cyclase and if all the guanylyl cyclase in the brain were associated with NO transmission, then guanylyl cyclase and NOS localizations should be closely similar. However, they differ markedly, indicating that NO may act in other ways than via guanylyl cyclase and/or guanylyl cyclase may be regulated by other transmitters besides NO.

The discovery that NOS is identical to neuronal NADPH diaphorase (Bredt *et al.*, 1991a; Hope *et al.*, 1991; see Chapter 5) suggested a link between NO and neurotoxic processes. NADPH diaphorase is a histochemical stain which selectively marks neurons which are resistant to brain ischaemia and in neurodegenerative diseases such as Huntington's disease (Ferrante *et al.*, 1985; Koh *et al.*, 1986; Uemura *et al.*, 1990), suggesting a possible neuroprotective role for NO. Conversely, excessive synthesis of NO may be neurotoxic since NOS inhibitors block glutamate-mediated cell death in primary neuronal cultures (Dawson *et al.*, 1991). Why NOS neurons are resistant to NMDA-mediated damage despite their own generation of NO remains a central question, and one of considerable controversy (Choi, 1993).

A potential physiologic function for NO in the brain relates to synaptic plasticity (see Chapter 7). NO has been implicated in long-term potentiation (LTP), a model of spatial learning, in the hippocampus. Application of nitroarginine, a NOS inhibitor, to hippocampal slices blocks LTP formation (Böhme *et al.*, 1991; O'Dell *et al.*, 1991). Injection of nitroarginine into pyramidal cells of the hippocampus also inhibits LTP, suggesting that NO might act as a retrograde messenger for LTP passing from pyramidal cells to Schaffer collateral terminals (Schuman and Madison, 1991). A major problem with the notion that NO mediates LTP is the lack of bNOS in hippocampal pyramidal neurons (Bredt and Synder, 1992).

Because NO freely diffuses down its concentration gradient without regard for synaptic membranes, it represents an ideal candidate to regulate neuronal development. On theoretical grounds, NO has been suggested to function as a retrograde messenger subserving use-dependent modification of synaptic efficacy during the establishment of ordered cortical connections (Gally *et al.*, 1990). This would require a postsynaptic locus for NOS during development. An alternative role for NO in axonal guidance is suggested in experiments by Hess *et al.* (1993) showing that NO modulates neurite extension in regenerating dorsal root ganglion growth cones. These latter data are consistent with NO functioning as a presynaptic messenger which may segregate axons with temporally correlated activity such that those which 'fire together, wire together' (Edelman and Gally, 1992).

Consistent with the notion that NO plays a role during brain development, recent studies indicate that NOS is transiently expressed in discrete populations of neurons during neurogenesis (Kalb and Agonstini, 1993; Bredt and Snyder, 1994). In embryonic rat brain the most prominent staining occurs in the cerebral cortex (Figure 3). At all embryonic stages the entire cortical plate stains prominently, while the ventricular zone and the cortical subplate display diminished NOS immunoreactivity. At E14 staining is confined to the Cajal–Retzius cells, the earliest cortical neurons to develop with some of their processes stained migrating inward, while at E15, 17 and 19 the great majority of cortical cells stain (Figure 3). At E17 and 19, staining is also prominent in the intermediate zone. High power magnification establishes that the staining in the intermediate zone involves NOS fibres rather than cell bodies (data not shown). At E19, PO, and P-2 some of these processes extend through the striatum and into the thalamus. Thus, they appear to represent a corticothalamic projection. The absence of fibres staining in the ventral aspect of the medulla suggests that the corticospinal pathways are not involved.

The cortical subplate does not show prominent NOS staining. The vast majority of neurons in the subplate die during development. Interestingly, thymidine birth dating shows that somatostatin neurons of the cortex of adult rats derive from surviving subplate neurons. Cortical NOS neurons are all somatostatin containing (Chun and Shatz, 1989). If all surviving subplate neurons are NOS positive, this suggests that NO in these neurons has a protective effect reminiscent of the resistance of adult NOS neurons to neurotoxicity.

NOS is also transiently expressed in embryonic sensory neurons, including cells of the dorsal root, trigeminal, nodose and jugular ganglia. At E12 essentially all neuronal cells in the lumbar dorsal root ganglia stain for NOS (Figure 4). The percentage of cells staining for NOS diminishes gradually. Thus, at E15 only 50% of cells stain for NOS and the percentage further declines to 20% at P-3 and 1% in adults. The role of NOS in these developing sensory ganglia is unclear. Nerve growth factor (NGF) is crucial for development of cells in many sensory ganglia (Levi-Montalcini, 1982). Conceivably NGF and NO might interact somehow in the development of these cells. In support of this idea, NGF treatment of PC12 cells

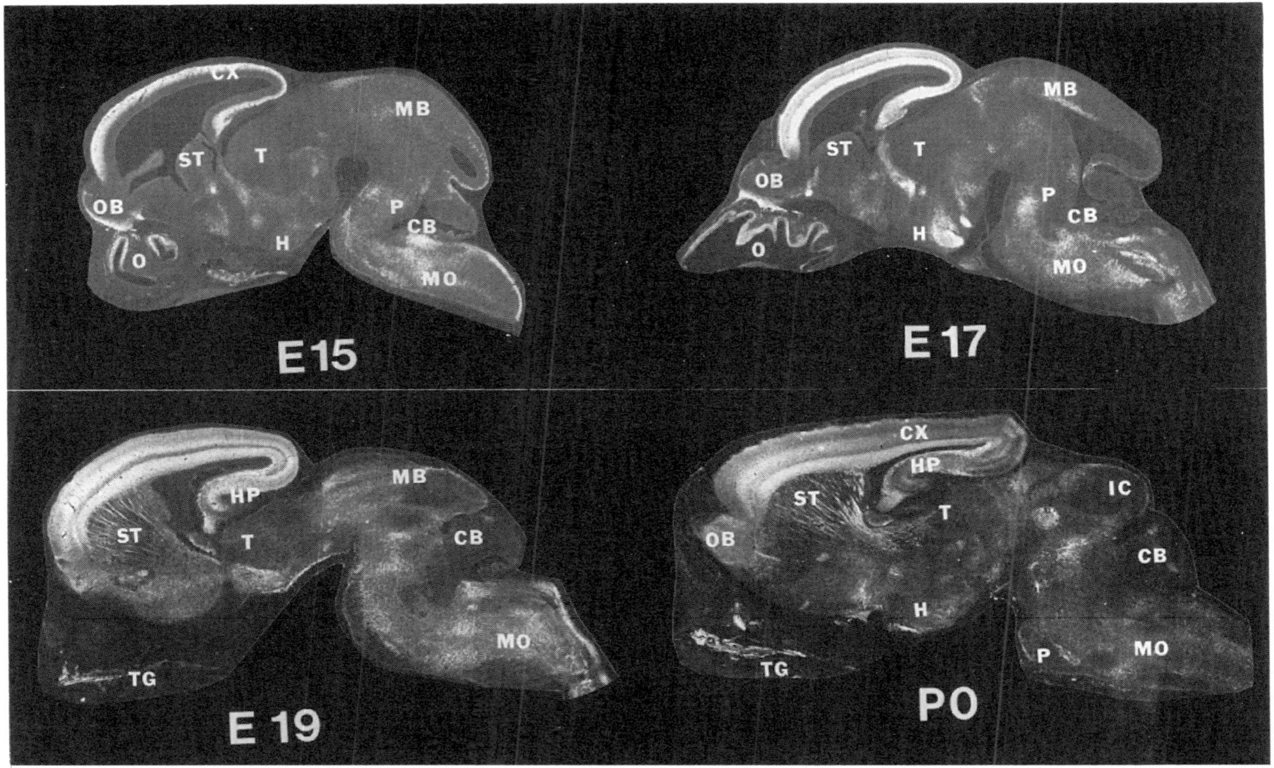

Figure 3 Localization of NOS in the brain during development. Sagittal sections of rat brain were processed for NOS immunohistochemistry. NOS-containing cells appear white in these dark field images. At early embryonic stages (E15–19), NOS is enriched in the neurons of the cortical plate (CX), olfactory epithelium (O) and medulla oblongata (MO). In the late embryonic and early postnatal stages, bNOS is most enriched in axons of the cortical neurons which can be seen coursing through the striatum (ST). In later postnatal stages, NOS expression disappears from the cortex and assumes its adult (AD) distribution, with highest levels in the cerebellum (CB), olfactory bulb (OB), and superior and inferior colliculi (SC, IC) (see Figure 2). Abbreviations: H, hypothalamus; HP, hippocampus; MB, midbrain; T, thalamus. In these dark-field images, NOS staining appears white.

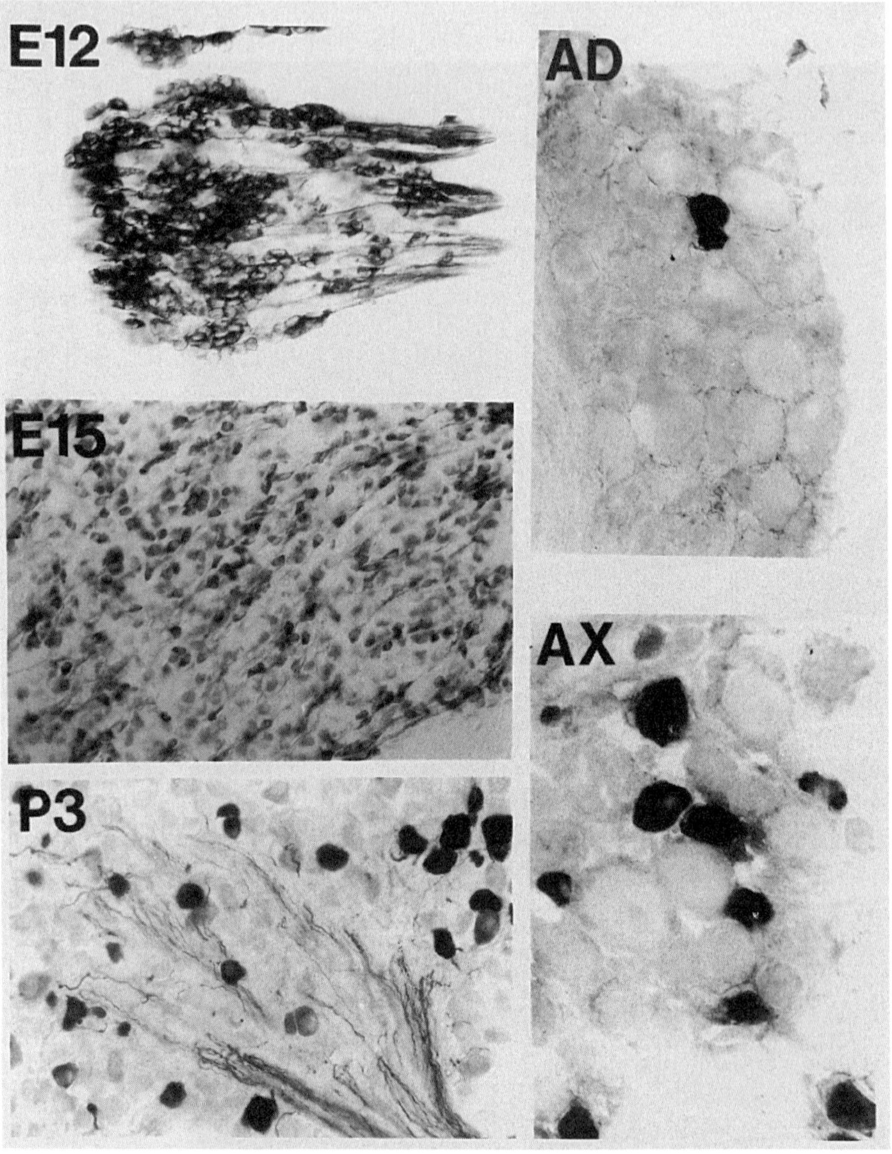

Figure 4 Transient expression of NOS in dorsal root ganglion neurons (×400). NOS is abundantly expressed in 100% of lumbar DRG neurons at E12. The percentage of NOS-positive L5 DRG neurons falls to 50% at E15, 20% at P3, and 1–2% in adult (AD). Thirty-six days following a crush injury of the sciatic nerve axon (AX) in the adult rat, NOS expression is present in 20% of ipsilateral L5 DRG neurons.

results in the expression of NOS in these cells (Hirsch *et al.*, 1993). NGF also prominently regulates NOS expression in cholinergic neurons of the basal forebrain (Holzmann *et al.*, 1994).

The effects of neuronal lesions on NOS expression may be relevant to considerations of NO function in development. Hökfelt and colleagues have observed a major increase in the percentage of NOS positive cells in sensory ganglia following peripheral axotomy (Verge *et al.*, 1992). In axotomized axons NOS inhibitors selectively suppress ongoing neuronal activity suggesting that NO may be involved in the generation of spontaneous discharges in deafferented sensory neurons (Wiesenfield-Hallin *et al.*, 1993). Spontaneous waves of neuronal activity are essential for the establishment of certain somatotopic maps (Shatz, 1990). The role of NO in neuronal activity in developing neurons has not yet been explored.

It is alternatively possible that developmental and lesion-induced NO play a role in eliciting cell death. Following ventral root avulsion, motor neurons of the spinal cord exhibit prominent NOS staining, whereas these cells do not stain for NOS in unlesioned animals (Wu and Li, 1993). In the brain, lesions of the medial forebrain bundle and mammillothalamic tract are followed by the appearance of NOS staining in cell bodies of these fibre bundles with the staining persisting for up to 150 days (Herdegen *et al.*, 1993). The persistence of staining in these cells has been suggested to establish that NO has a neuroprotective effect in these neurons. On the other hand, in the spinal cord, the NOS-positive motor neurons following ventral root avulsion ultimately die. Moreover, treatment of these animals with the NOS inhibitor nitroarginine prior to ventral root avulsion protects the cells from death (Wu and Li, 1993).

One powerful technique for exploring the functions of a gene product is to construct null mutants. Homologous recombination techniques have now been employed to disrupt the gene for bNOS, resulting in homozygous bNOS 'knockout' mice (Huang *et al.*, 1993). NOS catalytic activity is depleted from the brain, and NOS staining in undetectable in central and peripheral neurons. Yet in most respects these animals appear normal. Microscopic examination fails to reveal morphologic abnormalities in the brain or most peripheral tissues. The stomachs of bNOS-deficient mice are greatly distended compared to age-matched control mice. Histologic examination reveals circular muscular hypertrophy, especially in the pyloric region, which is likely the result of chronic muscle contraction. The pathology of the stomach resembles that observed in hypertrophic pyloric stenosis suggesting a role for nitric oxide in this disorder. This conclusion is supported by studies showing a lack of NADPH diaphorase activity in the myenteric neurons of human newborns with pyloric stenosis (Vanderwinden *et al.*, 1992). In these patients diaphorase staining is normal outside the pyloric region so that generalized bNOS deficiency is not likely the cause of the disorder.

The overall normality of the architecture of the brain in these transgenics implies that either (a) compensatory mechanisms are induced in the knockouts to replace the lack of bNOS or (b) NO is not required for morphologic development, but perhaps is important in higher level processes in neurogenesis. An important challenge for the future centres on defining specific physiologic functions in brain for the free radical nitric oxide.

References

Abu-Soud, H.M. & Stuehr, D.J. (1993) *Proc. Natl. Acad. Sci. USA* **90**, 10769–10772.

Arnold, W.P., Mittal, C.K., Katsuki, S. & Murad, F. (1977) *Proc. Natl. Acad. Sci. USA* **74**, 3203–3207.

Assreuy, J., Cunha, F.Q., Liewm, F.Y. & Moncada, S. (1993) *Br. J. Pharmacol.* **108**, 833–837.

Böhme, G.A., Bon, C., Stutzmann, J.-M., Doble, A. & Blanchard, J.-C. (1991) *Eur. J. Pharmacol.* **199**, 379–381.

Bredt, D.S. & Snyder, S.H. (1989) *Proc. Natl. Acad. Sci. USA* **86**, 9030–9033.

Bredt, D.S. & Snyder, S.H. (1990) *Proc. Natl. Acad. Sci. USA* **87**, 682–685.

Bredt, D.S. & Snyder, S.H. (1992) *Neuron* **8**, 3–11.

Bredt, D.S. & Snyder, S.H. (1994) *Neuron* **13**, 301–313.

Bredt, D.S., Hwang, P.M. & Snyder, S.H. (1990) *Nature* **347**, 768–770.

Bredt, D.S., Hwang, P.H., Glatt, C., Lowenstein, C., Reed, R.R. & Snyder, S.H. (1991a) *Nature* **351**, 714–718.

Bredt, D.S., Glatt, C.E., Hwang, P.M., Fotuhi, M., Dawson, T.M. & Snyder, S.H. (1991b) *Neuron* **7**, 615–624.

Bredt, D.S., Ferris, C.D. & Snyder, S.H. (1992) *J. Biol. Chem.* **267**, 10976–10981.

Brüne, B. & Lapetina, E.G. (1991) *Biochem. Biophys. Res. Commun.* **181**, 921–926.

Burnett, A.L., Lowenstein, C.J., Bredt, D.S., Chang, T.S.K. & Snyder, S.H. (1992) *Science* **257**, 401–403.

Buscoini, L. & Michel, T. (1993) *J. Biol. Chem.* **268**, 8410–8413.

Charles, I.G., Palmer, M.J., Hickery, M.S., Bayliss, M.T., Chubb, A.P., Adl, V.S., Moss, D.W. & Moncada, S. (1993) *Proc. Natl. Acad. Sci. USA* **90**, 11419–11423.

Cho, H.J., Xie, Q.-W., Calaycay, J., Mumford, R.A., Swiderek, K.M., Lee, T.D. & Nathan, C. (1992) *J. Exp. Med.* **176**, 599–604.

Choi, D.W. (1993) *Proc. Natl. Acad. Sci. USA* **90**, 9741–9743.

Chun, J.J.M. & Shatz, C.J. (1989) *J. Neurosci.* **9**, 1648–1667.

Costa, M., Furness, J.B., Pompolo, S., Brookes, S.J.H., Bornstein, J.C., Bredt, D.S. & Snyder, S.H. (1992) *Neurosci. Lett.* **148**, 121–125.

Dawson, T.M., Steiner, J.P., Dawson, V.L., Dinersman, J.L., Uhl, G.R. & Snyder, S.H. (1993) *Proc. Natl. Acad. Sci. USA* **90**, 9808–9812.

Dawson, V.L., Dawson, T.M., London, E.D., Bredt, D.S. & Snyder, S.H. (1991) *Proc. Natl. Acad. Sci. USA* **88**, 6368–6371.

Denis, M. (1993) *J. Leukocyte Biology* **49**, 380–387.

Desai, K.M., Sessa, W.C. & Vane, J.R. (1991) *Nature* **351**, 477–479.

Edelman, G.M. & Gally, J.A. (1992) *Proc. Natl. Acad. Sci. USA* **89**, 11651–11652.

Evans, T., Carpenter, A. & Cohen, J. (1992) *Proc. Natl. Acad. Sci. USA* **89**, 5361–5365.

Ferrante, R.J., Kowall, N.W., Beal, M.F., Richardson Jr., E.P., Bird, E.D. & Martin, J.B. (1985) *Science* **230**, 561–563.

Furchgott, R.F. & Zawadzki, J.V. (1980) *Nature* **288**, 373–376.

Gally, J.A., Montague, P.R., Reeke Jr., G.N. & Edelman, G.M. (1990) *Proc. Natl. Acad. Sci. USA* **87**, 3547–3551.

Garthwaite, J., Charles, S.L. & Chess-Williams, R. (1988) *Nature* **336**, 385–388.

Garthwaite, J., Garthwaite, G., Palmer, R.M.J. & Moncada, S. (1989) *Eur. J. Pharmacol.* **172**, 413–416.

Geller, D.A., Lowenstein, C.J., Shapiro, R.A., Nassler, A.K., Di Silvio, M., Wang, S.C., Nakayama, D.K., Simmons, R.L., Snyder, S. & Billiar, T.M. (1993) *Proc. Natl. Acad. Sci. USA* **90**, 3491–3495.
Giovanelli, J., Campos, K.L. & Kaufman, S. (1991) *Proc. Natl. Acad. Sci. USA* **88**, 7091–7095.
Green, L.C., Wagner, D.A., Ruiz-le-Luzuriaga, K., Istfan, N., Young, V. & Tannenbaum, S.R. (1981a) *Proc. Natl. Acad. Sci. USA* **78**, 7764–7768.
Green, L.C., Tannenbaum, S.R. & Goldman, P. (1981b) *Science* **212**, 56–58.
Herdegen, T., Brecht, S., Mayer, B., Leah, J., Kummer, W., Bravo, R. & Zimmerman, M. (1993) *J. Neurosci.* **13**, 4130–4145.
Hess, D.T., Patterson, S.I., Smith, D.S. & Pate Skene, J.H. (1993) *Nature* **366**, 562–565.
Hevel, J.M. & Marletta, M.A. (1992) *Biochemistry* **31**, 7160–7165.
Hevel, J.M., White, K.A. & Marletta, M.A. (1991) *J. Biol. Chem.* **266**, 22789–22791.
Hibbs Jr., J.B., Taintor, R.R. & Vavrin, Z. (1987) *Science* **235**, 473–476.
Hiki, K., Hattori, R., Kawai, C. & Yui, Y. (1992) *J. Biochem.* **111**, 556–558.
Hirsch, D.B., Steiner, J.P., Dawson, T.M., Mammen, A., Hayek, E. & Snyder, S.H. (1993) *Cur. Biol.* **13**, 749–754.
Holtzmann, D.M., Kilbridge, J., Bredt, D.S., Black, S.M., Li, Y., Clary, D.O., Reichardt, L.F. & Mobley W.C. (1994) NOS Induction by NGF in Basal Forebrain Cholinergic Neurons. Suggests Novel Mechanism for NGF Effects on Learning and Memory. In *Neurobiology of Disease* (in press).
Hope, B.T., Michael, G.J., Knigge, K.M. & Vincent, S.R. (1991) *Proc. Natl. Acad. Sci. USA* **88**, 2811–2814.
Huang, P.L., Dawson, T.M., Bredt, D.S., Snyder, S.H. & Fishman, M.C. (1993) *Cell* **75**, 1276–1285.
Ignarro, L.J., Lippton, H., Edwards, J.C., Baricos, W.H., Hyman, A.L., Kadowitz, P.J. & Gruetter, C.A. (1981) *J. Pharmacol. Exp. Ther.* **218**, 739–749.
Ignarro, L.J., Buga, G.M., Wood, K.S., Byrns, R.E. & Chaudhuri, G. (1987) *Proc. Natl. Acad. Sci. USA* **84**, 9265–9269.
Janssens, S.P., Shimouchi, A., Quertermous, T., Bloch, D.B. & Bloch, K.D. (1992) *J. Biol. Chem.* **267**, 14519–14522.
Kalb, R.G. & Agonstini, J. (1993) *Neuroscience* **57**, 1–8.
Klatt, P., Schmidt, K., Uray, G. & Mayer, B. (1993) *J. Biol. Chem.* **268**, 14781–14787.
Knowles, R.G., Palacios, M., Palmer, R.M.J. & Moncad, S. (1989) *Proc. Natl. Acad. Sci.* **86**, 5159–5162.
Kobzik, L., Bredt, D.S., Lowenstein, C.J., Drazen, J., Gaston, B., Sugarbaker, D. & Stamler, J.S. (1993) *Am. J. Respir. Cell Mol. Biol.* **9**, 371–377.
Kobzik, L., Reid, M.B., Bredt, D.S. & Stamler, J.S. (1994) *Nature* **372**, 546–548.
Koh, J.-Y., Peters, S. & Choi, D.W. (1986) *Science* **234**, 73–76.
Kwon, N.S., Nathan, C.F. & Stuehr, D.J. (1989) *J. Biol. Chem.* **264**, 20496–20501.
Lamas, S., Marsden, P.A., Li, G.K., Tempst, P. & Michel, T. (1992) *Proc. Natl. Acad. Sci. USA* **89**, 6348–6352.
Levi-Montalcini, R. (1982) *Ann. Rev. Neurosci.* **5**, 341–361.
Lowenstein, C.J., Glatt, C.S., Bredt, D.S. & Snyder, S.H. (1992) *Proc. Natl. Acad. Sci. USA* **89**, 6711–6715.
Lowenstein, C.J., Alley, E.W., Raval, P., Snowman, A.M., Synder, S.H., Russell, S.W. & Murphy, W.J. (1993) *Proc. Natl. Acad. Sci. USA* **90**, 9730–9734.
Lyons, C.R., Orloff, G.J. & Cunningham, J.M. (1992) *J. Biol. Chem.* **267**, 6370–6374.
Marletta, M.A., Yoon, P.S., Iyengar, R., Leaf, C.D. & Wishnok, J.S. (1988) *Biochemistry* **27**, 8706–8711.
Marsden, P.A., Heng, H.H.Q., Scherer, S.W., Stewart, R.J., Hall, A.V., Shi, X.-M., Tsui, L.-C. & Shappert, K.T. (1993) *J. Biol. Chem.* **268**, 17478–17488.
Mayer, B., John, M. & Böhme, E. (1990) *FEBS Lett.* **277**, 215–219.

Mayer, B., John, M., Heinzel, B., Werner, E.R., Wachter, H., Schultz, G. & Böhme, E. (1991) *FEBS Lett.* **288**, 187–191.
McMillan, K. & Masters, B.S.S. (1993) *Biochemistry* **32**, 9875–9880.
McMillan, K., Bredt, D.S., Hirsch, D.J., Snyder, S.H., Clark, J.E. & Masters, B.S.S. (1992) *Proc. Natl. Acad. Sci. USA* **89**, 11141–11145.
Michel, T., Li, G.K. & Busconi, L. (1993) *Proc. Natl. Acad. Sci. USA* **90**, 6252–6256.
Nakane, M., Mitchell, J., Förstermann, U. & Murad, F. (1991) *Biochem. Biophys. Res. Commun.* **180**, 1396–1402.
Nakane, M., Schmidt, H.H.H.W., Pollock, J.S., Förstermann, U. & Murad, F. (1993) *FEBS Lett.* **316**, 175–180.
Narhi, L.O. & Fulco, A.J. (1986) *J. Biol. Chem.* **261**, 7160–7169.
Nozaki, K., Moskowitz, M.A., Maynard, K.I., Koketsu, N., Dawson, T.M., Bredt, D.S. & Snyder, S.H. (1993) *J. Cereb. Blood Flow Metab.* **13**, 70–79.
O'Dell, T.J., Hawkins, R.D., Kandel, E.R. & Arancio, O. (1991) *Proc. Natl. Acad. Sci. USA* **88**, 11285–11289.
Ogura, T., Yokoyama, T., Fujisawa, H., Kurshima, Y. & Esumi, H. (1993) *Biochem. Biophys. Res. Commun.* **193**, 1014–1022.
Palmer, R.M.J. & Moncada, S. (1989) *Biochem. Biophys. Res. Commun.* **158**, 348–352.
Palmer, R.M., Ferrige, A.G. & Moncada, S. (1987) *Nature* **327**, 524–526.
Pollock, J.S. Förstermann, U., Mitchell, J.A., Warner, T.D., Schmidt, H.H.H.W., Nakane, M. & Murad, F. (1991) *Proc. Natl. Acad. Sci. USA* **88**, 10480–10484.
Pou, S., Surichamorn, W., Bredt, D.S., Snyder, S.H. & Rosen, G.M. (1992) *J. Biol. Chem.* **267**, 24173–24176.
Rajfer, J., Aronson, W.J., Bush, P.A., Dorey, F.J. & Ignarro, L.J. (1992) *New Eng. J. Med.* **326**, 90–94.
Rengasamy, A. & Johns, R.A. (1993) *Mol. Pharmacol.* **44**, 124–128.
Robinson, L.I., Weremowicz, S., Morton, C. & Michel, T. (1994) *Genomics* **19**, 350–357.
Rogers, N.E. and Ignarro, L.J. (1992) *Biochem. Biophys. Res. Commun.* **189**, 242–249.
Schmidt, H.H.W., Pollock, J.S., Nakane, M., Gorsky, L.D., Förstermann, U. & Murad, F. (1991) *Proc. Natl. Acad. Sci. USA* **88**, 365–369.
Schmidt, H.H.W., Smith, R.M., Nakane, M. & Murad, F. (1992) *Biochemistry* **31**, 3243–3249.
Schuman, E.M. & Madison, D.V. (1991) *Science* **254**, 1503–1506.
Sessa, W.C., Harrison, J.K., Barber, C.M., Zeng, D., Durieux, M.E., D'Angelo, D.D., Lynch, K.R. & Peach, K.J. (1992) *J. Biol. Chem.* **267**, 15274–15276.
Shatz, C.J. (1990) *Neuron* **5**, 745–756.
Stuehr, D.J. & Ikeda-Saito, M. (1992) *J. Biol. Chem.* **267**, 20547–20550.
Stuehr, D.J., Gross, S.S., Sukuma, I., Levi, R. & Nathan, C.F. (1989) *J. Exp. Med.* **169**, 1011–1020.
Stuehr, D.J., Cho, H.J., Kwon, N.S., Weise, M.F. & Nathan, C.F. (1991) *Proc. Natl. Acad. Sci. USA* **88**, 7773–7777.
Tayeh, M.A. & Marletta, M.A. (1989) *J. Biol. Chem.* **264**, 19654–19658.
Toda, N. (1981) *Br. J. Pharmacol.* **72**, 281–283.
Uemura, Y., Kowall, N.W. & Beal, M.F. (1990) *Ann. Neurol.* **27**, 620–625.
Vanderwinden, J.-M., Mailleux, P., Schiffmann, S.N., Vanderhaeghen, J.-J. & DeLaet, M.H. (1992) *New Eng. J. Med.* **327**, 511–515.
Verge, V.M.K., Xu, Z., Xu, X.-J., Wiesenfeld-Hallin, Z., Hökfelt, T. (1992) *Proc. Natl. Acad. Sci. USA* **89**, 11617–11621.
Vorherr, T., Knopfel, L., Hofmann, F., Mollner, S., Pfeuffer, T. & Carafoli, E. (1993) *Biochemistry* **32**, 6081–6088.
White, L.A. & Marletta, M.A. (1992) *Biochemistry* **31**, 6627–6631.
Wiesenfeld-Hallin, Z., Hao, J.X., Xu, X.J. & Hökfelt, T. (1993) *J. Neurophys.* **70**, 2350–2353.
Wilcox, C.S., Welch, W.J., Murad, F., Gross, S.S., Taylor, G. & Levi, R. (1992) *Proc. Natl. Acad. Sci. USA* **89**, 11993–11997.

Wu, W. & Li, L. (1993) *Neurosci. Lett.* **153**, 121–124.
Xie, Q.-W., Cho, H.J., Calaycay, J., Mumford, R.A., Swiderek, K.M., Lee, T.D., Ding, A., Troso, T. & Nathan, C. (1992) *Science* **256**, 225–228.
Xie, Q.W., Whisnant, R. & Nathan, C. (1993) *J. Exp. Med.* **177**, 1779–1784.
Yui, Y., Hattori, R., Kosuga, K., Eizawa, H., Hiki, K. & Kawai, C. (1991) *J. Biol. Chem.* **266**, 12544–12547.

CHAPTER 2

BIOCHEMISTRY AND MOLECULAR PHARMACOLOGY OF NITRIC OXIDE SYNTHASES

Bernd Mayer

Institut für Pharmakologie und Toxikologie, Universität Graz, Universitätsplatz 2, A-8010 Graz, Austria

Table of Contents

2.1 Introduction 21
2.2 Biochemical properties of NO synthases 22
  2.2.1 Detection of enzymatic NO formation 22
  2.2.2 Purification and characterization of NO synthases 23
2.3 Identification of cofactors 25
2.4 Mechanisms of NO synthesis 26
  2.4.1 L-Arginine oxidation 27
  2.4.2 Uncoupled reduction of molecular oxygen 28
  2.4.3 Function of tetrahydrobiopterin 29
2.5 Regulation of NOS activity 31
  2.5.1 Availability of L-arginine and tetrahydrobiopterin 31
  2.5.2 Phosphorylation 32
  2.5.3 Feedback inhibition by NO 33
  2.5.4 Gene expression 33
2.6 Molecular pharmacology of NOS inhibition 34
  2.6.1 N^G-Methyl-L-arginine 35
  2.6.2 N^G-Nitro-L-arginine 36
  2.6.3 7-Nitro indazole 37

Acknowledgements 38
References 38

2.1 Introduction

Since the first reports on mammalian NO synthesis in 1987 (Ignarro *et al.*, 1987; Palmer *et al.*, 1987), our knowledge about this novel biochemical pathway has rapidly grown. So far three distinct NO synthase (NOS; EC 1.14.13.39) isozymes

Nitric Oxide in the Nervous System
ISBN 0-12-721985-4

have been purified, cloned and biochemically characterized. Brain and endothelial NOSs are constitutively expressed and regulated by the intracellular concentration of free Ca^{2+}, whereas a Ca^{2+}-independent isozyme is cytokine inducible in macrophages and many other cells. In the course of a complex, only partially understood redox reaction, NOSs catalyse the conversion of L-arginine to L-citrulline and NO. We know that substrate oxidation occurs at a cytochrome P450-like haeme iron and requires NADPH as donor of electrons.

In this chapter, several topics of NOS enzymology will be reviewed, with a special focus on the brain enzyme. Purification and properties of NOS isozymes will be described in Section 2.2; Section 2.3 is devoted to the discussion of mechanistic aspects of NO biosynthesis. Selected topics of enzyme regulation, including substrate and cofactor availability, protein phosphorylation, feedback inhibition by NO, and transcriptional control of NOS activity will be covered in Section 2.4. Finally, several inhibitors of NOS and their mechanisms of action will be described in Section 2.5. The molecular biology of NOS is described in detail in Chapter 1 of this volume and will not be addressed here.

2.2 Biochemical properties of NO synthases

2.2.1 Detection of enzymatic NO formation

Enzymatic NO formation was shown first for the macrophage cell line RAW 264.7 (Marletta *et al.*, 1988). Addition of L-arginine and NADPH to cytosols prepared from γ-interferon/LPS (lipopolysaccharide)-activated cells resulted in formation of L-citrulline and NO. Production of NO was assayed either directly by NO chemiluminescence or as production of the stable inactivation products NO_2^-/NO_3^-. Using $^{15}N^G$-labelled L-arginine as substrate, mass spectrometric analysis allowed an unequivocal identification of the guanidino group of L-arginine as precursor of endogenously produced NO.

Expressed levels of constitutive NOSs seem to be considerably lower than those of the inducible enzyme, so more sensitive assay methods had to be established to characterize biosynthesis of NO in other tissues. Since NO elicits pronounced stimulation of soluble guanylyl cyclase, measurement of NO-induced cGMP accumulation was applied by several groups as a useful bioassay for NOS activity. The brain is capable of producing both NO and cGMP (Garthwaite *et al.*, 1988), so it was possible to monitor the stimulation of endogenous guanylyl cyclase in crude supernatants upon addition of L-arginine, NADPH and Ca^{2+} (Bredt and Snyder, 1989; Knowles *et al.*, 1989). A similar approach allowed the first characterization of constitutive NOS in adrenal gland (Palacios *et al.*, 1989), bovine lung (Mayer and Böhme, 1989) and blood platelets (Radomski *et al.*, 1990). Endothelial cells express comparably low levels of endogenous soluble guanylyl cyclase, so we have modified the bioassay for L-arginine-derived NO and measured cGMP formation by purified exogenous soluble guanylyl cyclase added to crude fractions of cultured endothelial cells (Mayer *et al.*, 1989). Alternatively, cGMP accumulation was measured in cultured fibroblast detector cells in response to NO enzymatically produced in a separate system (Gorsky *et al.*, 1990).

Albeit highly sensitive and specific for the biologically relevant metabolite of L-arginine, these bioassays are rather difficult to handle, require availability of an appropriate cyclase preparation and do not allow quantitative determination of NO production. Palmer and Moncada (1989) have assayed NOS activity in endothelial cytosols by HPLC determination of ^{3}H-citrulline produced from labelled L-arginine. However, radiochemical determination of NO formation became a widely used routine assay only after the introduction of a more convenient and rapid method for the separation of substrate and product on small anion exchange columns (Bredt and Snyder, 1990). This assay is a useful routine method allowing quick handling of a large number of samples. Sensitivity of the assay is rather low in the presence of saturating concentrations of L-arginine but can be increased by reduction of the substrate concentration. Comparison of the ^{3}H-citrulline assay with other methods, i.e. NO-induced stimulation of purified soluble guanylyl cyclase (Mayer *et al.*, 1990), NO-induced loss of oxyhaemoglobin (Mayer *et al.*, 1992), as well as production of NO_2^-/NO_3^- (Klatt *et al.*, 1993), gave similar results with various NOS preparations and clearly demonstrated stoichiometric production of NO and L-citrulline in the NOS reaction.

2.2.2 Purification and characterization of NO synthases

2.2.2.1 Brain NO synthase

Ca^{2+}/calmodulin-dependent brain NOS was first purified to homogeneity from rat cerebellum by sequential anion exchange chromatography over DEAE agarose and affinity chromatography over 2′,5′-ADP sepharose (Bredt and Snyder, 1990). The latter material selectively binds adenine nucleotide-dependent enzymes, and NOS could be eluted from the column with an excess of NADPH. The pure protein appeared as a single 150 kDa band on SDS PAGE gels, and the molecular mass of the native enzyme was estimated as 200 kDa by gel permeation chromatography, pointing to a monomeric conformation of native NOS. The isolated enzyme was found to be labile, with a specific activity of about 1 μmol L-citrulline min^{-1} mg^{-1} and a K_m for L-arginine of 1.5 μM.

We attempted to apply this method for the large-scale preparation of NOS from porcine cerebellum, but the enzyme was too labile in crude supernatants to allow time-consuming column chromatography as a first purification step. Instead, we have rapidly removed NOS from the supernatants by ammonium sulphate precipitation. Thereafter, the enzyme was rather stable, and by chromatography of the ammonium sulphate precipitate over a 2′,5′-ADP sepharose column we obtained a homogeneous preparation of a protein which appeared as a single 160 kDa band on SDS PAGE gels (Mayer *et al.*, 1990). Functional reconstitution with purified soluble guanylyl cyclase and measurement of NO formation by NO chemiluminescence allowed the positive identification of the isolated protein as NOS. (6*R*)-5,6,7,8-Tetrahydro-L-biopterin (H_4biopterin) stimulated the enzyme 4-fold from 0.2 to 0.8 μmol L-citrulline min^{-1} mg^{-1}, suggesting a role of the pteridine as a cofactor in neuronal NO synthesis.

Schmidt and coworkers (1991) used a different approach to purify brain NOS. Crude rat cerebellar supernatants were first subjected to affinity chromatography on 2′,5′-ADP sepharose, and NOS was subsequently purified to homogeneity from the eluate by chromatography over calmodulin agarose. The isolated enzyme exhibited a K_m for L-arginine of 2.2 μM, and v_{max} was about 0.1 μmol L-citrulline min^{-1} mg^{-1}. A molecular mass of 155 kDa was estimated from SDS PAGE gels, but detailed physical characterization of isolated NOS by gel permeation chromatography and velocity sedimentation revealed that the native protein may represent a dimer, consisting of two identical subunits.

Taken together, these results clearly show that the brain contains a soluble enzyme that converts L-arginine to L-citrulline and NO and presumably represents a homodimer with a subunit molecular mass of 150–160 kDa. More recent results suggest that brain NOS may be associated with membranes, too. Using essentially the same purification scheme as described by Schmidt *et al.* (1991), Hiki and coworkers (1992) have isolated a 150 kDa protein from Triton X-100-solubilized particulate fractions of rat cerebellum. The enzyme had a rather low specific activity of ~10 nmol L-citrulline min^{-1} mg^{-1}, required addition of Ca^{2+}/calmodulin, NADPH and FAD, and was stimulated 3-fold by H_4biopterin. According to these data, brain NOS may be loosely associated with membranes and become solubilized to variable degrees upon tissue homogenization. This is further suggested by histochemical studies demonstrating that NADPH diaphorase activity, reflecting NOS-catalysed reduction of nitro-blue tetrazolium (Bredt *et al.*, 1991a; Hope *et al.*, 1991), is predominantly localized at the endoplasmic reticulum of rat brain (Wolf *et al.*, 1992).

2.2.2.2 Endothelial NO synthase

Initial characterizations of endothelial NOS have been carried out with soluble fractions of cultured endothelial cells (Mayer *et al.*, 1989; Palmer and Moncada, 1989), but more recent work showed that the enzyme is predominantly localized in KCl-insoluble membrane fractions (Förstermann *et al.*, 1991; Mayer *et al.*, 1993b). This particulate NOS was purified from cultured bovine aortic endothelial cells by detergent solubilization from crude membranes, affinity chromatography on 2′,5′-ADP sepharose, and gel permeation chromatography as a final step (Pollock *et al.*, 1991). The enzyme had a molecular mass of 135 kDa on SDS gels, required Ca^{2+}/calmodulin for activity, was stimulated by H_4biopterin, and exhibited a K_m for L-arginine of 2.9 μM. As observed with membrane-associated brain NOS (Hiki *et al.*, 1992), specific activity was rather low (~15 nmol L-citrulline min^{-1} mg^{-1}), suggesting instability of NOS in detergent or removal of an essential cofactor during purification. Cloning and sequence analysis of endothelial NOS revealed a consensus motive for N-terminal myristoylation (Lamas *et al.*, 1992; Sessa *et al.*, 1992), and mutation of this site resulted in expression of a soluble enzyme (Busconi and Michel, 1993; Sessa *et al.*, 1993), demonstrating that membrane association of this isozyme is due to post- or co-translational modification.

2.2.2.3 Inducible NO synthase

Most of the work on inducible NOS has been done with the enzyme obtained from activated macrophages, but expression of this isozyme is induced by cytokines in many other mammalian cells, including brain microglia, suggesting an important role of inducible NOS in neurotoxicity (Boje and Arora, 1992; Chao *et al.*, 1992; Murphy *et al.*, 1993). Using basically the same purification protocols as described above, i.e. affinity chromatography on 2′,5′-ADP sepharose and gel permeation chromatography, inducible NOS was isolated from soluble fractions of activated rat peritoneal macrophages (Yui *et al.*, 1991) and the murine RAW 264.7 cell line (Hevel *et al.*, 1991; Stuehr *et al.*, 1991a). The native enzyme was reported to represent a dimer with a subunit molecular mass of 130 kDa (RAW 264.7) or 150 kDa (rat peritoneal macrophages). Enzyme activity was independent of Ca^{2+} and enhanced by H_4biopterin; v_{max} was 1–1.5 μmol L-citrulline min^{-1} mg^{-1}. K_m values for L-arginine were 2.8, 16 and 32.3 μM in the studies by Stuehr *et al.*, Hevel *et al.*, and Yui *et al.*, respectively.

Cloning of macrophage NOS (Lyons *et al.*, 1992; Xie *et al.*, 1992) revealed a consensus site for calmodulin binding similar to the motive found in the sequence of the brain enzyme (Bredt *et al.*, 1991b). Indeed, inducible NOS apparently contains calmodulin tightly bound as a subunit even in solutions nominally free of Ca^{2+} (<30 nM) (Cho *et al.*, 1992), suggesting that all NOS isoforms require bound calmodulin for activity. Accordingly, only the constitutive brain and endothelial enzymes seem to require micromolar concentrations of free Ca^{2+} for calmodulin binding, while calmodulin binds to inducible NOS in a Ca^{2+}-independent fashion.

2.3 Identification of cofactors

Biochemical characterization of brain and macrophage NOSs showed that they are rather complex enzymes which require various redox-active cofactors for optimal function (Marletta, 1993; Mayer, 1993). Requirement for NADPH (Marletta *et al.*, 1988), stimulation by FAD (Stuehr *et al.*, 1990), and inhibition by diphenyliodonium compounds (Stuehr *et al.*, 1991b) suggested that NOSs may be flavoproteins. Indeed, supernatants of heat-denaturated brain NOS exhibited excitation and emission fluorescence spectra typical for reduced flavins, and HPLC analysis of the supernatants revealed the presence of equimolar amounts of FAD and FMN (Mayer *et al.*, 1991). Subsequently, identification of FAD and FMN as prosthetic groups of NOSs was confirmed for the enzymes from rat brain (Bredt *et al.*, 1992; H.H.H.W. Schmidt *et al.*, 1992) and macrophages (Hevel *et al.*, 1991; Stuehr *et al.*, 1991a).

H_4Biopterin was shown to stimulate NO formation in soluble fractions of cytokine-activated macrophages (Kwon *et al.*, 1989; Tayeh and Marletta, 1989) and in purified preparations of neuronal NOS (Mayer *et al.*, 1990). We found that the purified brain enzyme contained residual amounts of tightly bound pteridines, suggesting that the basal enzyme activities seen in the absence of added cofactor may be due to residual endogenous H_4biopterin (Mayer *et al.*, 1991). Co-purification of the pteridine with NOSs was subsequently confirmed for the enzymes obtained from rat (H.H.H.W. Schmidt *et al.*, 1992) and human (Klatt *et al.*, 1992b) brain as well as murine macrophages (Hevel and Marletta, 1992).

Atomic absorption spectroscopic analysis of purified neuronal NOS revealed the presence of 1 mol of iron per mol of subunit (Mayer *et al.*, 1991), and more recently both brain and macrophage NOSs were identified as haem proteins (Klatt *et al.*, 1992c; McMillan *et al.*, 1992; Stuehr and Ikeda-Saito, 1992; White and Marletta, 1992). Shift of the Soret band to 450 nm upon binding of CO, inhibition of NO formation by miconazole, cyanide and CO as well as sequence similarities to other P450s indicate that NOS isozymes belong to the cytochrome P450 superfamily (for sequence alignment see Renaud *et al.*, 1993).

2.4 Mechanisms of NO synthesis

Mammalian microsomal P450-containing mono-oxygenase systems consist of two protein components: a cytochrome P450 haem protein which catalyses substrate hydroxylation and an FAD- and FMN-containing cytochrome P450 reductase, responsible for electron transfer from NADPH to the haem (Guengerich, 1991; Degtyarenko and Archakov, 1993). NOSs contain a haem moiety as well as FAD and FMN within one single peptide chain and exhibit sequence similarities to both P450 mono-oxygenases and the respective reductases (Bredt *et al.*, 1991b; Renaud *et al.*, 1993), suggesting that they are unique in mammalian cells and represent self-sufficient cytochrome P450s. A bacterial enzyme with similar properties has been purified and cloned from *Bacillus megaterium* (Fulco, 1991).

As illustrated in Figure 1, NOSs seem to combine previously evolved catalytic functions: flavin-catalysed transport of electrons from NADPH to haem iron and activation of molecular oxygen for the mono-oxygenase-like N-oxidation of L-arginine. The flavins apparently operate independently of the haem, since electrons can be transferred also to exogenously added electron acceptors, including cytochrome P450 purified from liver microsomes, cytochrome *c*, and the low molecular mass dye nitro blue tetrazolium (Klatt *et al.*, 1992a). Reduction of cytochrome *c* by the brain enzyme requires the presence of Ca^{2+}/calmodulin, is insensitive to superoxide dismutase and cyanide, and occurs at rates about 10-fold higher than NO for-

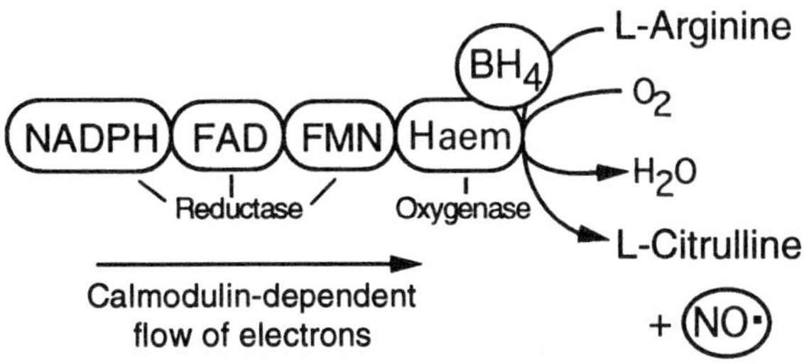

Figure 1 Coenzymes and catalytic domains of NO synthase. NO synthases consist of two catalytic domains: an FAD- and FMN-containing reductase and a haem-containing oxygenase domain. The flavins shuttle electrons from NADPH to haem iron which catalyses reductive activation of molecular oxygen coupled to NO formation in the presence of saturating concentrations of L-arginine and H_4biopterin (BH_4). Electron transfer requires enzyme-bound calmodulin (Klatt *et al.*, 1992a; Abou-Soud and Stuehr, 1993).

mation. Ca^{2+}/calmodulin dependence of NOS-catalysed electron shuttling was confirmed recently using spectrophotometrical techniques (Abu-Soud and Stuehr, 1993).

2.4.1 L-Arginine oxidation

The precise mechanism of NO formation is not known. The reaction involves a 5-electron oxidation of one of the chemically equivalent guanidino-nitrogens of L-arginine, leading to the concomitant production of L-citrulline and NO. It is accompanied by an NADPH-dependent reduction of molecular oxygen (Mayer *et al.*, 1991) which is incorporated into both reaction products (Kwon *et al.*, 1990; Leone *et al.*, 1991). Studies with purified macrophage and brain NOSs hint at a two step reaction involving formation of N^G-hydroxy-L-arginine as an intermediate which is normally not released in appreciable quantities from the purified enzymes but immediately transformed further to NO and L-citrulline (Stuehr *et al.*, 1991c; Klatt *et al.*, 1993) (see Figure 2). Yet, some tumour cell lines expressing inducible NOS were shown to release substantial amounts of N^G-hydroxy-L-arginine to the extracellular medium (Chenais *et al.*, 1991). These findings may be of particular importance because N^G-hydroxylated L-arginine represents a biologically active molecule (Zembowicz *et al.*, 1991; Chenais *et al.*, 1993), and is metabolized to NO and L-citrulline by various haem proteins other than NOS (Boucher *et al.*, 1992).

Incubation of the macrophage and brain enzymes with substoichiometrical amounts of NADPH revealed that 1.5 mol of the nucleotide are consumed for the formation of 1.0 mol of L-citrulline (Mayer *et al.*, 1991; Stuehr *et al.*, 1991c). Two of the three NADPH-derived reducing equivalents seem to be utilized for the initial N-hydroxylation of L-arginine, and oxidative cleavage of N^G-hydroxy-L-arginine to NO and L-citrulline may require one additional electron (Stuehr *et al.*, 1991c). Since NADPH can only transfer electrons two at a time, it has been suggested that

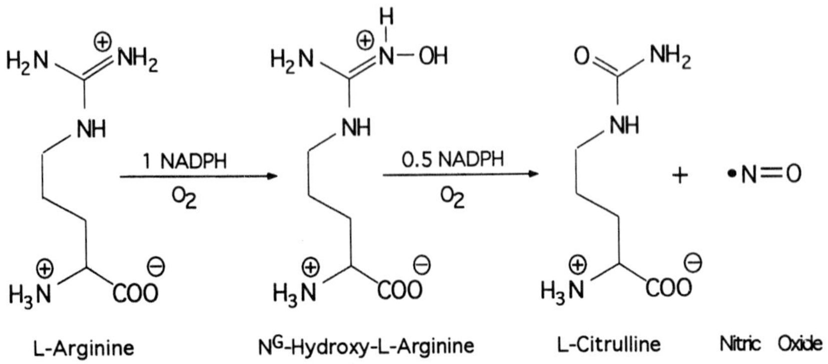

Figure 2 NO synthase-catalysed oxidation of L-arginine. NO synthases generate NO by catalysing NADPH-dependent 5-electron oxidation of the guanidino group of L-arginine. In the course of two sequential oxidation steps L-arginine is N-hydroxylated to N^G-hydroxy-L-arginine first. The intermediate remains bound to the enzyme and is oxidatively cleaved to L-citrulline and NO. The overall reaction involves reduction of two oxygen molecules requiring a total of eight electrons. Five reducing equivalents are derived from guanidino nitrogen oxidation; co-oxidation of NADPH provides three additional electrons. For a detailed discussion of reaction mechanisms, see Feldman *et al.* (1993b), Klatt *et al.* (1993) and Marletta (1993).

enzyme-bound flavins store the extra fourth electron for use in the next catalytic cycle. Accordingly, six electrons would be transferred to the haem from three molecules of NADPH in the course of two reaction cycles (Feldman *et al.*, 1993b).

2.4.2 Uncoupled reduction of molecular oxygen

When brain NOS is activated by Ca^{2+}/calmodulin it starts to oxidize NADPH and transfer reducing equivalents to the prosthetic haem group for reductive activation of molecular oxygen. As observed with other P450-containing mono-oxygenases, enzymatic oxygen reduction also takes place in the absence of substrate, leading to generation of superoxide anions as well as hydrogen peroxide instead of NO (Mayer *et al.*, 1991; Heinzel *et al.*, 1992; Pou *et al.*, 1992). Upon addition of L-arginine, oxygen activation proceeds along with substrate metabolism, resulting in increased NO formation and decreased production of hydrogen peroxide. Although H_4biopterin is not directly involved in oxygen reduction, coupling of the reaction to L-arginine oxidation requires the presence of the pteridine. Thus, only in the presence of saturating concentrations of both L-arginine and H_4biopterin, does brain NOS efficiently utilize molecular oxygen for NO formation without concomitantly producing reactive oxygen metabolites.

Inducible NOS apparently down-regulates uncoupled oxygen reduction, as it consumes only minute quantities of NADPH in the absence of L-arginine (Abu-Soud and Stuehr, 1993). Significant amounts of hydrogen peroxide are generated, however, in the presence of the NOS inhibitor N^G-methyl-L-arginine (Olken and Marletta, 1993), indicating that inducible NOS, in contrast to the brain enzyme, requires a ligand bound to its substrate site for reduction of molecular oxygen to occur.

2.4.3 Function of tetrahydrobiopterin

H_4Biopterin has been known for decades as a cofactor of enzymes which hydroxylate the aromatic amino acids phenylalanine, tyrosine and tryptophan (Nichol *et al.*, 1985). These amino acid hydroxylases are non-haem iron proteins and use H_4biopterin as donor of electrons for activation of molecular oxygen. Accordingly, H_4biopterin is co-oxidized with the amino acid substrates. The quinoid form of H_2biopterin (*q*-H_2biopterin) is produced as the initial oxidation product; its intramolecular rearrangement yields H_2biopterin. In intact cells or crude preparations, these oxidation products are cycled back to H_4biopterin at the expense of NAD(P)H by dihydropteridine (*q*-H_2biopterin) or dihydrofolate (H_2biopterin) reductases (see Figure 3).

H_4Biopterin stimulates all NOS isozymes known so far (Kwon *et al.*, 1989; Tayeh and Marletta, 1989; Mayer *et al.*, 1990; Pollock *et al.*, 1991). The pteridine seems to be obligatory for NO synthesis, since basal enzyme activities observed in the absence of exogenously added H_4biopterin correlate well with the amount of NOS-bound biopterins (Mayer *et al.*, 1991; Hevel and Marletta, 1992; H.H.H.W. Schmidt *et al.*, 1992). It may depend on homogenization and purification techniques how much of H_4biopterin remains tightly bound to NOS, perhaps accounting for the variable degrees of enzyme stimulation by added H_4biopterin reported by different groups (Dwyer *et al.*, 1991; Giovanelli *et al.*, 1991; Schmidt and Murad, 1991; Evans *et al.*, 1992; Klatt *et al.*, 1992b).

Identification of H_4biopterin as a cofactor of NOSs led several researchers to speculate that the pteridine may have a function in NO formation similar to its role in aromatic amino acid hydroxylation, and catalyse the initial N-oxygenation of L-arginine. However, it was not possible to assign the activity of H_4biopterin to a distinct step of L-arginine oxidation (Stuehr *et al.*, 1991c; Klatt *et al.*, 1993), and the pteridine is apparently not involved directly in the NOS-catalysed, L-arginine-independent reduction of molecular oxygen (Heinzel *et al.*, 1992). Moreover, H_4biopterin seems to be active as cofactor in NO synthesis at much lower than stoichiometric concentrations, since one molecule of the cofactor was found to be sufficient for the formation of up to 20 molecules of product (Giovanelli *et al.*, 1991; Mayer *et al.*, 1991). Accordingly, redox activity of H_4biopterin would implicate its continuous recycling at the cost of NADPH during NO synthesis, but in spite of several attempts we obtained no experimental evidence to support enzymatic redox cycling of H_4biopterin (E.R. Werner and B. Mayer, unpublished observations). However, the methods which are currently available test for redox cycling of free H_4biopterin only and therefore do not exclude the possibility that the enzyme-bound cofactor is redox active.

Negative results similar to ours led Giovanelli and coworkers to propose that

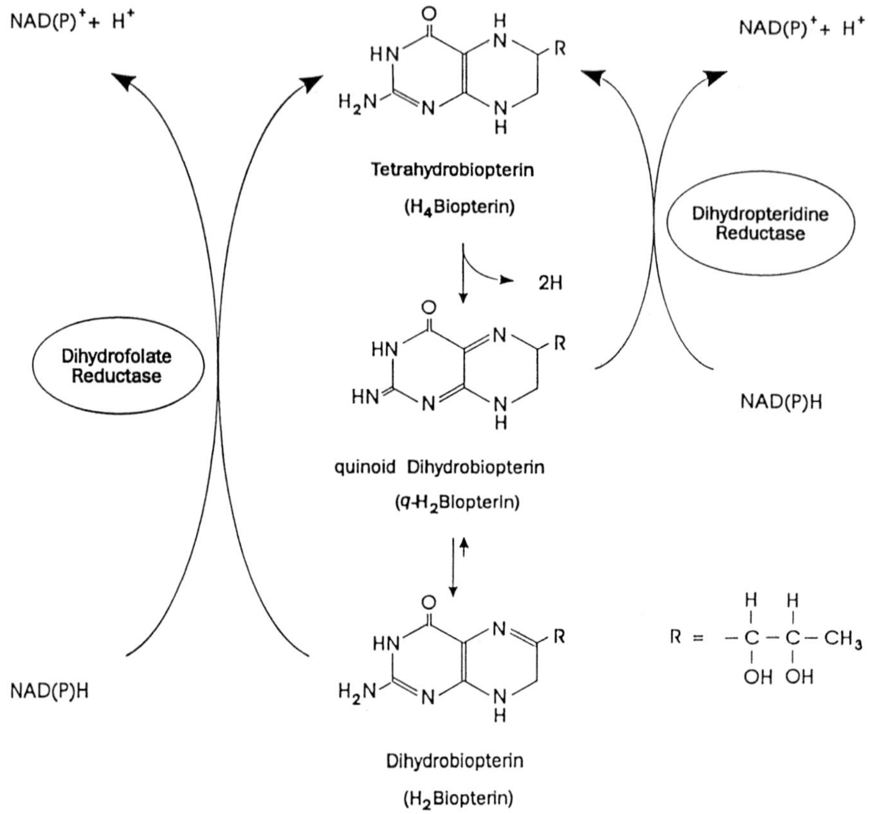

Figure 3 Redox cycling of reduced biopterins. The function of H_4biopterin in the NO synthase reaction is not known. As a redox-active cofactor of aromatic amino acid hydroxylation H_4biopterin is co-oxidized with the substrates to *q*-H_2biopterin and finally H_2biopterin. The oxidized products are reduced back to H_4biopterin by NAD(P)H-dependent dihydrofolate (H_2biopterin) and dihydropteridine (*q*-H_2biopterin) reductases.

H_4biopterin is not a reactant in NO synthesis but rather acts as an allosteric effector of the enzyme (Giovanelli *et al.*, 1991). In support of this hypothesis, it was shown that H_4biopterin is required to keep macrophage NOS in its active dimeric state and that dissociated inactive monomers only reassociate when haem, H_4biopterin and L-arginine are present at the same time (Baek *et al.*, 1993). Although neuronal NOS seems to behave differently and remains a dimer under conditions that induce dissociation of macrophage NOS (Schmidt *et al.*, 1991; Klatt *et al.*, 1994a), some kind of synergistic interaction between L-arginine and H_4biopterin binding sites may occur in the course of constitutive NO synthesis, too. Thus, L-arginine decreased the K_D of brain NOS for H_4biopterin from about 200 nM to 30 nM and pteridine binding apparently enhanced the affinity of the substrate site, suggesting an allosteric interaction between the two binding domains (Klatt *et al.*, 1994b).

However, the role of the pteridine may not be confined to its effect on NOS conformation. We found that the oxidized derivative H_2biopterin bound to the pteridine site of brain NOS with a fairly high affinity of about 2 μM but was inactive as a cofactor in NO synthesis (Klatt *et al.*, 1994b). These data may indicate a dual role of H_4biopterin in NO synthesis: binding of the pteridine may convert NOS into an active conformational state, and the bound cofactor may participate in a redox-active manner in L-arginine oxidation.

2.5 Regulation of NOS activity

Although regulation of NO formation in the nervous system occurs primarily through intracellular Ca^{2+}, several additional mechanisms may contribute to the modulation of NO-mediated cellular signalling. Here I will briefly focus on possible effects of substrate and cofactor availability, phosphorylation of NOS, feedback inhibition by NO, and regulation via *de novo* synthesis of the NOS protein.

2.5.1 Availability of L-arginine and tetrahydrobiopterin

Conceivably, generation of NO by Ca^{2+}/calmodulin-activated NOS is controlled by the intracellular availability of substrates or cofactors. The content of free L-arginine in total brain is ~10 μmol per 100 g wet weight (Clarke *et al.*, 1989), a concentration sufficiently high for saturation of brain NOS. Nevertheless, several *in vivo* and *in vitro* studies demonstrate biological effects on addition of exogenous L-arginine, indicating that NOS may not be substrate saturated in certain brain areas. For example, nociceptive processing in the mouse brain was affected in an N^G-nitro-L-arginine-sensitive manner by intracerebroventricular administration of L-arginine (Kawabata *et al.*, 1993), and release of dopamine from rat striatal slices was stimulated by superfusion with either sodium nitroprusside or exogenous L-arginine (Zhu and Luo, 1992). Inhibition of NMDA-mediated cGMP accumulation in rat cerebellum by nanomolar concentrations of N^G-nitro-L-arginine also suggests low intracellular levels of free L-arginine in neuronal cells (East and Garthwaite, 1990). Others observed release of L-arginine to the extracellular medium upon electrical stimulation of rat cerebellar slices (Hansel *et al.*, 1992), indicating that neurons may depend on transport of the NOS substrate from specific L-arginine pools. As observed with cultured endothelial cells and macrophages (Schmidt *et al.*, 1993, 1994), transport system y^+ is responsible for uptake of L-arginine into neuronal cells, and several basic amino acids utilizing the same transporter can reduce intracellular availability of L-arginine (K. Schmidt *et al.*, in press).

It will be recalled that L-arginine deficiency would not only decrease rates of NO formation but also induce generation of superoxide anions and hydrogen peroxide. Superoxide rapidly combines with NO to form peroxynitrite (Beckman *et al.*, 1990), a cytotoxic reaction product involved in lipid peroxidation and other cell degenerating processes (Radi *et al.*, 1991). Reports on the role of NO in glutamate neurotoxicity are controversial (Choi, 1993). Recent work suggests that NO may protect neurons against overstimulation by glutamate due to blockade of NMDA receptors, while neurotoxic effects may be mediated by peroxynitrite rather than NO itself (Lipton *et al.*, 1993). Thus, NO toxicity seems to be concurrent with production of superoxide, precisely the brain NOS products under conditions of reduced L-arginine availability.

The relation between H_4biopterin levels and NO synthesis in neuronal cells is not known but deserves investigation, since a number of neurodegenerative diseases seem to involve impaired pteridine metabolism (Duch and Smith, 1991). Studies with other cells, including cultured fibroblasts (Werner-Felmayer *et al.*, 1990), macrophages (Sakai *et al.*, 1993), smooth muscle (Gross and Levi, 1992), human tumour cells (Werner-Felmayer *et al.*, 1993a) and endothelial cells (K. Schmidt *et al.*, 1992; Werner-Felmayer *et al.*, 1993b), clearly demonstrate that NO release from intact cells depends on intracellular H_4biopterin levels. The rate-limiting enzyme of H_4biopterin biosynthesis, GTP cyclohydrolase I, and NOS are co-induced in many cytokine-inducible cells, suggesting that the NOS cofactor H_4biopterin is synthesized as required (Werner *et al.*, 1993). On the other hand, human umbilical vein endothelial cells, which contain a constitutive form of NOS, also express enhanced levels of GTP cyclohydrolase I upon cytokine treatment (Werner-Felmayer *et al.*, 1993b). Since the increased levels of H_4biopterin in the cytokine-activated cells were accompanied by enhanced bradykinin-induced NO release, the pteridine may be rate limiting for NO synthesis in untreated human endothelial cells.

2.5.2 Phosphorylation

The published data on NOS phosphorylation and its functional consequences are controversial. Several reports demonstrate that purified brain NOS is stoichiometrically phosphorylated by various protein kinases (PK), including cAMP-dependent PK (Brüne and Lapetina, 1991; Bredt *et al.*, 1992), cGMP-dependent PK (B. Mayer and F. Hofmann, unpublished), PKC (Nakane *et al.*, 1991; Bredt *et al.*, 1992) and Ca^{2+}/calmodulin-stimulated PK-II (Nakane *et al.*, 1991; Bredt *et al.*, 1992). Nakane *et al.* (1991) reported an incorporation of 2 and 9 mol of phosphate per mol of NOS subunit by PKC and Ca^{2+}/calmodulin-stimulated PK-II, respectively, whereas Bredt *et al.* (1992) found incorporation of about 1 mol of phosphate in each case. Phosphopeptide mapping indicated that each kinase phosphorylates a different serine (Bredt *et al.*, 1992), but threonine residues were also described as putative targets (Nakane *et al.*, 1991).

Various groups seem to concur that the activity of purified NOS is not affected when the enzyme is phosphorylated by cAMP- (Brüne and Lapetina, 1991; Bredt *et al.*, 1992) or cGMP-dependent PK (B. Mayer and F. Hofmann, unpublished), but the reports on the effects of PKC and Ca^{2+}/calmodulin-stimulated PK-II are controversial (Nakane *et al.*, 1991; Bredt *et al.*, 1992). Yet, studies with intact cells indicate a direct or indirect involvement of PKC in the regulation of NO synthesis. For instance, activation of NMDA receptors resulted in a PKC-dependent stimulation of NOS activity in mouse striatal neurons (Marin *et al.*, 1992), and more recently it was demonstrated that the immunosuppressant FK506 enhanced NOS phosphorylation and reduced nitrite levels in 293 kidney cells stably transfected with a rat brain NOS cDNA (Dawson *et al.*, 1993). Since FK506, an inhibitor of the Ca^{2+}-activated phosphatase calcineurin (Liu *et al.*, 1991), protected cultured cortical neurons against NMDA toxicity, a challenging model has been constructed, according to which neuronal NOS is a substrate of both PKC and calcineurin. PKC phosphorylation could result in reduced rates of NO formation, and this effect would be reversed by calcineurin-catalysed dephosphorylation. Calcineurin in turn is inhibited by FK506, resulting in a net decrease of neuronal NO production and hence reduced neurotoxicity (Dawson *et al.*, 1993).

Clearly, it is not possible to draw definitive conclusions about the physiological role of brain NOS phosphorylation at this stage. Functional effects may not simply be reflected by changed enzyme activities but could be more subtle and affect, for example, the protein's affinity for one of its coenzymes or its subcellular localization, as shown for endothelial NOS (Michel *et al.*, 1993b). Furthermore, effects seen in intact cells with phorbol esters or kinase and phosphatase inhibitors could reflect interference with complex signal transduction cascades rather than direct consequences of NOS phosphorylation (Hecker *et al.*, 1993).

2.5.3 Feedback inhibition by NO

Recent reports suggest that NOS is inhibited by its reaction product NO (Rogers and Ignarro, 1992; Assreuy *et al.*, 1993; Buga *et al.*, 1993; Griscavage *et al.*, 1993; Rengasamy and Johns, 1993). Because of the possible therapeutic potential of administration of exogenous NO (Ahlner *et al.*, 1991; Geggel, 1993; Rossaint *et al.*, 1993), feedback inhibition of NOS could be of considerable clinical relevance. NO completely blocks rat liver microsomal cytochromes P450 at threshold concentrations as low as 70 nM (Wink *et al.*, 1993). Since NOSs have been identified as P450-like haem proteins (Klatt *et al.*, 1992c; McMillan *et al.*, 1992; Stuehr and Ikeda-Saito, 1992; White and Marletta, 1992), it is conceivable that NO formation may be blocked by binding of NO to the prosthetic haem group of the synthases. However, the studies supporting feedback inhibition of NOS have been performed with intact cells or crude enzyme preparations and high concentrations of added NO donors, which may exert unspecific effects. We found that purified NOS was neither appreciably inhibited by NO donors nor stimulated by oxyhaemoglobin, a scavenger of endogenously produced NO (Mayer *et al.*, 1994). Accordingly, the effects of exogenously added NO seen with more complex systems may be indirect and require further investigation.

2.5.4 Gene expression

Inducible NOS activity is predominantly regulated via gene expression, but several reports suggest that synthesis of the constitutive NOSs is also under transcriptional control. Following axotomy, long-lasting increases in neuronal NOS protein and mRNA levels were observed in neurons of dorsal root ganglia (Verge *et al.*, 1992; Fiallos-Estrada *et al.*, 1993) and intrinsic neurons of the brain (Herdegen *et al.*, 1993). The physiological function of this up-regulation of NOS after nerve injury is not understood, but conceivably NO could counteract the deleterious effects of axotomy by increasing blood flow required for cell regeneration or act as a messenger molecule affecting synaptic activity of neighbouring neurons. Immunohistochemistry and *in situ* hybridization studies hint at developmental changes of neuronal NOS expression levels and subcellular distribution in the brain (Brüning, 1993; Matsumoto *et al.*, 1993). The physiological consequences of these changes are elusive, especially because knock-out of the gene encoding neuronal NOS had no dramatic effect on brain development in transgenic mice (Huang *et al.*, 1993; see Chapter 1).

Expression levels of constitutive endothelial NOS were reported to be modified in pathophysiological situations. For instance, mRNA levels were decreased by inflammatory cytokines (MacNaul and Hutchinson, 1993), whereas synthesis of the enzyme was apparently increased after focal cerebral ischaemia (Zhang *et al.*, 1993). It remains to be seen whether this up-regulation of endothelial NOS enhances cytotoxicity or protects neuronal cells against damage.

2.6 Molecular pharmacology of NOS inhibition

NOSs are inhibited by a wide variety of drugs which interfere with one of the multiple catalytic functions of the enzyme. As mentioned earlier, imidazole derivatives as well as cyanide and CO prevent redox cycling of the haem (Marletta, 1993). Oxidation of haem iron also may be responsible for enzyme inhibition by methylene blue, a drug formerly regarded as a specific inhibitor of soluble guanylyl cyclase (Mayer *et al.*, 1993a). Several dyes like nitro blue tetrazolium or dichlorophenol indophenol inhibit NO formation as parasitic acceptors for NADPH-derived electrons (Klatt *et al.*, 1992a), and diphenyl iodonium compounds similarly interfere with flavin-catalysed electron transport (Stuehr *et al.*, 1991b). Finally, calmodulin antagonists prevent binding of Ca^{2+}/calmodulin and therefore activation of the constitutive isozymes (Bredt and Snyder, 1990). However, these drugs cannot be regarded as specific NOS inhibitors, because they block other cytochrome P450s, flavin-containing reductases or calmodulin-dependent enzymes, too.

The affinities of NOSs for L-arginine are in the micromolar range and thus much higher than those of other arginine metabolizing enzymes, suggesting that substrate analogues may be NOS inhibitors with sufficient specificity. In the past few years several N^G-derivatives of L-arginine have been developed. Although they predominantly act via competition for L-arginine, recent work hints at a more complex mode of molecular action of some of these drugs. In the first part of this section, the mechanisms of action of the most commonly used compounds

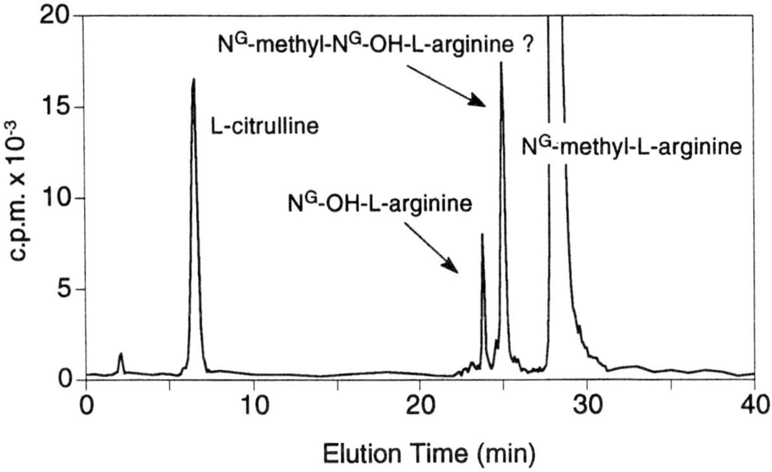

Figure 4 NO synthase metabolites of N^G-methyl-L-arginine. The NO synthase inhibitor L-NMA is slowly metabolized by brain NO synthase in a Ca^{2+}/calmodulin-dependent fashion. The final products of this pathway are L-citrulline and NO, indicating that L-NMA is an alternate substrate of the enzyme. HPLC analysis of radioactive metabolites produced from ^{14}C-labelled L-NMA suggested that L-NMA is first N-hydroxylated to N^G-methyl-N^G-hydroxy-L-arginine, which may be demethylated to the NO synthase substrate N^G-hydroxy-L-arginine. Data from Klatt *et al.* (1994a).

N^G-me/thyl-L-arginine (L-NMA) and N^G-nitro-L-arginine (L-NNA) will be described. 7-Nitro indazole may represent a novel NOS inhibitor (Babbedge *et al.*, 1993; Moore *et al.*, 1993a,b). It is structurally unrelated to L-arginine and exhibits a very interesting action profile, which will be discussed in the final part of this chapter.

2.6.1 N^G-Methyl-L-arginine

L-NMA was the first compound used for inhibition of NO production in tissues (Moncada *et al.*, 1991). L-NMA exhibits similar potencies for inhibition of constitutive and inducible NOSs and was initially described as a purely competitive inhibitor of the brain enzyme (Knowles *et al.*, 1990). However, more recent studies revealed that L-NMA induced an irreversible inactivation of inducible (Olken *et al.*, 1991) and brain (Feldman *et al.*, 1993a; Klatt *et al.*, 1994a) NOSs. Inactivation of NOSs is coupled to enzymatic turnover of the inhibitor resulting in the formation of L-citrulline and NO in a 1:1 stoichiometry (Olken and Marletta, 1993; Klatt *et al.*, 1994a). Metabolism of L-NMA by brain NOS requires the presence of Ca^{2+}/calmodulin and is around 50-fold slower than L-arginine oxidation. The routes of L-NMA metabolism are not fully understood. From the HPLC chromatogram shown in Figure 4 we have concluded that the inhibitor is first hydroxylated to N^G-hydroxy-N^G-methyl-L-arginine, followed by demethylation to the NOS substrate N^G-hydroxy-L-arginine Katt *et al.*, 1994a). Chemical modification of critical amino acid residues by formaldehyde, a co-product of the demethylation reaction, could account for the observed irreversibility of enzyme inhibition (Olken and Marletta, 1993). Thus, L-NMA is an alternate substrate and mechanism-based inhibitor ('partial agonist') of both NOS isozymes, possibly explaining reports on NO-mediated relaxation of isolated blood vessels by L-NMA (Archer and Hampl, 1992; Tseng *et al.*, 1993; Winn *et al.*, 1993).

Since L-NMA competes for L-arginine binding but does not block enzymatic oxygen reduction, NADPH oxidation uncouples from NO synthesis upon addition of the inhibitor, resulting in the generation of superoxide and hydrogen peroxide (Heinzel *et al.*, 1992; Pou *et al.*, 1992; Olken and Marletta, 1993). It has not been investigated yet whether L-NMA also induces formation of reactive oxygen species in intact cells, but such an effect could explain, for instance, the unexpected toxicity of the NOS inhibitor in hepatic tissue (Billiar *et al.*, 1990).

2.6.2 N^G-Nitro-L-arginine

There is experimental evidence that mechanisms of NOS inhibition by L-NNA and its methyl ester (L-NAME) differ from the mode of action of the methyl analogue. First, L-NMA blocks constitutive and inducible NOSs with about equal potency, whereas the nitro compounds exhibit a pronounced selectivity for the brain and endothelial isozymes (Furfine *et al.*, 1993; Mayer *et al.*, 1993b). Secondly, L-NNA and L-NAME inhibit uncoupled oxygen reduction, whereas L-NMA does not (Heinzel *et al.*, 1992; Pou *et al.*, 1992; Furfine *et al.*, 1993; Klatt *et al.*, 1993).

The apparently high potency of L-NNA as inhibitor of brain NOS was attributed to irreversible inhibition of the enzyme both *in vitro* and *in vivo* (Dwyer *et al.*, 1991). In accordance with these data we found that low concentrations of L-NNA (1 μM) induced a rapid inactivation of endothelial and brain NOSs, whereas the cytokine-induced macrophage enzyme was not inhibited under these conditions (see Figure 5). Further studies showed that enzyme inactivation was slowly reversed upon removal of the inhibitor due to slow dissociation of L-NNA from the substrate binding domain ($t_{1/2}$ ~10 min at 37°C). It is noteworthy that association of L-NNA was also remarkably slow ($k_{on}=3.6\times10^5$ M^{-1} min^{-1}), pointing to a steric hindrance of its access to the substrate site of NOS (Klatt *et al.*, 1994a).

The mechanisms of L-NNA-induced inactivation of brain NOS are not known. The effect is independent of L-NNA turnover and competitive with L-arginine, reversibility of inhibition correlates well with L-NNA dissociation kinetics, and the drug does not convert NOS homodimers into inactive monomers (Klatt *et al.*, 1994a). Together with the fact that haem-catalysed, L-arginine-independent reduction of oxygen is also blocked by L-NNA, these data suggest that the inhibitor induces a profound conformational change of the catalytic centre of constitutive NOSs (Klatt *et al.*, 1993).

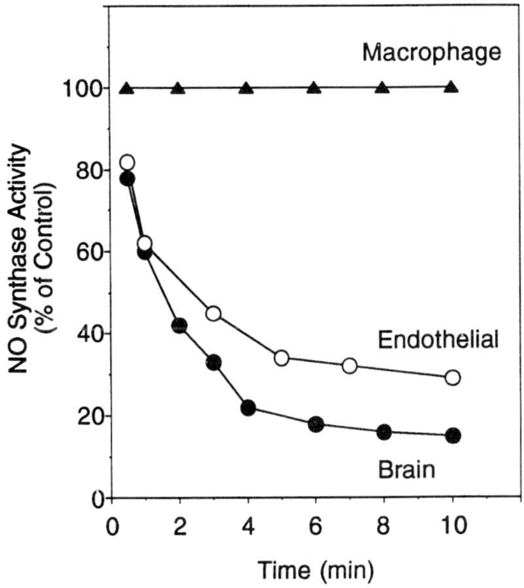

Figure 5 Inactivation of nitric oxide synthases by N^G-nitro-L-arginine. Time courses of L-citrulline formation revealed that 1 μM L-NNA rapidly inactivated the constitutively expressed endothelial and brain isozymes, whereas inducible NO synthase from activated RAW 264.7 cells was not inhibited.

2.6.3 7-Nitro indazole

Moore and colleagues have recently introduced 7-nitro indazole (7-NI) as a novel potent NOS inhibitor with an apparent selectivity for the brain enzyme (Babbedge *et al.*, 1993; Moore *et al.*, 1993a,b). This is particularly interesting because this rather simple compound does not exhibit structural similarities with L-arginine or any of the known NOS cofactors. Nevertheless, inhibition of rat cerebellar NOS by 7-NI was reported to be competitive with L-arginine (Babbedge *et al.*, 1993), and the drug was found to competitively antagonize L-[^{3}H]NNA binding to NOS in rat brain cytosol (Michel *et al.*, 1993a).

We investigated this issue with the purified brain enzyme and found that 7-NI was not only competitive with L-arginine but also with H_4biopterin binding (Klatt *et al.*, 1994b). These data suggest that binding of the drug occurs at a distinct site, which allosterically interacts with the binding domains for both L-arginine and H_4biopterin. 7-NI did not block enzymatic reduction of cytochrome *c* but was a potent inhibitor of uncoupled oxygen reduction (IC_{50}=0.3 μM), indicating that it interferes with haem redox cycling. 7-NI is structurally similar to imidazole, which binds to the prosthetic haem group of NOS (McMillan and Masters, 1993). Since we found that the action profile of imidazole is similar to that of 7-NI, we have proposed that the inhibitory effect of 7-NI is a consequence of its binding to the haem group of NOS, resulting in decreased affinities of the enzyme for L-arginine and H_4biopterin (Klatt *et al.*, 1994b).

The specificity of 7-NI as NOS inhibitor is yet unknown and needs to be investigated. Neither presence of a prosthetic haem group nor requirement for H_4biopterin is a unique feature of NOSs, but the combination of both is certainly without precedent in mammalian systems. Given that this unusual combination of functional domains is requisite for inhibition by 7-NI, future studies could reveal this drug as a highly selective inhibitor of NO synthesis.

Acknowledgements

I wish to thank Drs Friedrich Brunner and Peter Klatt for critical reading of the manuscript and Dr Burghard Heinzel for preparing Figure 3. Experimental work in my laboratory was supported by Grant P 8836 from the Fonds zur Förderung der Wissenschaftlichen Forschung in Österreich.

References

Abu-Soud, H.M. & Stuehr, D.J. (1993) *Proc. Natl. Acad. Sci. USA* **90**, 10769–10772.

Ahlner, J., Andersson, R.G.G., Torfgard, K. & Axelsson, K.L. (1991) *Pharmacol. Rev.* **43**, 351–423.

Archer, S.L. & Hampl, V. (1992) *Biochem. Biophys. Res. Commun.* **188**, 590–596.

Assreuy, J., Cunha, F.Q., Liew, F.Y. & Moncada, S. (1993) *Br. J. Pharmacol.* **108**, 833–837.

Babbedge, R.C., Bland-Ward, P.A., Hart, S.L. & Moore, P.K. (1993) *Br. J. Pharmacol.* **110**, 225–228.

Baek, K.J., Thiel, B.A., Lucas, S. & Stuehr, D.J. (1993) *J. Biol. Chem.* **268**, 21120–21129.

Beckman, J.S., Beckman, T.W., Chen, J., Marshall, P.A. & Freeman, B.A. (1990) *Proc. Natl. Acad. Sci. USA* **87**, 1620–1624.

Billiar, T.R., Curran, R.D., Harbrecht, B.G., Stuehr, D.J., Demetris, A.J. & Simmons, R.L. (1990) *J. Leukocyte Biol.* **48**, 565–569.

Boje, K.M. & Arora, P.K. (1992) *Brain Res.* **587**, 250–256.

Boucher, J.L., Genet, A., Vadon, S., Delaforge, M. & Mansuy, D. (1992) *Biochem. Biophys. Res. Commun.* **184**, 1158–1164.

Bredt, D.S. & Snyder, S.H. (1989) *Proc. Natl. Acad. Sci. USA* **86**, 9030–9033.

Bredt, D.S. & Snyder, S.H. (1990) *Proc. Natl. Acad. Sci. USA* **87**, 682–685.

Bredt, D.S., Glatt, C.E., Hwang, P.M., Fotuhi, M., Dawson, T.M. & Snyder, S.H. (1991a) *Neuron* **7**, 615–624.

Bredt, D.S., Hwang, P.M., Glatt, C.E., Lowenstein, C., Reed, R.R. & Snyder, S.H. (1991b) *Nature* **351**, 714–718.

Bredt, D.S., Ferris, C.D. & Snyder, S.H. (1992) *J. Biol. Chem.* **267**, 10976–10981.

Brüne, B. & Lapetina, E.G. (1991) *Biochem. Biophys. Res. Commun.* **181**, 921–926.

Brüning, G. (1993) *J. Neurosci. Res.* **36**, 580–587.

Buga, G.M., Griscavage, J.M., Rogers, N.E. & Ignarro, L.J. (1993) *Circ. Res.* **73**, 808–812.

Busconi, L. & Michel, T. (1993) *J. Biol. Chem.* **268**, 8410–8413.

Chao, C.C., Hu, S.X., Molitor, T.W., Shaskan, E.G. & Peterson, P.K. (1992) *Immunol.* **149**, 2736–2741.

Chenais, B., Yapo, A., Lepoivre, M. & Tenu, J.P. (1991) *J. Chromatogr.* **539**, 433–441.

Chenais, B., Yapo, A., Lepoivre, M. & Tenu, J.P. (1993) *Biochem. Biophys. Res. Commun.* **196**, 1558–1565.

Cho, H.J., Xie, Q.W., Calaycay, J., Mumford, R.A., Swiderek, K.M., Lee, T.D. & Nathan, C. (1992) *J. Exp. Med.* **176**, 599–604.

Choi, D.W. (1993) *Proc. Natl. Acad. Sci. USA* **90**, 9741–9743.

Clarke, D.D., Lajtha, A.L. & Maker, H.S. (1989) In *Basic Neurochemistry. Molecular, Cellular, and Medical Aspects* (eds Siegel, G.J., Agranoff, B.W., Albers, R.W., and Molinoff, P.B.), pp. 541–564. New York, Raven Press.

Dawson, T.M., Steiner, J.P., Dawson, V.L., Dinerman, J.L., Uhl, G.R. & Snyder, S.H. (1993) *Proc. Natl. Acad. Sci. USA* **90**, 9808–9812.

Degtyarenko, K.N. & Archakov, A.I. (1993) *FEBS Lett.* **332**, 1–8.

Duch, D.S. & Smith, G.K. (1991) *J. Nutr. Biochem.* **2**, 411–423.

Dwyer, M.A., Bredt, D.S. & Snyder, S.H. (1991) *Biochem. Biophys. Res. Commun.* **176**, 1136–1141.

East, S.J. & Garthwaite, J. (1990) *Eur. J. Pharmacol.* **184**, 311–313.

Evans, T., Carpenter, A. & Cohen, J. (1992) *Proc. Natl. Acad. Sci. USA* **89**, 5361–5365.

Feldman, P.L., Griffith, O.W., Hong, H. & Stuehr, D.J. (1993a) *J. Med. Chem.* **36**, 491–496.

Feldman, P.L., Griffith, O.W. & Stuehr, D.J. (1993b) *Chem. Eng. News* **71**, 26–38.

Fiallos-Estrada, C.E., Kummer, W., Mayer, B., Bravo, R., Zimmermann, M. & Herdegen, T. (1993) *Neurosci. Lett.* **50**, 169–173.

Förstermann, U., Pollock, J.S., Schmidt, H.H.H.W., Heller, M. & Murad, F. (1991) *Proc. Natl. Acad. Sci. USA* **88**, 1788–1792.

Fulco, A.J. (1991) *Annu. Rev. Pharmacol. Toxicol.* **31**, 177–203.

Furfine, E.S., Harmon, M.F., Paith, J.E. & Garvey, E.P. (1993) *Biochemistry* **32**, 8512–8517.

Garthwaite, J., Charles, S.L. & Chess-Williams, R. (1988) *Nature* **336**, 385–388.

Geggel, R.L. (1993) *J. Pediatr.* **123**, 76–79.

Giovanelli, J., Campos, K.L. & Kaufman, S. (1991) *Proc. Natl. Acad. Sci. USA* **88**, 7091–7095.

Gorsky, L.D., Förstermann, U., Ishii, K. & Murad, F. (1990) *FASEB J.* **4**, 1494–1500.

Griscavage, J.M., Rogers, N.E., Sherman, M.P. & Ignarro, L.J. (1993) *J. Immunol.* **151**, 6329–6337.

Gross, S.S. & Levi, R. (1992) *J. Biol. Chem.* **267**, 25722–25729.

Guengerich, F.P. (1991) *J. Biol Chem.* **266**, 10019–10022.

Hansel, C., Batchelor, A., Cuenod, M., Garthwaite, J., Knopfel, T. and Do, K.Q. (1992) *Neurosci. Lett.* **142**, 211–214.

Hecker, M., Lückhoff, A. & Busse, R. (1993) *J. Cell. Physiol.* **156**, 571–578.

Heinzel, B., John, M., Klatt, P., Böhme, E. & Mayer, B. (1992) *Biochem. J.* **281**, 627–630.

Herdegen, T., Brecht, S., Kummer, W., Mayer, B., Leah, J., Bravo, R. & Zimmermann, M. (1993) *J. Neurosci.* **13**, 4130–4146.

Hevel, J.M. & Marletta, M.A. (1992) *Biochemistry* **31**, 7160–7165.

Hevel, J.M., White, K.A. & Marletta, M.A. (1991) *J. Biol. Chem.* **266**, 22789–22791.

Hiki, K., Hattori, R., Kawai, C. & Yui, Y. (1992) *J. Biochem.* **111**, 556–558.

Hope, B., Michael, G., Knigge, K. & Vincent, S. (1991) *Proc. Natl. Acad. Sci. USA* **88**, 2811–2814.

Huang, P.L., Dawson, T.M., Bredt, D.S., Snyder, S.H. & Fishman, M.C. (1993) *Cell* **75**, 1273–1286.

Ignarro, L.J., Buga, G.M., Wood, K.S., Byrns, R.E. & Chaudhuri, G. (1987) *Proc. Natl. Acad. Sci. USA* **84**, 9265–9269.

Kawabata, A., Umeda, N. & Takagi, H. (1993) *Br. J. Pharmacol.* **109**, 73–79.

Klatt, P., Heinzel, B., John, M., Kastner, M., Böhme, E. & Mayer, B. (1992a) *J. Biol. Chem.* **267**, 11374–11378.

Klatt, P., Heinzel, B., Mayer, B., Ambach, E., Werner-Felmayer, G., Wachter, H. & Werner, E.R. (1992b) *FEBS Lett.* **305**, 160–162.
Klatt, P., Schmidt, K. & Mayer, B. (1992c) *Biochem. J.* **288**, 15–17.
Klatt, P., Schmidt, K., Uray, G. & Mayer, B. (1993) *J. Biol. Chem.* **268**, 14781–14787.
Klatt, P., Schmidt, K., Brunner, F. & Mayer, B. (1994a) *J. Biol. Chem.* **269**, 1674–1680.
Klatt, P., Schmid, M., Leopold, E., Schmidt, K., Werner, E.R. & Mayer, B. (1994b) *J. Biol. Chem.* **269**, 13861–13866.
Knowles, R.G., Palacios, M., Palmer, R.M.J. & Moncada, S. (1989) *Proc. Natl. Acad. Sci. USA* **86**, 5159–5162.
Knowles, R.G., Palacios, M., Palmer, R.M.J. & Moncada, S. (1990) *Biochem. J.* **269**, 207–210.
Kwon, N.S., Nathan, C.F. & Stuehr, D.J. (1989) *J. Biol. Chem.* **264**, 20496–20501.
Kwon, N.S., Nathan, C.F., Gilker, C., Griffith, O.W., Matthews, D.E. & Stuehr, D.J. (1990) *J. Biol. Chem.* **265**, 13442–13445.
Lamas, S., Marsden, P.A., Li, G.K., Tempst, P. & Michel, T. (1992) *Proc. Natl. Acad. Sci. USA* **89**, 6348–6352.
Leone, A.M., Palmer, R.M.J., Knowles, R.G., Francis, P.L., Ashton, D.S. & Moncada, S. (1991) *J. Biol. Chem.* **266**, 23790–23795.
Lipton, S.A., Choi, Y.B., Pan, Z.H., Lei, S.Z.Z., Chen, H.S.V., Sucher, N.J., Loscalzo, J., Singel, D.J. & Stamler, J.S. (1993) *Nature* **364**, 626–632.
Liu, J., Farmer, J.D., Lane, W.S., Friedman, J., Weissman, I. & Schreiber, S.L. (1991) *Cell* **66**, 807–815.
Lyons, C.R., Orloff, G.J. & Cunningham, J.M. (1992) *J. Biol. Chem.* **267**, 6370–6374.
MacNaul, K.L. & Hutchinson, N.I. (1993) *Biochem. Biophys. Res. Commun.* **196**, 1330–1334.
Marin, P., Lafon-Cazal, M. & Bockaert, J. (1992) *Eur. J. Neurosci.* **4**, 425–432.
Marletta, M.A. (1993) *J. Biol. Chem.* **268**, 12231–12234.
Marletta, M.A., Yoon, P.S., Iyengar, R., Leaf, C.D. & Wishnok, J.S. (1988) *Biochemistry* **27**, 8706–8711.
Matsumoto, T., Pollock, J.S., Nakane, M. & Förstermann, U. (1993) *Devel. Brain Res.* **73**, 199–203.
Mayer. B. (1993) *Semin. Neurosci.* **5**, 197–205.
Mayer, B. & Böhme, E. (1989) *FEBS Lett.* **256**, 211–214.
Mayer, B., Schmidt, K., Humbert, P. & Böhme, E. (1989) *Biochem. Biophys. Res. Commun.* **164**, 678–685.
Mayer, B., John, M. & Böhme, E. (1990) *FEBS Lett.* **277**, 215–219.
Mayer, B., John, M., Heinzel, B., Werner, E.R., Wachter, H., Schultz, G. & Böhme, E. (1991) *FEBS Lett.* **288**, 187–191.
Mayer, B., Klatt, P., Böhme, E. & Schmidt, K. (1992) *J. Neurochem.* **59**, 2024–2029.
Mayer, B., Brunner, F. & Schmidt, K. (1993a) *Biochem. Pharmacol.* **45**, 367–374.
Mayer, B., Schmid, M., Klatt, P. & Schmidt, K. (1993b) *FEBS Lett.* **333**, 203–206.
Mayer, B., Werner, E.R., Leopold, E., Klatt, P. & Schmidt, K. (1994) In *Biology of Nitric Oxide* (eds Feelisch, M., Busse, R. and Moncada, S.). Colchester, Portland Press (in press).
McMillan, K. & Masters, B.S.S. (1993) *Biochemistry* **32**, 9875–9880.
McMillan, K., Bredt, D.S., Hirsch, D.J., Snyder, S.H., Clark, J.E. & Masters, B.S.S. (1992) *Proc. Natl. Acad. Sci. USA* **89**, 11141–11145.
Michel, A.D., Phul, R.K., Stewart, T.L. & Humphrey, P.P.A. (1993a) *Br. J. Pharmacol.* **109**, 287–288.
Michel, T., Li, G.K. & Busconi, L. (1993b) *Proc. Natl. Acad. Sci. USA* **90**, 6252–6256.
Moncada, S., Palmer, R.M.J. & Higgs, E.A. (1991) *Pharmacol. Rev.* **43**, 109–142.
Moore, P.K., Babbedge, R.C., Wallace, P., Gaffen, Z.A. & Hart, S.L. (1993a) *Br. J. Pharmacol.* **108**, 296–297.
Moore, P.K., Wallace, P, Gaffen, Z., Hart, S.L. & Babbedge, R.C. (1993b) *Br. J. Pharmacol.* **110**, 219–224.

Murphy, S., Simmons, M.L., Agullo, L., Garcia, A., Feinstein, D.L., Galea, E., Reis, D.J., Minc-Golomb, D. & Schwartz, J.P. (1993) *Trends Neurosci.* **16**, 323–328.
Nakane, M., Mitchell, J., Förstermann, U. & Murad, F. (1991) *Biochem. Biophys. Res. Commun.* **180**, 1396–1402.
Nichol, C.A., Smith, G.K. & Duch, D.S. (1985) *Annu. Rev. Biochem.* **54**, 729–764.
Olken, N.M. & Marletta, M.A. (1993) *Biochemistry* **32**, 9677–9685.
Olken, N.M., Rusche, K.M., Richards, M.K. and Marletta, M.A. (1991) *Biochem. Biophys. Res. Commun.* **177**, 828–833.
Palacios, M., Knowles, R.G., Palmer, R.M.J. & Moncada, S. (1989) *Biochem. Biophys. Res. Commun.* **165**, 802–809.
Palmer, R.M.J. & Moncada, S. (1989) *Biochem. Biophys. Res. Commun.* **158**, 348–352.
Palmer, R.M.J., Ferrige, A.G. & Moncada, S. (1987) *Nature* **327**, 524–526.
Palmer, R.M.J., Ashton, D.S. & Moncada, S. (1988) *Nature* **333**, 664–666.
Pollock, J.S., Förstermann, U., Mitchell, J.A., Warner, T.D., Schmidt, H.H.H.W., Nakane, M. & Murad, F. (1991) *Proc. Natl. Acad. Sci. USA* **88**, 10480–10484.
Pou, S., Pou, W.S., Bredt, D.S., Snyder, S.H. & Rosen, G.M. (1992) *J. Biol. Chem.* **267**, 24173–24176.
Radi, R., Beckman, J.S., Bush, K.M. & Freeman, B.A. (1991) *Arch. Biochem. Biophys.* **288**, 481–487.
Radomski, M.W., Palmer, R.M.J. & Moncada, S. (1990) *Proc. Natl. Acad. Sci. USA* **87**, 5193–5197.
Renaud, J.P., Boucher, J.L., Vadon, S., Delaforge, M. & Mansuy, D. (1993) *Biochem, Biophys. Res. Commun.* **192**, 53–60.
Rengasamy, A. & Johns, R.A. (1993) *Mol. Pharmacol.* **44**, 124–128.
Rogers, N.E. & Ignarro, L.J. (1992) *Biochem. Biophys. Res. Commun.* **189**, 242–249.
Rossaint, R., Falke, K.J., Lopez, F., Slama, K., Pison, U. & Zapol, W.M. (1993) *New Engl. J. Med.* **328**, 399–405.
Sakai, N., Kaufman, S. & Milstien, S. (1993) *Mol. Pharmacol.* **43**, 6–10.
Schmidt, H.H.H.W. & Murad, F. (1991) *Biochem. Biophys. Res. Commun.* **181**, 1372–1377.
Schmidt, H.H.H.W., Pollock, J.S., Nakane, M., Gorsky, L.D., Förstermann, U. & Murad, F. (1991) *Proc. Natl. Acad. Sci. USA* **88**, 365–369.
Schmidt, H.H.H.W., Smith, R.M., Nakane, M. & Murad, F. (1992) *Biochemistry* **31**, 3243–3249.
Schmidt, K., Werner, E.R., Mayer, B., Wachter, H. & Kukovetz, W.R. (1992) *Biochem. J.* **281**, 297–300.
Schmidt, K., Klatt, P. & Mayer, B. (1993) *Mol. Pharmacol.* **44**, 615–621.
Schmidt, K., Klatt, P. & Mayer, B. (1994) *Biochem. J.* **301**, 313–316.
Schmidt, K., List, B., Klatt, P. & Mayer, B. (1994) *J. Neurochem.* (in press).
Sessa, W.C., Harrison, J.K., Barber, C.M., Zeng, D., Durieux, M.E., Dangelo, D.D., Lynch, K.R. & Peach, M.J. (1992) *J. Biol. Chem.* **267**, 15274–15276.
Sessa, W.C., Barber, C.M. & Lynch, K.R. (1993) *Circ. Res.* **72**, 921–924.
Stuehr, D.J. & Ikeda-Saito, M. (1992) *J. Biol. Chem.* **267**, 20547–20550.
Stuehr, D.J., Kwon, N.S. & Nathan, C.F. (1990) *Biochem. Biophys. Res. Commun.* **168**, 558–565.
Stuehr, D.J., Cho, H.J., Kwon, N.S., Weise, M.F. & Nathan, C.F. (1991a) *Proc. Natl. Acad. Sci. USA* **88**, 7773–7777.
Stuehr, D.J., Fasehun, O.A., Kwon, N.S., Gross, S.S., Gonzales, J.A., Levi, R. & Nathan, C.F. (1991b) *FASEB J.* **5**, 98–103.
Stuehr, D.J., Kwon, N.S., Nathan, C.F., Griffith, O.W., Feldman, P.L. & Wiseman, J. (1991c) *J. Biol. Chem.* **266**, 6259–6263.
Tayeh, M.A. & Marletta, M.A. (1989) *J. Biol. Chem.* **264**, 19654–19658.
Tseng, C.M., Goodman, L.W., Rubin, L.J. & Tod, M.L. (1993) *J. Appl. Physiol.* **74**, 549–558.
Verge, V.M.K., Xu, Z., Xu, X.J., Wiesenfeld-Hallin, Z. & Hökfelt, T. (1992) *Proc. Natl. Acad. Sci. USA* **89**, 11617–11621.

Werner, E.R., Werner-Felmayer, G. & Wachter, H. (1993) *Proc. Soc. Exp. Biol. Med.* **203**, 1–12.
Werner-Felmayer, G., Werner, E.R., Fuchs, D., Hausen, A., Reibnegger, G. & Wachter, H. (1990) *J. Exp. Med.* **172**, 1599–1607.
Werner-Felmayer, G., Werner, E.R., Fuchs, D., Hausen, A., Mayer, B., Reibnegger, G., Weiss, G. & Wachter, H. (1993a) *Biochem. J.* **289**, 357–361.
Werner-Felmayer, G., Werner, E.R., Fuchs, D., Hausen, A., Reibnegger, G., Schmidt, K., Weiss, G. & Wachter, H. (1993b) *J. Biol. Chem.* **268**, 1842–1846.
White, K.A. & Marletta, M.A. (1992) *Biochemistry* **31**, 6627–6631.
Wink, D.A., Osawa, Y., Darbyshire, J.F., Jones, C.R., Eshenaur, S.C. & Nims, R.W. (1993) *Arch. Biochem. Biophys.* **300**, 115–123.
Winn, M.J., Asante, N.K. & Ku, D.D. (1993) *J. Pharmacol. Exp. Ther.* **264**, 265–270.
Wolf, G., Würdig, S. & Schunzel, G. (1992) *Neurosci. Lett.* **147**, 63–66.
Xie, Q.W., Cho, H.J., Calaycay, J., Mumford, R.A., Swiderek, K.M., Lee, T.D., Ding, A.H., Troso, T. & Nathan, C. (1992) *Science* **256**, 225–228.
Yui, Y., Hattori, R., Kosuga, K., Eizawa, H., Hiki, K. & Kawai, C. (1991) *J. Biol. Chem.* **266**, 12544–12547.
Zembowicz, A., Hecker, M., MacArthur, H., Sessa, W.C. & Vane, J.R. (1991) *Proc. Natl. Acad. Sci. USA* **88**, 11172–11176.
Zhang, Z.G., Chopp, M., Zaloga, C., Pollock, J.S. & Förstermann, U. (1993) *Stroke* **24**, 2016–2021.
Zhu, X.Z. & Luo, L.G. (1992) *J. Neurochem.* **59**, 932–935.

CHAPTER 3

THE NO RECEPTOR: CHARACTERIZATION AND REGULATION OF SOLUBLE GUANYLYL CYCLASE

Doris Koesling, Peter Humbert and Günter Schultz

Institut für Pharmakologie, Freie Universität Berlin, Thielallee 67-73, D-14195 Berlin, Germany

Table of Contents

3.1 Introduction 43
3.2 Primary structure of soluble guanylyl cyclase 44
3.3 Catalytic activity 46
3.4 Isoforms 47
3.5 Regulation 48

References 50

3.1 Introduction

NO sensitivity of soluble guanylyl cyclase has been known for almost 20 years, but the physiological significance of the activation by NO was unclear until a few years ago. Guanylyl cyclases catalyse the conversion of GTP to cGMP; the latter is an important intracellular signal molecule involved in the regulation of various cellular events (Waldman and Murad, 1987). The role of cGMP has not been elucidated in most systems, but it is an established mediator of smooth muscle relaxation, an inhibitor of platelet aggregation and involved in retinal signal transduction (Walter, 1989; Lincoln and Cornwell, 1993). In the vision system, cGMP acts by directly operating a cation channel, which is kept open in the darkness by high cGMP levels and which closes as a consequence of the drop of the intracellular cGMP concentration during visual excitement leading to hyperpolarization (Fesenko *et al.*, 1985). The occurrence of the cGMP-operated ion channels appears not to be limited to the retina as there is accumulating evidence for the existence of related channels in a variety of other tissues (Biel *et al.*, 1993). Therefore, in addition to the cGMP-dependent protein kinases and the cGMP-regulated phosphodiesterases, which have

Nitric Oxide in the Nervous System
ISBN 0-12-721985-4

been known as effector systems for cGMP for quite a while, cGMP-operated ion channels represent important target molecules for cGMP as well.

In contrast to the formation of cAMP, which is restricted to the plasma membrane in most mammalian cells, cGMP is formed by membrane-bound and soluble enzymes (Figure 1). The membrane-bound guanylyl cyclases belong to the group of receptor-linked enzymes, which differ in their extracellular ligand binding domains but share a common intracellular catalytic domain. The enzymes, which have one membrane-spanning region, are activated by different peptide hormones (Garbers, 1992). So far four different membrane-bound guanylyl cyclases have been identified in mammals (GC-A, GC-B, GC-C, GC-ret). GC-A is stimulated by the atrial natriuretic peptide ANP, and GC-B displays the highest affinity for the C-type natriuretic peptide CNP. GC-C, which is mainly found in the intestine, is stimulated by an endogenous peptide termed guanylin and by the heat-stable enterotoxin of *Escherichia coli*, which is known to cause diarrhoea. Recently GC-ret, a membrane-bound guanylyl cyclase of the retina, was cloned and sequenced (Shyjan *et al.*, 1992); nothing is known about the physiological ligand of this cyclase. In addition to these known isoforms of membrane-bound guanylyl cyclases, probably more members of this family or receptor-linked enzymes exist. Mainly because of the progress in molecular biological methods, knowledge about membrane-bound and soluble guanylyl cyclases has increased a lot within the last years, and the following chapter tries to give an overview on recent developments on the soluble enzyme, emphasizing structural features.

3.2 Primary structure of soluble guanylyl cyclase

Soluble guanylyl cyclase was purified as a heterodimer consisting of an α and a β subunit with molecular masses of 73 and 70 kDa on SDS PAGE gels (Humbert *et al.*, 1990). The purified enzyme was shown to contain haem as a prosthetic group which has been proposed to act as the acceptor molecule for NO (Gerzer *et al.*, 1981). Using peptide sequences derived from the purified protein, both subunits were cloned and sequenced (Koesling *et al.*, 1988, 1990; Nakane *et al.*, 1988, 1990). So far, two isoforms of soluble guanylyl cyclase subunits have been identified using the polymerase chain reaction. One of the subunits shows more homologies to the β_1 subunit than to the α_1 subunit and was designated β_2 (Yuen *et al.*, 1990), the other one exhibits more similarities to the α_1 than to the β_1 subunit and was termed α_2 (Harteneck *et al.*, 1991). The overall homologies of N-terminal parts of the guanylyl cyclase subunits reveal some but relatively low homology. Noticeable is a stretch of about 100 amino acids in the N-terminal region which reveals a significantly higher degree of identical amino acids between α_1 and α_2 and between β_1 and β_2 than between α_1 and β_1 (Figure 2). One can speculate that this region is responsible for the respective properties of α and β subunits. All subunits share a C-terminal region of about 250 amino acids which is also conserved in the membrane-bound guanylyl cyclases and which is also found in the cytosolic parts of all adenylyl

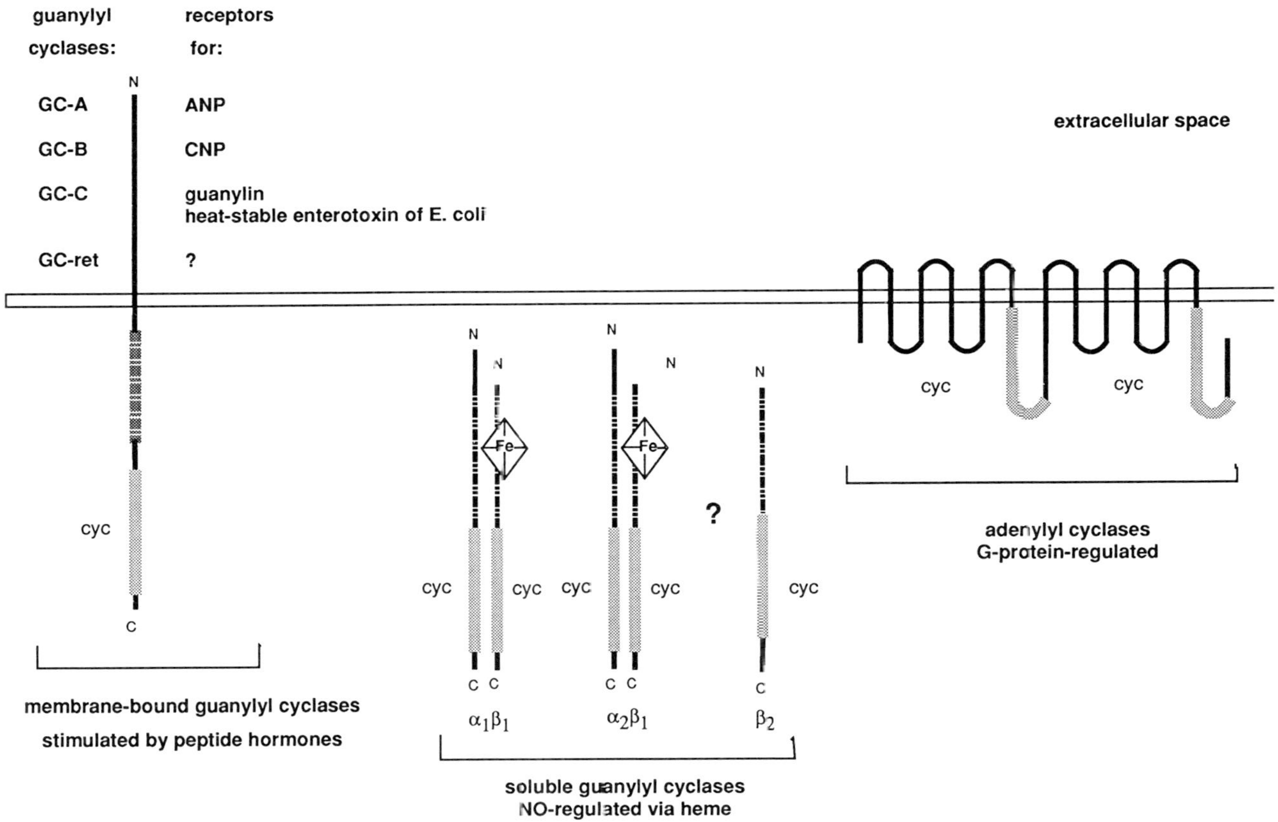

Figure 1 Schematic representation of the overall structures of guanylyl cyclases and adenylyl cyclases. Shown are the different heterodimers of soluble guanylyl cyclase, a representative membrane-bound guanylyl cyclase with different receptor domains and a representative membrane-bound adenylyl cyclase. Labelled boxes indicate sequence homologies. GC-A, membrane-bound guanylyl cyclase type A; GC-B, membrane-bound guanylyl cyclase type B; GC-C, membrane-bound guanylyl cyclase type C; GC-ret, membrane-bound guanylyl cyclase of the retinae type; ANP, natriuretic peptide type A; CNP, natriuretic peptide type C; cyc, putative catalytic domain of guanylyl cyclases and adenylyl cyclases.

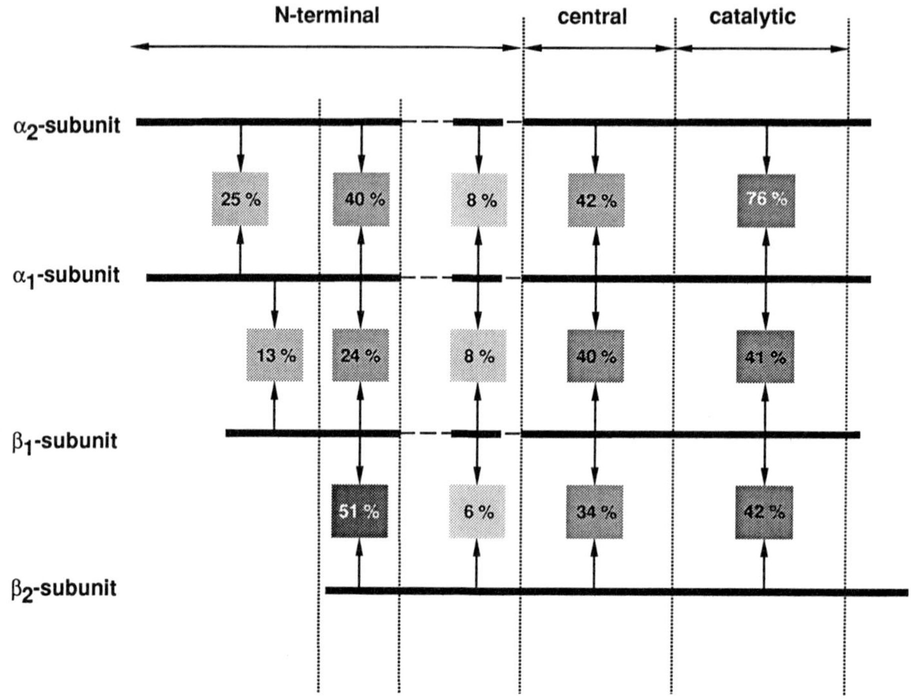

Figure 2 Comparison of the subunits of soluble guanylyl cyclase. The polypeptide chains are shown as a bold line, gaps are indicated by a dotted line. The numbers give the amount of identical amino acids shared between the respective subunits in the regions indicated by vertical lines. The C-termini of the subunits of soluble guanylyl cyclase are unrelated. The regions chosen for comparison are: amino acids 1–161 of the α_2, 1–126 of the α_1, and 1–56 of the β_1 subunit; 162–268 of the α_2, 127–234 of the α_1, 57–164 of the β_1, and 1–101 of the β_2 subunit; 269–321 of the α_2, 235–282 of the α_1, 165–212 of the β_1, and 102–208 of the β_2 subunit; 322–483 of the α_2, 283–441 of the α_1, 213–383 of the β_1, and 205-368 of the β_2 subunit; 484–698 of the α_2, 442–660 of the α_1, 384–605 of the β_1, and 369–581 of the β_2 subunit.

cyclases sequenced so far (see Figure 1). There are some weak similarities between the putative catalytic domains of guanylyl cyclases and the corresponding regions of adenylyl cyclase from *Saccharomyces cerevisiae* (Kataoka *et al.*, 1985) and *Schizosaccharomyces pombe* (Yamawaki-Kataoka *et al.*, 1989). In contrast, the catalytic domains of the adenylyl cyclase from *Rhizobium meliloti* are highly homologous to the guanylyl cyclase catalytic domain (Beuve *et al.*, 1990).

3.3 Catalytic activity

Expression experiments with a truncated form of the membrane-bound GC-A just containing the putative catalytic domain demonstrated that the region conserved in all cyclases indeed is capable of cyclic nucleotide formation (Thorpe and Morkin, 1990). Expression experiments of the soluble enzyme revealed that an α and β subunit are needed for a catalytically active enzyme, as co-expression of the α_1 and β_1 subunits and of the α_2 and β_1 subunits yielded enzymes with catalytic activity (Harteneck *et al.*, 1990). These results suggest that at least two putative catalytic domains are essential for enzymatic activity. The requirement of two catalytic domains is supported by the existence of two putative catalytic domains in the adenylyl cyclases and by the finding that the membrane-bound guanylyl cyclases exist at least as a dimer even in the absence of the ligand (Chinkers and Wilson, 1992). In addition, the catalytically active truncated form of the membrane-bound guanylyl cyclases mentioned above was shown to exist as a dimer. Therefore, dimerization of two catalytic domains appears to be a prerequisite for cyclase activity, but it is still unknown whether both catalytic domains participate in the formation of one catalytic centre or if they both are able to bind and convert the nucleotide, although kinetic studies with the purified enzyme did not indicate different active sites.

3.4 Isoforms

Two isoforms of guanylyl cyclase subunits, α_2 and β_2, have been identified on the cDNA level using homology screening, but to date the corresponding proteins have not been detected. The so called α_3 and β_3 subunits (Giuili *et al.*, 1992) represent the human variants of the α_1 and β_1 subunits rather than new isoforms of the subunits, and shifts in the reading frame probably account for differences between the α_3 and α_1 subunits. Co-expression of the α_2 subunit with the β_1 subunit yields a cGMP-forming enzyme and, as the $\alpha_2\beta_1$ heterodimer is stimulated by NO, it is very likely that this heterodimer also contains haem as a prosthetic group, although the haem content of this heterodimer has not been proven. The finding that co-expressed α_2 and β_1 subunit exhibit enzymatic activity shows the interchangeability of the α subunits, although it is not known whether the $\alpha_2\beta_1$ heterodimer physiologically exists, and the physiological meaning of two similarly regulated heterodimers is unclear. Possibly, the α_1 and α_2 subunits are expressed in tissue-specific manner and both serve the same function, i.e. dimerization with the β_1 subunit to form a catalytically active enzyme. Because of low levels of mRNA coding for the guanylyl cyclase subunits, the tissue distribution of the isoforms is difficult to investigate. Relatively high levels of mRNA coding for the β_1 and α_1 subunits are found in brain and lung.

In contrast to the α_2 subunit, the β_2 subunit may differ from the other guanylyl cyclase subunits concerning cellular localization as it contains an isoprenylation consensus sequence; although isoprenylation of the protein has not been demonstrated, it appears to be possible that the β_2 subunit is attached to the membrane via the isoprenoid group. Compared to the other subunits, the β_2 subunit shows some structural differences since it lacks the first 62 amino acids of the β_1 subunit but extends 86 amino acids beyond the C-terminus of the β_1 subunit. For unknown reasons, the β_2 subunit does not form a catalytically active enzyme with either one of the α subunits. The β_2 subunit may need another not yet identified subunit to form an active heterodimer, or the β_2 subunit may represent in inactive subunit, which for example may dimerize with the α_1 subunit, thereby inhibiting the formation of catalytically active $\alpha_1\beta_1$ heterodimers.

3.5 Regulation

Since the early seventies, regulation of intracellular cGMP concentration has been demonstrated by various hormones and neurotransmitters in intact cells. The failure of hormonal guanylyl cyclase stimulation in broken cell preparations led to the proposal of the existence of an intracellular stimulator of the soluble enzyme, and intermediates of eicosanoid metabolism were considered as good candidates for a long time (Waldman and Murad, 1987). In addition, reagents modulating the oxidative/reductive state of thiols were found to influence the activity of the enzyme (Böhme *et al.*, 1983).

Recently, carbon monoxide (CO), like NO a normally noxious gas, which is known to stimulate the soluble guanylyl cyclase several-fold (Brüne *et al.*, 1992), has been suggested to act as a physiological activator, mainly because haem oxygenase-2, a CO-forming enzyme, co-localizes with the enzyme throughout the brain (Verma *et al.*, 1993). Haem oxygenase releases CO in the process of degrading haem to biliverdin. This appears to be a rather costly way to produce a signal molecule, but other biological pathways of CO synthesis may be discovered.

In the late seventies, NO-containing compounds were shown to be potent activators of soluble guanylyl cyclase (Böhme *et al.*, 1978; Murad *et al.*, 1978) although the physiological implication of this stimulation was unclear until a few years ago, when endogenous enzymatically produced NO was found (see Chapters 1 and 2). Soluble guanylyl cyclase contains haem as a prosthetic group, and the finding that the haem-free enzyme exhibits basal activity but is not stimulated by NO led to the proposal that enzyme-bound haem acts as the receptor molecule for NO, transducing the binding signal to activation of the cyclase catalytic domain (Gerzer *et al.*, 1981). The haem is apparently not covalently bound as it can be removed from the haem-containing enzyme and also can be reinserted into a haem-free enzyme, thereby restoring the stimulation by NO (Ignarro *et al.*, 1982a). The stimulation of soluble guanylyl cyclase is initiated by NO binding to the sixth coordination position of the haem iron. Binding of NO is supposed to pull the iron out of the plane of the porphyrin ring, leading to the loss of the protein-iron bond and resulting in activation of the enzyme (Traylor and Sharma, 1992). The finding that protoporphyrin IX, a haem without the central iron, is a potent NO-haem-independent activator of the enzyme (Ignarro *et al.*, 1982b) is in good accordance

α haemoglobin chain	L	S	A	L	S	D	L	H	A	man
	L	S	E	L	S	D	L	H	A	cow
	L	S	A	L	S	D	L	H	A	platypus
	L	S	K	L	S	D	L	H	A	chicken
β haemoglobin chain	F	A	T	L	S	E	L	H	C	man
	F	A	A	L	S	E	L	H	C	cow
	F	A	K	L	S	E	L	H	C	platypus
	F	S	Q	L	S	E	L	H	C	chicken
myoglobin	I	K	P	L	A	Q	S	H	A	man
	V	K	H	L	A	E	S	H	A	cow
	L	K	P	L	A	Q	S	H	A	platypus
	L	K	P	L	A	Q	T	H	A	chicken
soluble guanylyl cyclase	L	Q	N	L	D	A	L	H	D	β_1-subunit
	I	E	N	L	D	A	L	H	S	β_2-subunit

Figure 3 Vicinities of the proximal histidines in different haemo- and myoglobin chains. Shown are the amino acids preceding the proximal histidine of the α and β haemo- and myoglobin chains of the indicated species. Marked are the proximal histidines and the leucines in the −4 position, which are conserved in all vertebrate haemo- and myoglobin types. The last two lines show the amino acids of the β_1 (99–106) and β_2 subunits (36–44) with the putative proximal histidine 105 of the β_1 subunit which is conserved in the β_2 subunit. For further explanations see text.

with the proposed mechanism of activation. Recently, it was shown that the histidine 105 of the β_1 subunit is essential for stimulation by NO as the substitution of this residue by phenylalanine yielded a catalytically active guanylyl cyclase which was not stimulated by NO (Wedel *et al.*, 1994). Investigations of the haem content of the mutant revealed a loss of enzyme-bound haem and, thus, histidine 105 is apparently involved in binding of the haem moiety. Structural analogies of the His 105 of the β_1 subunits to the proximal histidine of the haemo- and myoglobin chains, i.e. the leucine in the −4 position (Figure 3), suggest that this histidine possibly represents the 'proximal histidine' of soluble guanylyl cyclase, forming the linkage to the fifth coordination position of the haem iron. Future studies have to confirm the rupture the histidine–iron bond during activation of the enzyme and have to show whether the histidine plays an important role in the activation, e.g. by interacting with the substrate or with the catalytic centre of the enzyme.

References

Beuve, A., Boesten, B., Cranier, M., Danchin, A. & O'Gara, F. (1990) *J. Bacteriol.* **172**, 2614–2621.

Biel, M., Altenhofen, W., Hullin, R., Ludwig, J., Freidel, M., Flockerzi, V., Dascal, N., Kaupp, U.B. & Hofmann, F. (1993) *FEBS Lett.* **329**, 134–138.

Böhme, E., Graf, H. & Schultz, G. (1978) *Adv. Cycl. Nucl. Res.* **9**, 131–143.

Böhme, E., Gerzer, R., Grossmann, G., Herz, J., Mülsch, A., Spies, C. & Schultz, G. (1983) *Horm. Cell Regul.* **7**, 147–161.

Brüne, B., Schmidt, K.-U. & Ulrich, V. (1990) *Eur. J. Biochem.* **192**, 683–688.

Chinkers, M. & Garbers, D.L. (1991) *Annu. Rev. Biochem.* **60**, 553–575.

Chinkers, M. & Wilson, E.M. (1992) *J. Biol. Chem.* **267**, 18589–18597.

Fesenko, E.E., Kolesnikov, S.S. & Lyubarsky, A.L. (1985) *Nature* **313**, 310–313.

Garbers, D.L. (1992) *Cell* **72**, 1–4.

Gerzer, R., Böhme, E., Hofmann, F. & Schultz, G. (1981) *FEBS Lett.* **132**, 71–74.

Giuili, G., Scholl, U., Bulle, F. & Guellaen (1992) *FEBS Lett.* **304**, 83–88

Harteneck, Ch., Koesling, D., Söling, A., Schultz, G. & Böhme, E. (1990) *FEBS Lett.* **272**, 221–223.

Harteneck, Ch., Wedel, B., Koesling, D., Malkewitz, J., Böhme, E. & Schultz, G. (1991) *FEBS Lett.* **292**, 217–222.

Humbert, P., Niroomand, F., Fischer, G., Mayer, B., Koesling, D., Hinsch, K.-H., Gausepohl, H., Frank, R., Schultz, G. & Böhme, E. (1990) *Eur. J. Biochem.* **190**, 273–278.

Ignarro, L.J., Degnan, J.N., Baricos, W.H., Kadowitz, P.J. & Wolin, M.S. (1982a) *Biochim. Biophys. Acta* **781**, 49–59.

Ignarro, L.J., Wood, K.S. & Wolin, M.S. (1982b) *Proc. Natl. Acad. Sci. USA* **79**, 2870–2873.

Kataoka, T., Broek, D. & Wigler, M. (1990) *Cell* **43**, 493–505.

Koesling, D., Herz, J., Gausepohl, H., Niroomand, F., Hinsch, K.-D., Mülsch, H., Böhme, E., Schultz, G. & Frank, R. (1988) *FEBS Lett.* **239**, 29–34.

Koesling, D., Harteneck, Ch., Humbert, P., Bosserhoff, A., Frank, R., Schultz, G. & Böhme, E. (1990) *FEBS Lett.* **266**, 128–132.

Lincoln, T.M. & Cornwell, T.L. (1993) *FASEB J.* **7**, 328–338.

Maines, M.D. (1988) *FASEB J.* **2**, 2557–2568.

Murad, F., Mittal, C.K., Arnold, W.P., Katsuki, S. & Kimura, H. (1978) *Adv. Cycl. Nucleotide Res.* **9**, 175–204.

Nakane, M., Saheki, S., Kuno, T., Ishii, K. & Murad, F. (1988) *Biochem. Biophys. Res. Commun.* **157**, 1139–1147.

Nakane, M., Arai, K., Saheki, S., Kuno, T., Buechler, W. & Murad, F. (1990) *J. Biol. Chem.* **265**, 16841–16845.

Shyjan, A.W., de Sauvage, F.J., Gillett, N.A., Goeddel, D.V. & Lowe, D.G. (1992) *Neuron* **9**, 727–737.

Thorpe, D.S. & Morkin, E. (1990) *J. Biol. Chem.* **265**, 14717–14720.

Traylor, T.G. & Sharma, V.S. (1992) *Biochemistry* **31**, 2847–2849.

Verma, A., Hirsch, D.J., Glatt, C.E., Ronnett, G.V. & Snyder, S.H. (1993) *Science* **259**, 381–384.

Waldman, S.A. & Murad, F. (1987) *Pharmacol. Rev.* **39**, 163–189.

Walter, U. (1989) *Rev. Physiol. Biochem. Pharmacol.* **113**, 42–88.

Wedel, B., Humbert, P., Harteneck, C., Foerster, J., Malkewitz, J., Böhme, E., Schultz, G. & Koesling, D. (1994) *Proc. Natl. Acad. Sci. USA* **91**, 2592–2596.

Yamawaki-Kataoka, Y., Tamaoki, T., Choe, H.R., Tanaka, H. & Kataoka, T. (1989) *Proc. Natl. Acad. Sci. USA* **86**, 5693–5697.

Yuen, P.S.T., Potter, L.R. & Garbers, D.L. (1990) *Biochemistry* **29**, 10872–10878.

CHAPTER 4

CYCLIC GMP RECEPTOR PROTEINS: ROLE IN NERVOUS SYSTEM AND OTHER TISSUES

Thomas M. Lincoln

Department of Pathology, Division of Molecular and Cellular Pathology, University of Alabama at Birmingham, Birmingham, AL 35294, USA

Table of Contents

4.1 Cyclic GMP-dependent protein kinases 51
4.1.1 Structure of cGMP kinases 52
4.1.2 Regulation of activity of cGMP kinase 58
4.1.3 Distribution of cGMP kinases 58
4.1.4 Substrates for cGMP kinase 59
4.1.5 Role of kinase and substrate localization in the cell 61
4.1.6 Biologic roles of the cGMP kinase 62
4.1.7 Cross-activation of cyclic nucleotide-dependent protein kinases 63
4.2 Cyclic GMP-regulated ion channels 64
4.2.1 The rod photoreceptor cGMP-gated channel 65
4.2.2 Renal cGMP-regulated ion channels 67
4.2.3 Cyclic GMP-regulated olfactory epithelium ion channels 68
4.3 Cyclic GMP-regulated phosphodiesterases 69
4.3.1 Phosphodiesterases that are regulated by cGMP 69
4.3.2 Properties of cGMP-binding phosphodiesterases 73
4.4 Role of cyclic GMP kinase in the nervous system 74
4.4.1 Long-term potentiation and depression 75
4.4.2 Cyclic GMP kinase in neurons 76
4.4.3 Effects of cGMP on neuronal ion fluxes 77

References 78

4.1 Cyclic GMP-dependent protein kinases

The cellular effects of cyclic GMP (cGMP) appear to be mediated by several types of cGMP receptor proteins (Lincoln and Cornwell, 1993). Perhaps the best

Nitric Oxide in the Nervous System
ISBN 0-12-721985-4

characterized cGMP receptor protein is the cGMP-dependent protein kinase (cGMP kinase), which is in reality a class of closely related enzymes. The two other major classes of receptor proteins for cGMP are the cGMP-regulated ion channels and the cGMP-binding phosphodiesterases. Each receptor protein has been shown to mediate some of the physiologic actions of cGMP in specific cell types. It is also possible that a specific cell type may have more than one type of cGMP receptor protein expressed. The capacity for cGMP to interact with each of these different types of receptor proteins raises the interesting possibilities that cGMP may *increase* protein phosphorylation in the cell by activating cGMP kinases, *decrease* protein phosphorylation in the cell by activating cAMP hydrolysis, or perhaps even have *no effect* on protein phosphorylation in the cell by regulating monovalent cation fluxes. Even the cyclic nucleotide-dependent protein kinases are not specific for their cyclic nucleotides; thus, cGMP can activate the cAMP kinase in some circumstances and vice versa.

4.1.1 Structure of cGMP kinases

Cyclic GMP kinases are serine/threonine protein kinases which belong to the very large protein kinase family (see Hofmann *et al.*, 1992; Lincoln and Cornwell, 1993, for recent reviews). The enzyme has been isolated from several mammalian sources, arthropods and unicellular organisms including *Paramecium*, *Tetrahymena* and *Dictyostelium*. In addition, two cGMP kinase genes have been cloned from *Drosophila* that together encode at least six different cDNAs (Kalderon and Rubin, 1989). There exist in mammalian cells two types of cGMP kinase: a Type I kinase which is comparatively abundant in tissues such as those containing smooth muscle as well as the Purkinje cells of the cerebellum and platelets (Wernet *et al.*, 1989; Sandberg *et al.*, 1989), and a Type II kinase which is more limited in distribution and is expressed primarily in the intestinal epithelium and brain (Uhler, 1993). The Type I kinase is expressed as two isoforms: an α and a β isoform (Lincoln *et al.*, 1988; Wolfe *et al.*, 1989). These two forms appear to be derived by alternate splicing of mRNA encoding the amino-terminal region (Wernet *et al.*, 1989; Francis *et al.*, 1989). Thus, the first 89 residues of the Type Iα kinase and the first 104 residues of the Type Iβ kinase are different. The rest of the sequence of the Type I kinase is identical. The Type II kinase, which is slightly larger than the Type I kinase, is highly homologous with the latter particularly in the carboxyl-terminal catalytic domain and the cGMP binding domain (Figure 1). The variability in the sequences of the cGMP kinases is most evident at the amino-terminal domain. The role for the amino-terminal domain in cGMP kinase function will be discussed below. It appears that cGMP kinase may be one of the most highly conserved proteins inasmuch as the human and bovine Type Iβ enzymes differ by only a few amino acid residues.

The overall structure of the cGMP kinase shown in Figure 1 can be subdivided into an amino-terminal domain (A), cGMP binding domains (B and C) and a catalytic domain (D). The cGMP kinases from mammalian sources exist as homod-

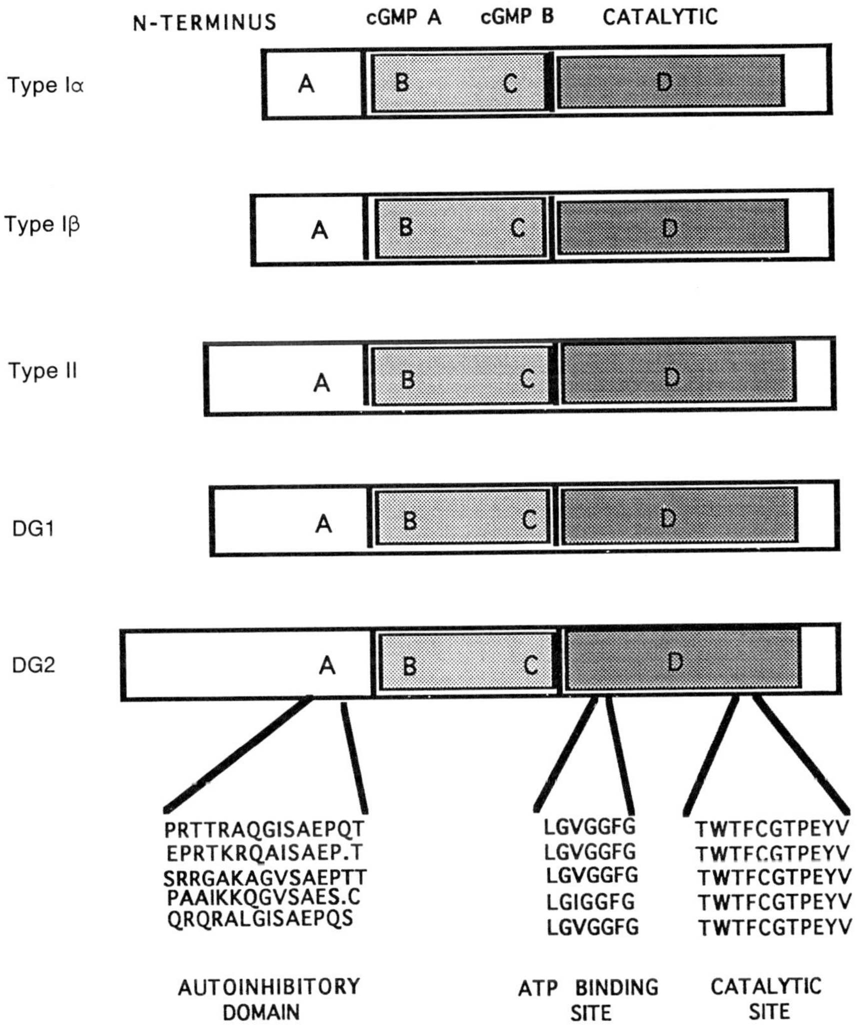

Figure 1 Structure of cyclic GMP-dependent protein kinases. The enzymes isolated from mammalian sources (Types Iα and Iβ, Type II) and the *Drosophila* enzymes (DG1 and DG2) have their domains aligned and homology indicated. The positions of the autoinhibitory domains, ATP binding sites, and catalytic sites are shown. Substantial sequence homology is observed in the N-terminus of the Types I and II enzymes in the region of the leucine zipper motif (residues 1–30 of the Iα kinase ≈50%, but not with the rest of the N-terminal domain). Types Iα and β are identical after residue 104. Reproduced with permission from the R.G. Landes Company, Austin.

imers, presumably linked together through a leucine/isoleucine zipper motif located within approximately the first 30–50 residues in the amino-terminal domain (Landgraf *et al.*, 1990; Atkinson *et al.*, 1991). Two moles of cGMP are bound per one mole of monomer, although the sequences for the two binding sites are not

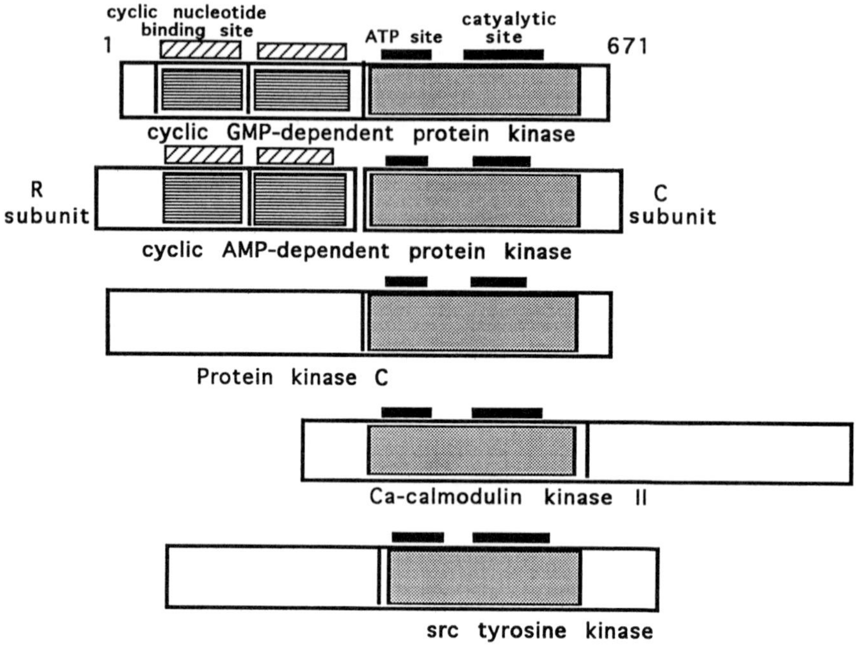

Figure 2 Schematic of structures of several different protein kinases based on cDNA cloning. The conserved portions are noted with similar boxed patterns. Sites for the cyclic nucleotide binding sites, ATP binding sites, and catalytic sites for these different protein kinases are noted. The numbering scheme (1–671) is based on the structure of cyclic GMP-dependent protein kinase Iα. Reproduced with permission from the R.G. Landes Company, Austin.

identical within the monomer. Several studies have indicated that the dissociation rates for cGMP from the two binding sites are quite different, suggesting that they act in concert with each other such that occupation of the higher affinity site (B) increases the affinity for cGMP binding to the lower affinity site (A) (Corbin *et al.*, 1986; Doskeland *et al.*, 1987). This positive cooperativity in cGMP binding has physiologic significance in that the kinase appears 'primed' to be activated by elevations in intracellular cGMP due to the occupation of site B by basal levels of cGMP.

As illustrated in Figure 2, the cGMP kinases display a high degree of homology with the cAMP kinase regulatory (R) and catalytic (C) subunits when the sequences are compared by aligning the R and C subunits along the axis of the cGMP kinase subunit. Similarities between these two kinases were noted even before the amino acid sequences were determined and the cDNAs encoding the proteins cloned (Gill, 1977; Lincoln and Corbin, 1977). Thus, the cyclic nucleotide-dependent protein kinases were the first kinases shown to be homologous proteins and part of a much larger family of enzymes. It is likely that the cGMP kinase has evolved by the fusion of the R and C subunit genes followed by specific mutations that allowed for the selective binding of cGMP substrate specificity. If this situation is correct, then cGMP kinase would not appear to be the more ‘primitive’ of the two kinases as suggested in some of the older literature. The highest degrees of homology between the cyclic nucleotide-dependent protein kinases are seen in the catalytic domains (≃70%) and in the regulatory domains where the cyclic nucleotide binding sites are located (≃50%). The structures of other representative protein kinases deduced from cDNA cloning are also aligned with the cyclic nucleotide-dependent enzymes, and all demonstrate a significant amount of homology with the cGMP kinase in the catalytic domain (Figure 2). As shown in Figure 3, several invariant residues have been identified in all protein kinase catalytic domains, as have others that are highly conserved substitutions. In particular the lysyl residue at position 546 in the Type Iα cGMP kinase is conserved in all protein kinases and appears to be critical for the phosphoryl transfer. The crystal structure for the C subunit of the cAMP kinase has recently been determined (Knighton *et al.*, 1991a,b). Based on the coordinates of the cAMP kinase, the cGMP kinase consists of two lobes: an upper lobe which contains the ATP binding site within a ‘glycine loop’ (^{366}LGVGGFG^{371}) and a lower lobe containing binding sites for the protein or peptide substrate for the kinase. Modelling of other protein kinases based on this structure suggests that all these enzymes share a similar overall topology.

In contrast to the catalytic domains, the regulatory domains of the cyclic nucleotide-dependent protein kinases share little or no sequence similarities with the regulatory domains of the other representative protein kinases. This is not unexpected due to the different modes of regulation of the various protein kinases by ligands and other effector molecules. The cyclic nucleotide binding domains of both the cyclic nucleotide-dependent protein kinases, on the other hand, demonstrate a significant degree of homology with the cAMP-binding catabolite activator protein (CAP) from *Escherichia coli* (Weber *et al.*, 1987). As shown in Figure 4, the cAMP binding sites in the R subunit and the cGMP binding sites in the regulatory domain of cGMP kinase have sequences conserved with those in the CAP protein. Based on the crystal structure of the CAP protein, it has been possible to determine which residues are critical for the interaction of cyclic nucleotides with cyclic nucleotide binding sites (Weber *et al.*, 1989). The high affinity binding of cGMP to the cGMP kinase has been proposed to involve a major substitution of threonyl residues (177 and 301 in the Type Iα isoform of the cGMP kinase for alanyl residues 210 and 334 in the bovine Type I R subunit of the cAMP kinase). Shabb *et al.* (1990, 1991) have suggested that threonine-177 and -301 of the Type Iα cGMP kinase hydrogen-bond to the 2-amino group of the guanine ring of cGMP. Cyclic AMP, which lacks a 2-amino group on the adenine ring, would not be able to form a hydrogen bond with threonine. When the corresponding alanine in the R subunit is mutated to a threonine, the R subunit binds cGMP with a similar affinity to that of cAMP. These results suggest that threonyl residues in this position of cyclic nucleotide binding sites on proteins predict selectivity for cGMP binding.

```
cGK    SDFNIIDTLGVGGFGRVELV QLKSEESKTFAMKILKKRHI VDTRQQEHIRSEKQIMQGAH -SDFIVRLYRTFKDS
cAK    DQFERIKTLGIGSFGRVMLV K-HMETGNHYAMKILDKQKV VKLKQIEHTLNEKRILQAVN -FPFLVKLEFSFKDN
PKC    TDFNFLMVLGKGSFGKVMLA D-RKGTEELYAIKILKKDVV IQDDDVECTMVEKRVLALLD KPPFLTQLHSCFQTV
CAM    EEYQLFEELGKGAFSVVRRC V-KVLAGQEYAAKIINTKK- LSARDHQKLEREARICRLLK -HPNIVRLHDSISEE
SRC    ESLRLEVKLGQGCFGEVWMG T-WNGTTRV-AIKTLKP--- -GTMSPEAFLQEAQVMKKLR -HEKLVQLYAVVSEE

cGK    KYLYMLMEACLGGELWTILR -DRGSF-EDSTTRFYTACVV EAFAYLHSKGIIYRDLKPEN LILDHR---GYAKLV
cAK    SNLYMVMEYVPGGEMFSHLR -RIGRF-SEPHARFYAAQIV LTFEYLHSLDLIYRDLKPEN LLIDQQ---GYIQVT
PKC    DRLYFVMEYVNGGDLMYHIQ -QVGKF-KEPQAVFYAAEIS IGLFFLHKRGIIYRDLKLDN VMLDSE---GHIKIA
CAM    GHHYLIFDLVTGGELFEDIV -AREYY-SEADASHCIQQIL EAVLHCHQMGVVHRDLKPEN LLLASKLKGAAVKLA
SRC    P-IYIVTEYMSKGSLLDFLK GEMGKYLRLPQLVDMAAQIA SGMAYVERMNYVHRDLRAAN ILVGENL---VCKVA

cGK    DFGFAK-KIGFGKKTWTFCG TP-EYVAPEIILNKGHDISA DYWSLGILMYELLT-GSPPF SGPDPMKTYNIILRG
cAK    DFGFAK--RVKG-RTWTLCG TP-EYLAPEIILSKGYNKAV DWWALGVLIYEMAA-GYPPF FADQPIQIYEKIVSG
PKC    DFGMCKEHMMDGVTTRTFCG TP-DYIAPEIIAYQPYGKSV DWWAYGVLLYEMLA-GQPPF DGEDEDELFQSIMEH
CAM    DFGLAIEVEGEQQAWFGFAG TP-GYLSPEVLRKDPYGKPV DLWACGVILYILLV-GYPPF WDEDQHRLYQQIKAG
SRC    DFGLARLIEDNEYTARQGAK FPIKWTAPEAALYGRFTIKS DVWSFGILLTELTTKGRVPY PGMVNREVLDQVERG

cGK    IDMIEFP--KKIAKNAANLI KKLCRDNPSERL-------- ------------------G NLK--NGVKDIQKHK
cAK    --KVRFP--SHFSSDLKDLL RNLLQVDLTKRF-------- ------------------G NLK--DGVNDIKNHK
PKC    --NVSYP--KSLSKEAVSIC KGLMTKHPAKRL-------- ------------------G CGP--EGERDVREHA
CAM    AYDFPSPEWDTVTPEAKDLI NKMLTINPSKRITAAEALKH PWISHRSTVASCMHRQETVD CLKKFNARRKLKGAI
SRC    -YRMPCP--PECPESLHDLM CQCWRKDPEERP-------- ----------------TFE YLQ--AFLEDYFTST

cGK    W--------FEGFNW----- ------------------E GLRKGTL------------T PPIIP--SV--AS--
cAK    W--------FATTDW----- ------------------I AIYQRKV------------E APFIP--KF--KG--
PKC    F--------FRRIDW----- ------------------E KLENREI------------Q PPFKP--KVCGKG--
CAM    LTTMLATRNFSGGKSGGNKK NDGVKESSESTNTTIEDEDT KVRKQEIIKVTEQLIEAISN GDFESYTKMCDPGMT
SRC    E--------PPVPAW----- -----------------RE PIGLELL------------L APEAS---LWGTG--

cGK    ---PT---------DTSNF- --D----------S---FPE -----DN------DEPP--- ----------PDDNS
cAK    ---PG---------DTSNF- --D----------D---YEE -----EE------IRVS--- ----------INEKC
PKC    ---AE---------NFDKFF TRG----------QPVLTPP -----DQ------LVIA--- ----------NIDQS
CAM    AFEPEALGNLVEGLDFHRFY FENLWSRNSKPVHTTILNPH IHLMGDESACIAYIRITQYL DAGGIPRTAQSEETR
SRC    AWLRA---------EGPRF- --G----------E---QPQ -----SR------MWHG--- ----------EVSGA
```

Figure 3 Sequence alignments of representative protein kinases illustrating conserved residues. The catalytic domains are illustrated in this figure for the following representative kinases: cGK, cyclic GMP-dependent protein kinase (bovine Type I), cAK, catalytic subunit of cyclic AMP-dependent protein kinase (bovine α), PKC, protein kinase C (bovine α), CAM, calmodulin-dependent kinase II (rat α subunit), src (bovine c-src protein tyrosine kinase). Reproduced with permission from the R.G. Landes Company, Austin.

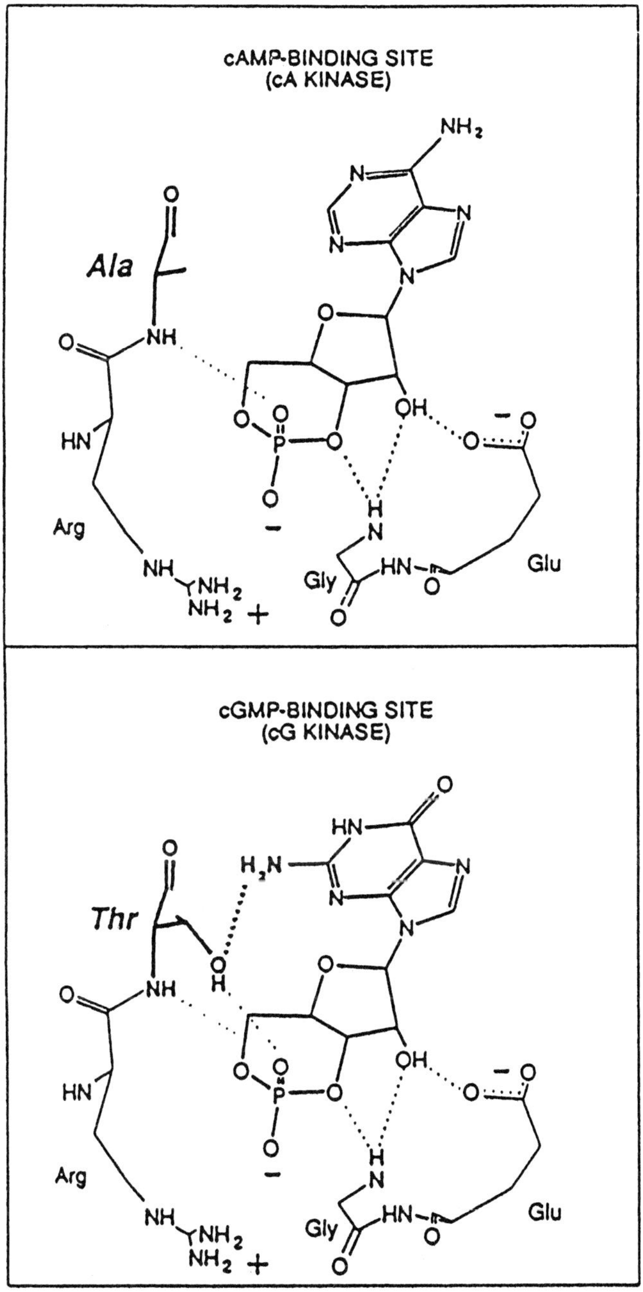

Figure 4 Structure of cyclic AMP binding sites (top panel) and cyclic GMP binding sites (bottom panel) in cyclic nucleotide receptor proteins. The model is predicted on the three-dimensional structure of the binding pocket of the bacterial CAP protein. All cyclic GMP-selective binding pockets have threonyl residues located in the position occupied by alanine in cyclic AMP binding proteins. From Shabb *et al.* (1991).

4.1.2 Regulation of activity of cGMP kinase

Several components in the regulation of cGMP kinase activity have been described. Cyclic GMP kinase contains within the amino-terminal domain an autoinhibitory site whose function it is to bind to and inhibit the catalytic domain (Lincoln *et al.*, 1978). The autoinhibitory domain is believed to 'compete' with substrates at the catalytic domain and thus has also been dubbed the substrate analogue region. Inspection of the autoinhibitory sequences for both the Type Iα and Iβ isoforms and Type II demonstrates that the sequences ^{59}RAQGISAEP, ^{74}KRQAISAEP and ^{118}RRGAKAGVSAEP, respectively, partially mimic phosphorylation sites in protein substrates. Some of the earliest studies on the mode of activation of cyclic nucleotide-dependent protein kinases demonstrated that alterations of residues in the autoinhibitory domain produced constitutive activation of both kinases (Corbin *et al.*, 1978; Lincoln *et al.*, 1978). Indeed from these early studies with the cAMP and cGMP kinases was derived the whole concept of autoinhibitory domains of all protein kinases including protein kinase C and protein tyrosine kinases.

One of the major differences between the Type Iα and Iβ isoforms of cGMP kinase is the concentration of cGMP needed to activate the enzymes. The K_A values for cGMP are 0.1 and 1.2 μM for Iα and Iβ, respectively. Presumably this shift in the K_A is due exclusively to the amino-terminal domain differences in the protein kinases since cells transfected with cDNAs encoding these kinase isoforms retain the characteristic shifts in K_A for cGMP (Ruth *et al.*, 1991).

The cGMP kinase is also phosphorylated by an intramolecular mechanism at serine and threonine residues located within or near the autoinhibitory domain (Lincoln *et al.*, 1978; Landgraf *et al.*, 1986). Phosphorylation produces a partially activated enzyme that is 'primed' for activation by cyclic nucleotide binding. Several laboratories have shown that the phosphorylated cGMP kinase is more readily activated by cyclic AMP than by cyclic GMP (Foster *et al.*, 1981; Landgraf *et al.*, 1986). This suggests a physiologic mechanism by which the cGMP kinase can be activated by the higher cellular concentrations of cAMP that exist in cells. Indeed, recent data by Jiang *et al.* (1992) have shown that elevations in intracellular cAMP activate cGMP kinase in intact vascular smooth muscle.

In addition to autoinhibitory sites and intramolecular phosphorylation, critical cysteines may be involved in the activation of the cGMP kinase. Hofmann and co-workers have proposed a mechanism of activation of cGMP kinase whereby binding of cGMP to the regulatory domain promotes the formation of a cysteine 'cross-bridge' between cys-312 in the cGMP binding domain (site B) and cys-518 in the catalytic domain (Landgraf *et al.*, 1991). This 'cross-bridge' might then function to open the catalytic domain allowing peptide binding sites access to substrates.

4.1.3 Distribution of cGMP kinases

Tissues that contain the highest concentrations of the cGMP kinase include vascular smooth muscle, visceral smooth muscle (e.g. uterine, intestinal), tracheal smooth muscle, platelets and Purkinje cells of the cerebellum (Walter, 1981; Lincoln and Corbin, 1983). The high concentration of cGMP kinase in the vasculature probably accounts for the nominal levels found in other organs such as liver, kidney and heart. Joyce *et al.* (1986), for example, have found that the highest concentrations of cGMP kinase in kidney are localized to the glomerulus and the other vasculature. Little or no enzyme could be detected in kidney tubules. Likewise, Ecker *et al.* (1989) have reported that cardiac cGMP kinase increases as vascularization of the heart is increased, suggesting that myocytes *per se* probably contain little cGMP kinase relative to the vasculature. The high concentrations of cGMP kinase found in lung, from which the enzyme was first purified, probably reflect the enormous vascularization and airway network in this organ. On the other hand, other cells are known to contain cGMP kinase at low, and perhaps nearly undetectable, levels. These include both neutrophils (Pryzwansky *et al.*, 1990) and endothelial cells (MacMillan-Crow *et al.*, 1994). In each of these cell types, the cGMP kinase is highly localized to the particulate fractions of the cells, most likely the cytoskeleton, and presumably in close association with substrate proteins.

4.1.4 Substrates for cGMP kinase

Few if any selective substrates for cGMP kinase have been well characterized, a problem that has hampered investigations into the role of the enzyme in cell function. Perhaps part of the problem is the fact that cAMP and cGMP kinases recognize similar substrates and phosphorylation sequences within substrates *in vitro* as well as in the intact cell. Therefore, selective cGMP-dependent protein phosphorylation has often been difficult to demonstrate. In addition, many of the selective substrate proteins that have been identified for the cGMP kinase are proteins whose functions are not known. A list of many of the potential physiologic substrates for cGMP kinase that have been reported in the literature is shown in Table 1. It is possible that some of these proteins do not have any physiologic relevance to cGMP actions in cells.

The first proteins identified as selective substrates for the cGMP kinase had been found in particulate fractions of smooth muscle membranes (Casnellie *et al.*, 1980; Ives *et al.*, 1980). These proteins, termed G_0, G_1 and G_2 based on their migration in polyacrylamide gels (M_r=240, 130 and 100 kDa, respectively) and their specificity for phosphorylation by cGMP, have since been determined to be cytoskeletal proteins and probably linked to the plasma membrane through the cytoskeleton network (Baltensperger *et al.*, 1990). Although selectively phosphorylated using cGMP kinase, these proteins are also substrates for the cAMP kinase and are phosphorylated on the same residues by each kinase (Parks *et al.*, 1987). The G_0 protein (M_r=240 kDa) has been suggested to be involved in the activation of the plasma membrane Ca^{2+}-ATPase pump in response to cGMP kinase (Yoshida *et al.*, 1991). The G_1 protein (M_r=130 kDa), and its proteolytic fragment, G_2 (M_r=100 kDa), which have been shown to be associated with the cytoskeleton, have not been identified. It has been suggested that the G_1 protein may be an integrin-like protein,

Table 1 Substrates for cyclic GMP-dependent protein kinase

Protein	Possible function
G_0	Activate Ca^{2+}-ATPase
G_1	Integrin-like function
Autophosphorylation	Activate enzyme; select for cAMP activation
G-substrate (cerebellum)	Phosphatase inhibitor
Phospholamban	Stimulate Ca^{2+}-ATPase
IP_3 receptor	Inhibit Ca^{2+} mobilization
VASP	Inhibit Ca^{2+} mobilization; reorganize cytoskeleton
Ryanodine receptor	Ca^{2+} release
Vimentin	Target kinase to substrates
Type V phosphodiesterase	Regulate cyclic GMP hydrolysis
DARPP-32	Phosphatase inhibitor

based on several of its physical properties, but it does not appear to represent any of the known integrin subunits (Baltensperger *et al.*, 1990).

Walter and co-workers have characterized a protein substrate for cGMP kinase from platelets and other tissues (Waldman *et al.*, 1986, 1987). This protein, known as VASP (for vasodilator-specific phosphoprotein), is another cytoskeletal and actin-binding protein (Reinhard *et al.*, 1992). Unlike most of the currently described cGMP kinase substrates, VASP has been well characterized with respect to stoichiometry of phosphorylation by cGMP kinase both *in vitro* and *in vivo*. As with other protein substrates for cGMP kinase, VASP is also an excellent substrate for the cAMP kinase *in vitro* and in the intact cell (Halbrugge *et al.*, 1990). Phosphorylation of VASP by either cAMP or cGMP kinase may regulate intracellular Ca^{2+} levels in platelets resulting in decreased adhesive and aggregatory properties of the cell (Geiger *et al.*, 1992).

The sarcoplasmic reticulum Ca^{2+}-ATPase regulatory protein, phospholamban, has also been shown to be a substrate for cGMP kinase *in vitro* and in the intact cell (Raeymaekers *et al.*, 1988; Sarcevic *et al.*, 1989; Cornwell *et al.*, 1991; Karczewski *et al.*, 1992). Phosphorylation of phospholamban by either the cAMP or cGMP kinase has been shown to promote the activation of the sarcoplasmic reticulum Ca^{2+}-pumping Ca^{2+}-ATPase (Cornwell *et al.*, 1991). Phospholamban is only expressed, however, in striated muscle and smooth muscle. Because of the potential role for cGMP kinase in the regulation of smooth muscle intracellular Ca^{2+} levels, several groups have suggested that phosphorylation of phospholamban by cGMP kinase may contribute to the relaxant effects of cGMP. Another sarcoplasmic reticulum protein, namely the receptor protein for inositol 1,4,5-trisphosphate (IP_3) has been shown to be phosphorylated *in vitro* using both cAMP and cGMP kinase (Komalavilas and Lincoln, 1994; Koga *et al.*, 1994). The role of phosphorylation of the IP_3 receptor is not known however.

From the previous discussion, it is obvious that many proteins are substrates for both cyclic nucleotide-dependent protein kinases. However, studies performed over several years have clearly indicated that these two protein kinases do indeed catalyse the phosphorylation of the same substrates *in vitro* and in the intact cell. The mechanism by which practically all serine/threonine protein kinases recognize substrate proteins has been thoroughly investigated in a number of laboratories. It has been established that primary sequence motifs in putative substrate proteins for the cAMP kinase are critical for the recognition of the substrate by the catalytic subunit (Krebs and Beavo, 1979). In particular, the sequence RRXS, where X is an uncharged residue, contains the basic structural requirements for recognition of the protein substrate by the catalytic subunit. Lincoln and Corbin (1977) have found that the cGMP kinase recognizes similar peptide sequences in which two basic amino acid residues are amino-terminal to the phosphorylatable serine/threonine. Thus, the same substitutions for arginyl residues that decrease cAMP kinase activity towards peptide substrates also decrease cGMP kinase activity towards these peptides. Glass and Krebs (1979) have suggested that certain amino acid residues within the phosphorylation sites of protein kinase substrates are better tolerated by the cGMP kinase than by the cAMP kinase, suggesting that substrate specificity does exist for the cGMP kinase. More recently, work from Corbin's laboratory has shown that the Type V phosphodiesterase from lung is phosphorylated selectively by the cGMP kinase (Thomas *et al.*, 1990b). The amino acid residue requirements for the phosphorylation site, ^{89}RKIS[**P**]ASEFDR, have been studied using amino acid substitutions within this sequence. As expected, replacement of the basic residues amino-terminal to the phosphorylated serine dramatically reduces activity of the enzyme towards the peptide substrate. However, the phenylalanyl residue at position 96 appears to enhance the activity of cGMP kinase towards the peptide but inhibits the activity of the cAMP kinase towards the peptide (Colbran *et al.*, 1992). Therefore, carboxyl-terminal aromatic residues appear to be well tolerated by cGMP kinase but not by cAMP kinase in certain cases. On the other hand, the phosphorylation site of the IP_3 receptor, which both cyclic nucleotide-dependent protein kinases recognize, has been determined to be ^{1752}RRES[**P**]LTSF (Ferris *et al.*, 1991; Komalavilas and Lincoln, 1994). A phenylalanyl residue in the IP_3 receptor is located in a similar position relative to the seryl residue as was described for the peptide derived from the phosphorylation site of the Type V phosphodiesterase. Whereas this phenylalanine may enhance cGMP kinase activity toward the peptide, it does not appear to inhibit cAMP kinase activity. Therefore, at this point one can only conclude that cGMP kinase recognizes some substrates in a unique fashion, but that the specificity determinants are not well defined.

4.1.5 Role of kinase and substrate localization in the cell

The exact manner by which apparently specific cGMP-dependent phosphorylation occurs for non-specific protein substrates in the cell is the subject of much speculation. One potential explanation was advanced by our laboratory, and was based on information available from the protein phosphatase field. Protein serine/threonine phosphatases are known to dephosphorylate a variety of phosphoproteins. In many cases, the catalytic subunits of phosphatases have been shown to be bound to other proteins or subunits that localize or 'target' the enzyme to the substrate (Hubbard and Cohen, 1993). Thus, in many cases, the phosphatase is thought to catalyse the dephosphorylation of the 'closest' phosphoprotein. The targeting mechanism may help explain the manner by which cGMP kinase catalyses the phosphorylation of protein substrates that are substrates for other kinases as well. Cornwell *et al.* (1991) have demonstrated that cGMP kinase is localized near the sarcoplasmic reticulum in rat aortic smooth muscle cells where it presumably has easy access to substrates in this structure, such as phospholamban and the IP_3 receptor protein. The association or targeting of cGMP kinase to structures that contain substrates for the enzyme would also be predicted to promote rapid, specific and efficient phosphorylation of these substrate proteins in response to elevations in intracellular cGMP. The targeting of cGMP kinase to regions in the cell containing potential substrates is a flexible mechanism that allows the kinase to recognize rather selectively many different substrate proteins in the cell, some of which might well be substrates for other protein kinases. Thus, by virtue of the binding or anchoring of the kinase near potential substrates in different regions of cells and in different types of cells, cGMP kinase could regulate a variety of cellular processes. Perhaps the best example of cGMP kinase targeting to a specific protein substrate is seen with the intermediate filament protein vimentin. Vimentin is phosphorylated by a number of protein kinases in cells. However, vimentin contains no true consensus phosphorylation sequences for cAMP or cGMP kinases. Studies by Wyatt *et al.* (1991) and MacMillan-Crow and Lincoln (1994) have shown that cGMP kinase binds to vimentin with high affinity and selectivity in neutrophils and smooth muscle cells, respectively. The binding of cGMP kinase to vimentin occurs primarily at the dimerization domain located at the amino-terminus of the kinase, since removal of the domain by proteolysis results in the dissociation of the cGMP kinase monomer from vimentin. In such a way, the catalytic domain of the kinase may be aligned near the phosphorylation site on vimentin (Figure 5). Upon activation of the cGMP kinase, rapid phosphorylation of vimentin occurs. Kinase that is dissociated from vimentin by removing the dimerization domain from the enzyme only slowly catalyses the phosphorylation of vimentin. Thus the only way that cGMP kinase may be able to catalyse the phosphorylation of this protein is through high affinity binding to the protein at a non-catalytic site. In summary, then, cGMP kinases catalyse the phosphorylation of protein substrates either by recognizing selective phosphorylation sites in the proteins or by being targeted to the proteins where the close association between the kinase, or its catalytic domains, and the phosphorylation site on the substrate allows for rapid phosphorylation.

4.1.6 Biologic roles of the cGMP kinase

Cyclic GMP kinase mediates many of the actions of cGMP in cells. These include the regulation of vascular smooth muscle relaxation and the inhibition of platelet

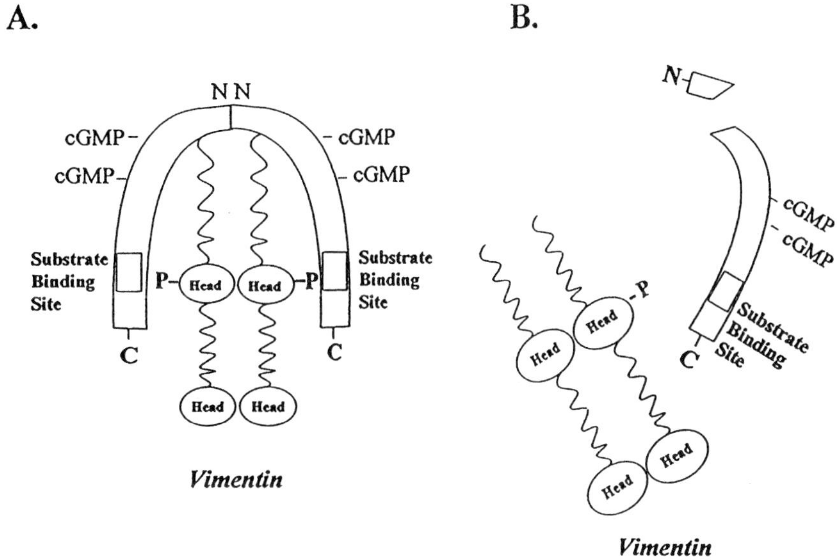

Figure 5 Model for targeting of cGMP to vimentin. The dimerization site at the N-terminus of cGMP kinase is proposed to interact with vimentin to align the catalytic domain with the phosphorylation site on vimentin. Removal of the dimerization domain to produce the active catalytic monomer prevents cGMP kinase binding to vimentin and phosphorylation of the protein. From MacMillan-Crow and Lincoln (1994).

adhesion and aggregation. Mechanisms for cGMP kinase regulation of contractile activity in vascular smooth muscle have been proposed, including the regulation of phospholipase C activity and IP_3 production, the regulation of Ca^{2+}-dependent K^+ channels, voltage-gated Ca^{2+} channels, and contractile protein activity (Lincoln and Cornwell, 1993; Lincoln *et al.*, 1994). Unfortunately, in many of these cases the physiologic substrate(s) for cGMP kinase is not known. The high levels of cGMP kinase (Lohmann *et al.*, 1981) and a specific substrate for the kinase (i.e. the G-substrate) (Schlichter *et al.*, 1980) in Purkinje cells in the cerebellum suggests that this enzyme mediates most, if not all, of the actions of cGMP in these cells. The possible roles of cGMP kinase in the nervous system will be addressed below.

4.1.7 Cross-activation of cyclic nucleotide-dependent protein kinases

From the previous discussions, it has become apparent that in many instances cyclic nucleotide-dependent protein kinases catalyse the phosphorylation of the same proteins both *in vitro* and in the intact cell. The situation is even more complex than this, however. It has long been known that cyclic nucleotide-dependent protein kinases are not *specifically* activated by their own cyclic nucleotides, but rather demonstrate only *selectivity* for cyclic nucleotides. In most instances, cAMP levels are significantly higher in cells than cGMP levels. Therefore, it can be predicted that cAMP can activate cGMP kinase *in situ*. Reports by Francis *et al.* (1988) and Lincoln *et al.* (1990) indicate that cAMP may mediate smooth muscle relaxation by activating cGMP kinase. More recently, Jiang *et al.* (1992) have shown that isoproterenol- or forskolin-evoked elevations in cAMP activate the cGMP kinase in intact coronary arteries. These findings have led to suggestions that cAMP relaxes vascular smooth muscle by activating cGMP kinase. On the other hand, cGMP cross-activates the cAMP kinase in certain cases. In particular, bacterial heat-stable entertoxin (STa) elevates cGMP in colonic cell lines resulting in Cl^- and water secretion. These effects appear to be mediated by activation of the cAMP kinase (Forte *et al.*, 1992). In addition to the potential physiologic significance of these findings, a salient point to be made is that the injudicious use of pharmacologic agents known to activate protein kinases could lead to erroneous interpretations. The 8-substituted analogues of cyclic nucleotides (8-bromo- or 8-para-chlorophenylthio derivatives) may cross-activate protein kinases in the intact cell, especially when used at concentrations greater than 100 μM. It would appear necessary for investigators to construct concentration–response curves for cyclic nucleotide derivatives in order to characterize the specific kinase that mediates the response.

4.2 Cyclic GMP-regulated ion channels

The role of cGMP in regulating ion channels has received much attention over the past several years. On the one hand, the regulation of ion channels by phosphorylation catalysed by cGMP kinase is of much current interest. Yet another mode of regulation of ion currents by cGMP has been established primarily as a result of studies from the vertebrate photoreceptor cell. In this instance, cGMP has been demonstrated to regulate non-selective ion channels by binding directly to the channel protein. This phenomenon underlies the mechanism of signal transduction in the photoreceptor cell.

The first evidence suggesting that cGMP regulated ion currents was obtained by Miller and colleagues working with the vertebrate rod outer segment membrane (Miller and Nicol, 1979). These investigators demonstrated that cGMP produced rapid and transient depolarization of the rod outer segment membrane. The cGMP-stimulated current is carried by a number of cations suggesting that the ion channel is non-selective. Upon hydrolysis of cGMP by cGMP phosphodiesterases, the current is rapidly terminated leading to the hyperpolarization of the photoreceptor cell (see Hurley, 1987, for a review). Unlike the case for other cell types, however, these effects of cGMP are not associated with cGMP kinase activation, suggesting that cGMP itself is regulating the channel protein (Fesenko *et al.*, 1985). The cGMP-gated non-selective ion channel has been identified in rod outer segment membranes and purified (Cook *et al.*, 1986, 1987), and the cDNA encoding the protein has been cloned and expressed in the oocyte system (Kaupp *et al.*, 1989). After the cloning of the rod outer segment of cGMP-gated channel, several investigators identified other cGMP-gated channels in cells and tissues (Nakamura and Gold, 1987; Ludwig *et al.*, 1990; Light *et al.*, 1990; Ahmad *et al.*, 1992; Biel *et al.*, 1993). It appears therefore that there exists an entire family of cGMP-regulated ion channels, some of which display subtle differences in regulation by cGMP, ion selectivity and sensitivity to drugs. The best characterized of the cGMP-regulated ion channels is, however, the photoreceptor channel, and the properties of this of channel will be discussed first.

4.2.1 The rod photoreceptor cGMP-gated channel

The rod photoreceptor cGMP-gated channel is expressed as a protein having a subunit M_r=79 000 (Kaupp *et al.*, 1989). The cDNA encodes a protein containing 690 amino acids (Figure 6). The topology of this channel is similar to that of several other monovalent cation channels such as K^+ and Na^+ channels, and contains six membrane spanning regions (Guy *et al.*, 1991). The pore region is presumed to lie between the S5 and S6 segment. Among the channel membrane spanning domains, homology between the rod outer segment channel and the others is observed mainly in the S4 segment (Guy *et al.*, 1991). The carboxyl-terminal domain contains the cGMP binding site within residues 498–577. There is significant homology between the cGMP binding site of the photoreceptor gated channel and the cGMP binding domains of the cGMP kinases suggesting that the channel binding site was evolutionarily derived from the primitive cyclic nucleotide binding proteins expressed in prokaryotes.

Despite the homology between the cGMP kinase binding domains and the photoreceptor channel cGMP binding site, these two binding sites demonstrate very different properties. Unlike the case for the cGMP kinase, only one cGMP binding site per channel monomer is present. Perhaps most significant is the fact that the affinity of the ion channel for cGMP is comparatively low, having a K_D for cGMP of approximately 20–50 μM. Even the affinity for cGMP at the lower affinity site A on the cGMP kinase is more than an order of magnitude greater than that for the photoreceptor site. Thus, while submicromolar concentrations of cGMP are capable of producing complete activation of the cGMP kinase, the photoreceptor ion channel requires greater than micromolar concentrations of cGMP to stimulate ion flux through the channel. The rod photoreceptor cGMP-gated ion channel demonstrates a high degree of selectivity for activation by cGMP. Neither cAMP nor non-cyclic nucleotide derivatives activate the channel (Cook *et al.*, 1986). The cGMP-regulated ion channels retain the specific threonyl residues which predict a high degree of selectivity towards the binding of cGMP as opposed to cAMP. Also, predictably, the photoreceptor channels, as well as other cGMP-regulated channels, are activated by 8-substituted cGMP analogues such as 8-bromo-cGMP similar to the case for cGMP kinases (Shabb and Corbin, 1992).

Injection of the mRNA for the rod outer segment channel in *Xenopus* oocytes

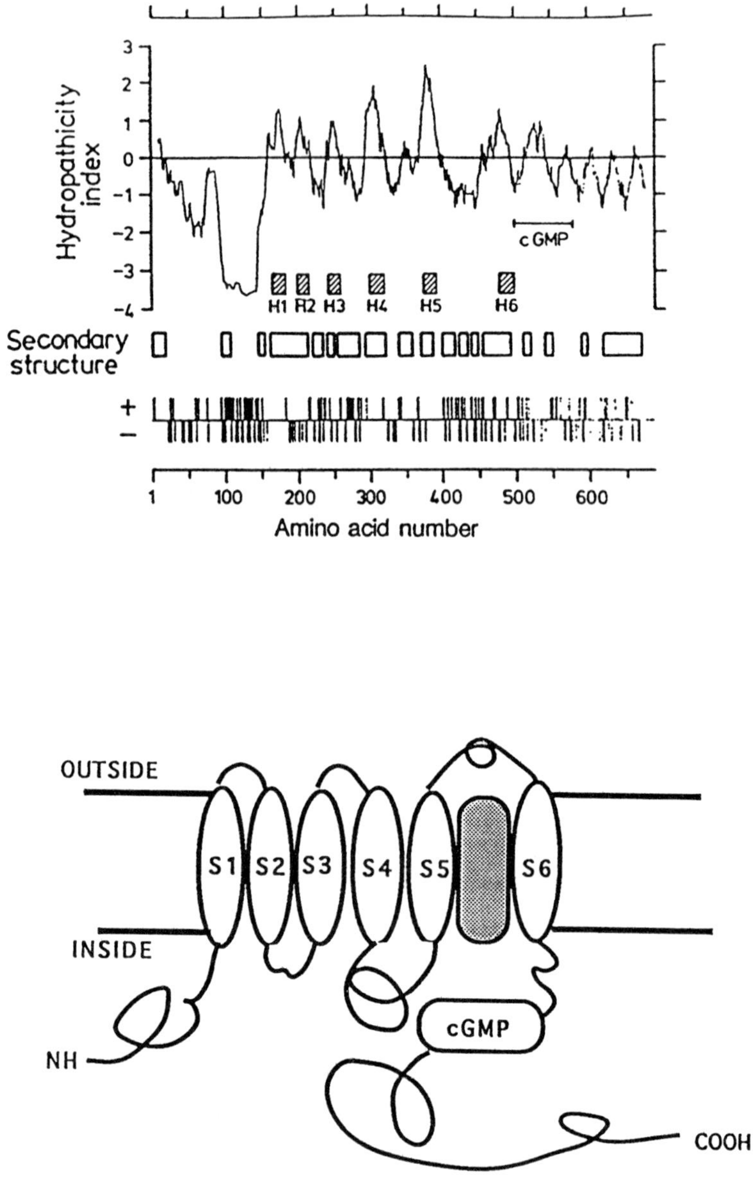

Figure 6 Structure of the cyclic GMP-gated cation channel from bovine rod outer segment membranes. The top panel is the hydropathy plot of the protein indicating six transmembrane segments (H1 through H6) and a hydrophilic cyclic GMP binding site. The bottom panel represents a schematic representation of the transmembrane domains based on the structure of the cloned potassium channels. The shaded area is the pore region located between segments S5 and S6. From Kaupp *et al.* (1989). Reproduced with permission from the R.G. Landes Company, Austin.

results in the expression of a protein whose properties closely resemble those of the native channel found in the photoreceptor membranes (Kaupp *et al.*, 1989). Cyclic GMP promotes a large increase in monovalent ion conductance, primarily carried by Na^+. The current carried in the presence of cGMP is quite substantial (2–5 nA). The number of channel subunits required to comprise the native protein is believed to be four, based on its similarity in topology with other monovalent cation channels. It is tempting to speculate that partial occupation of the cGMP binding sites on the channel complex results in a graded conductance of current. Thus, the degree of depolarization, action potential volley, etc. might well be related to the number of cGMP molecules associated with the channel protein complex.

4.2.2 Renal cGMP-regulated ion channels

A significant amount of circumstantial evidence had been generated over the past several years in several laboratories suggesting that cGMP regulates monovalent cation transport in renal collecting duct cells. In one study by Light *et al.* (1990) cGMP has been found to inhibit Na^+ transport in cultured renal inner medullary collecting duct cells by one mechanism that appears to involve cGMP kinase, but also by another mechanism that is unrelated to protein kinase. The interesting aspect of this study is that cGMP directly inhibits putative ion channels in this cell type. Furthermore, this cGMP-regulated ion channel is selective for Na^+ and is inhibited by amiloride. Work from other laboratories has now substantiated the presence of perhaps several types of cGMP-*inhibited* cation channels in the kidney. Ahmad *et al.* (1992) have found that a renal collecting duct cell line expresses a cGMP-inhibited non-selective cation channel that appears very similar if not identical to that found in the photoreceptor cell. The fact that the renal channel is inhibited, rather than activated, by cGMP may indicate that other components of the renal channel can regulate the gating activity by cGMP. Unlike the channel studied by Stanton's laboratory, this cGMP-inhibited channel is non-selective for monovalent cations and is not inhibited by amiloride. More recently, another cDNA encoding a cGMP-activated ion channel has been cloned using probes that hybridize with the photoreceptor channel (Biel *et al.*, 1993). This ion channel appears to be preferentially activated by cGMP and is expressed in kidney and testis. This particular cGMP-regulated ion channel is cation non-selective and quite homologous to the olfactory channel discussed below. Like the photoreceptor and other renal channels, only one binding site for cGMP is present on the channel subunit. However, the binding affinity for cGMP is rather high (K_D of approximately 1 μM) relative to the photoreceptor channel, but is still of low affinity compared with the cGMP kinase. Therefore, it is probable that cGMP regulates cation fluxes only when the intracellular cGMP levels are greatly elevated, or perhaps when the levels of cGMP in the immediate membrane environment where the channels are located are high.

Although photoreceptor cells and renal epithelium represent the major tissues expressing cGMP-regulated ion channels, it is conceivable that such channels exist in other tissues as well including cardiovascular tissue, brain and testis. In spermatozoa, for example, cGMP activates a K^+ channel in response to speract (Babcock *et al.*, 1992; Cook and Babcock, 1993). Spermatozoa have high levels of cGMP, but appear to be devoid of cGMP kinase (Garbers and Kopf, 1980). On the other hand, binding proteins for cGMP have been detected in sperm suggesting that the cGMP-regulated ion channels observed in testis come from the gamete itself. Further experiments are clearly needed to address this issue. Whether all tissues express cGMP-regulated channel proteins has not been verified at this time. If this is found to be the case, these findings could suggest a more widespread role for cGMP-gated channels in signal transduction than previously surmised based on information obtained from the photoreceptor work. It is necessary, however, to point out a significant caveat with the assumption that cGMP-regulated ion channels mediate the physiologic effects of cGMP in other tissues. The affinity for most of the well-described cGMP-regulated ion channels for cGMP is very low, with K_Ds in the order of 20 μM or greater. Cellular cGMP levels in most cells (with the notable exceptions of photoreceptor cells and spermatozoa) are below 100 nM. Therefore, it is difficult to conceive how basal or even elevated levels of cGMP could accumulate in most cells to high enough levels to bind to cGMP-gated channels. Perhaps local 'membrane environment' cGMP may be more important than the total cellular levels of the nucleotide.

4.2.3 Cyclic GMP-regulated olfactory epithelium ion channels

A cyclic nucleotide-regulated ion channel has been described in olfactory epithelium that appears to be gated by either cAMP or cGMP (Ludwig *et al.*, 1990). The cDNA encoding the olfactory ion channel has been cloned and there is a high degree of homology between the olfactory channel cGMP binding site and that of the photoreceptor channel (Figure 7). The affinity of the channel is higher for cGMP than for cAMP, but relatively high concentrations are required to activate the channel. The olfactory channel is somewhat unique in having the capability to bind cAMP with reasonable affinity since the channel has a threonyl residue at position 537 which would predict high affinity for cGMP relative to cAMP. On the other hand, sequences adjacent to or surrounding the olfactory channel binding site differ from that of the photoreceptor protein, for example, which could introduce determinants that allow for cAMP binding.

The physiologic regulation of the olfactory channels appears to be closely linked to the activation of adenylate cyclase rather than guanylate cyclase. Several specific odorant receptors have now been identified through cDNA cloning approaches, and almost invariantly these receptors couple to heterotrimeric G proteins to activate adenylate cyclase (Buck and Axel, 1991). The anatomic location of the receptors in the olfactory epithelium apparently plays an important role in the selectivity for odorant sensory perception. It should be kept in mind, especially with the widespread role for NO in cellular signalling, that these channels could be modulated by elevations in intracellular cGMP as well. For example, cGMP may be an

```
ROS  485/LLVELVLKLQPQVYSPGDYICKKGDIGREMYIIKEGKLAVVAD..........D
OLF  462/LLVELVLKLRPQVFSPGDYICRKGDIGKEMYIIKEGKLAVVAD..........D
GKA  110/QIQEIVDCMYPVEYGKDSCIIKEGDVGSLVYVMEDGKVEVTKE...........
GKB  228/ILSKLADVLEETHYENGEYIIRQGARGDTFFIISKGKVNVTREDSPN......E

ROS  GITQFVVLSDGSYFGEISILNIKGSKAGNRRTANIKS..IGYSDLFCLSKDDLMEALT
OLF  GVTQYALLSAGSCFGEISILNIKGSKMGNRRTANIRS..LGYSDLFCLSKDDLMEAVT
GKA  .GVKLCTMGPGKVFGELAILY......NCTRTATVKT..LVNVKLWAIDRQCFQTIMM
GKB  DPVFLRTLGKGDWFGEKALQG......EDVRTANVIA..AEAVTCLVIDRDSFKHLIG

ROS  EYPDAKGMLEEKGKQILMKDGLL/607
OLF  EYPDAKRVLEERGREILMKEGLL/584
GKA  RTGLIKHTEYMEFLKSVPTFQLP/225
GKB  GLDDVSNKAYEDAEAK/352
```

Figure 7 Sequence alignment of the rod outer segment cyclic GMP-regulated ion channel cyclic GMP binding site (ROS), the olfactory cyclic GMP-regulated ion channel (OLF), and the cyclic GMP binding sites from Type I bovine cyclic GMP-dependent protein kinase (GKA and GKB). The residues beginning with the binding pocket are indicated by the numbers preceding the sequences. Note the conserved identities (boxes) and conservative substitutions (dots). Reproduced with permission from the R.G. Landes Company, Austin.

important co-stimulant along with cAMP in olfactory epithelial cells. Conceivably, olfactory cells may respond to NO released from olfactory neuronal terminals to modulate olfactory sensory perception.

4.3 Cyclic GMP-regulated phosphodiesterases

The types of phosphodiesterases that bind cGMP, and therefore have the *potential* for cellular signalling, are divided into three classes (Beavo and Reifsnyder, 1990): (1) the cGMP-stimulated cyclic nucleotide phosphodiesterases (Type II phosphodiesterase); (2) the cGMP inhibitable cyclic AMP phosphodiesterases (Type III phosphodiesterases, which are the cilostamide-sensitive enzymes); and (3) the cGMP-binding cGMP specific phosphodiesterases (Type V phosphodiesterases which include the photoreceptor enzymes). The role of the Type II, Type III and Type V cGMP-binding cGMP phosphodiesterases in *mediating* cGMP signalling is of potential importance.

4.3.1 Phosphodiesterases that are regulated by cGMP

In addition to their roles in cGMP hydrolysis, cyclic nucleotide phosphodiesterases regulated by cGMP binding also perform critical roles in second messenger signal transduction.

4.3.1.1 The Type II cyclic nucleotide phosphodiesterases

The Type II cGMP-stimulated phosphodiesterase is actually one of the first cyclic nucleotide phosphodiesterase species characterized, having first been described in rat liver extracts in the early 1970s. The cGMP-stimulated phosphodiesterase is widely expressed in vertebrate cells where it has been implicated in such cyclic GMP-mediated events as the regulation of cardiac contractility and steroid hormone secretion. The Type II cGMP-stimulated phosphodiesterase is found predominantly in the soluble fraction of tissues, although a significant amount of the enzyme in brain appears to localize to particulate material (Whalin *et al.*, 1988). Kinetically, the enzyme displays little preference for the hydrolysis of either cAMP or cGMP. Given the relatively low affinities of the enzyme for cyclic nucleotides (K_m>10 μM) and the similarities in the V_{max} (120 units/mg), it is probable that the enzyme displays selectivity for cAMP hydrolysis in the intact cell. This has been shown to be the case in mammalian tissues (Manganiello *et al.*, 1990) and pheochromocytoma (PC12) cells (Whalin *et al.*, 1991). The unique feature of this enzyme, however, is that submicromolar concentrations of cGMP stimulate the enzyme by up to 10-fold to hydrolyse cAMP. Studies with the purified enzyme indicate that the effect of cGMP to stimulate cAMP hydrolysis is due primarily to an increase in the V_{max} of the enzyme. As with the other phosphodiesterases discussed above, the catalytic domain resides in the carboxyl-terminal half of the enzyme (Figure 8) (Charbonneau, 1990). Molecular cloning of the cDNA encoding the cardiac form of the Type II phosphodiesterase reveals substantial homology with other classes of cyclic nucleotide phosphodiesterases including those selective for cGMP hydrolysis (Sonnenburg *et al.*, 1991). The amino-terminal half of the enzyme contains the non-catalytic cGMP binding region consisting of tandem repeats of sequences and is homologous to that seen on the Type V phosphodiesterases (Figure 9). Biochemical studies aimed at determining the subunit structure of the native protein indicate that the cGMP-stimulated enzyme from liver, brain and heart exists as higher molecular weight species, perhaps homodimers.

Cyclic GMP signalling through activation of the Type II cGMP-stimulated phosphodiesterase has been described in bovine adrenal tissue, PC12 cells and amphibian heart. The characteristic effect of cGMP signalling by the activation of the Type II phosphodiesterase is an attenuation in the response of the tissue or cell to cAMP. In adrenal glomerulosa, for example, cAMP is a well-known second messenger for the stimulation of aldosterone secretion. MacFarland *et al.* (1991) have demonstrated that ANF-mediated inhibition of aldosterone secretion from isolated bovine zona glomerulosa cells is associated with a decrease in cAMP levels. It is interesting to note that these investigators found that the adrenal zona glomerulosa is one of the richest sources for the Type II cGMP-stimulated phosphodiesterase. Other studies by Fischmeister and co-workers (Hartzell and Fischmeister, 1986; Mery *et al.*, 1993) indicate that cGMP inhibits the β-adrenergic inward Ca^{2+} current in isolated frog cardiac myocytes. 8-Substituted derivatives of cGMP, which do not bind to cyclic nucleotide phosphodiesterases, have no demonstrable effect on the current.

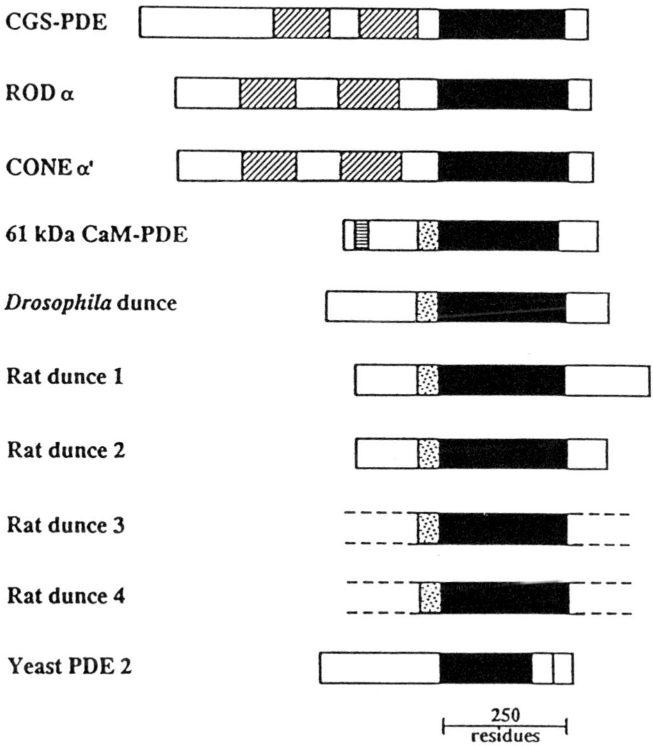

Figure 8 Cyclic nucleotide phosphodiesterase structures based on cDNA cloning. Areas of homology are indicated by similar shading. The catalytic domain is indicated with solid black boxes and boxes containing diagonal lines designate non-catalytic cGMP binding domains. From Charbonneau (1990).

In addition, inhibitors of phosphodiesterases block the effects of cGMP on the inward Ca^{2+} current. Collectively, these data suggest that cGMP inhibits cAMP accumulation in the cardiac myocytes thereby decreasing cAMP-regulated Ca^{2+} currents. The physiologic consequences of inhibitory β-adrenergic-regulated cardiac Ca^{2+} currents is a decrease in contractile force of the heart. The negative inotropic response to cGMP is an early and well-characterized cardiac action of the nucleotide. It is likely that the mechanism underlying this effect is the stimulation of cAMP hydrolysis by cGMP. In PC12 cells, elevations in cGMP by ANF and sodium nitroprusside result in a decrease in the levels of cAMP (Whalin *et al.*, 1991). These effects are due to an increase in the decay rate for cAMP, and can be prevented by phosphodiesterase inhibitors. Signalling roles for the Type II cGMP-stimulated phosphodiesterase have also been suggested for hepatocytes, endothelial cells and brain.

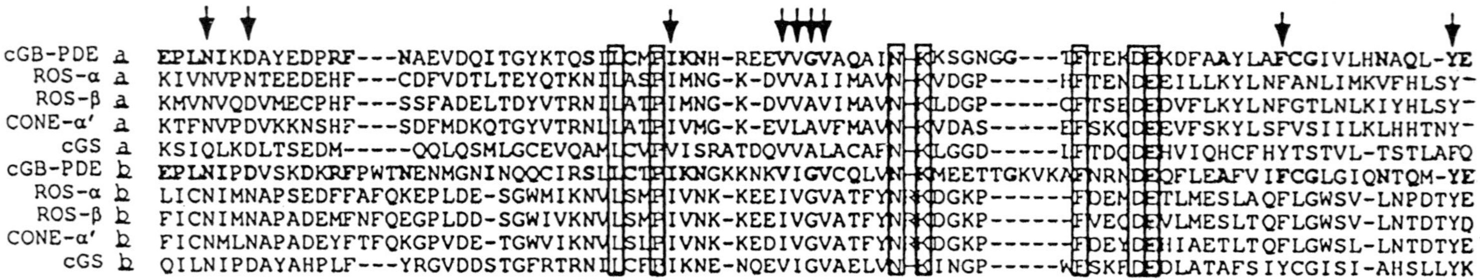

Figure 9 Sequence alignment of the non-catalytic cGMP binding sites from cGMP binding phosphodiesterases. Identical residues are indicated with boxes and arrows represent conservative amino acid substitutions. The abbreviations are cGB-PDE, lung Type V phosphodiesterase residues 228–311 in the *a* repeat and 410–500 in the *b* repeat; ROS-α, rod outer segment Type V α subunit residues 149–229 in the *a* repeat and residues 355–438 in the *b* repeat; ROS-β rod outer segment Type V β submit residues 147–227 in the *a* repeat and residues 353–436 of the *b* repeat; cone-α′, cone Type Vα′ subunit residues 146–226 in the *a* repeat and residues 352–435 in the *b* repeat; cGS, Type II cGMP-stimulated phosphodiesterase residues 284–364 in the *a* repeat and residues 454–534 of the *b* repeat. From McAllister-Lucas *et al.* (1993).

4.3.2 Properties of the cGMP-binding phosphodiesterases

There are at least four different Type V phosphodiesterases expressed in vertebrate tissues (Beavo and Reifsnyder, 1990). Three of these are expressed exclusively in photoreceptor cells and include the rod membrane-associated isozyme (V_{B1}), the rod soluble isozyme (V_{B2}), and the cone isozyme (V_{C1}). The rod and cone enzymes appear to be separate gene products whereas the rod soluble form appears to be a variant in subunit structure. The fourth isozyme in this class is the 'lung/platelet' enzyme (V_A) which appears to be more widely expressed in the cardiovascular system. The photoreceptor enzymes exist as ≃190 kDa heteromeric proteins containing smaller regulatory 15 kDa subunits (γ) bound to the native enzyme complex. The lung enzyme, however, appears to exist as a homodimer of approximately 170 kDa (McAllister-Lucas *et al.*, 1993). The properties of the Type V phosphodiesterases which distinguish them from all others are (a) they specifically hydrolyse cGMP and (b) they bind cGMP at an allosteric site.

The allosteric binding sites of Type V phosphodiesterases are homologous with those of the Type II cGMP-stimulated enzymes, but not those of the protein kinase family (Charbonneau *et al.*, 1990; Shabb and Corbin, 1992). This suggests that these sites had arisen independently of the cyclic nucleotide binding site of the bacterial CAP protein. It is conceivable, however, that the allosteric site arose from the catalytic site of the phosphodiesterase by gene duplication, since there appears to be some hint of structural similarity between these two sites (Charbonneau, 1990). An interesting property of the allosteric site on the Type V enzymes is that cGMP binding is stimulated several-fold by drugs which bind to and inhibit the catalytic site (i.e. competitive inhibitors) (Francis *et al.*, 1980). The order of potency for stimulating cGMP binding exactly parallels that of inhibition of catalytic activity (Thomas *et al.*, 1990a). For example, the most potent inhibitory methylxanthines like isobutyl-methylxanthine are also the most potent activators of cGMP binding. This property is not related to 'protection' of cGMP from hydrolysis since increased binding has been observed under conditions where hydrolysis can be completely prevented (EDTA, 4°C). Thus, conformational changes in the catalytic site appear to affect the binding to the allosteric site. Unfortunately, however, it has been difficult to demonstrate the reverse – that is, binding of cGMP to the allosteric site affects the activity of the catalytic site.

The fact that Type V phosphodiesterases all bind cGMP at a non-catalytic, allosteric site suggests that these enzymes also function as 'receptor' proteins for cGMP. Studies using the lung enzyme demonstrate that removal of the amino-terminal portion of the enzyme containing the allosteric binding site results in activation of the enzyme (Thomas *et al.*, 1990a). These findings may suggest that the amino-terminal domain serves as a regulatory domain of sorts to constrain catalytic activity. A natural corollary of this hypothesis is that cGMP binding would serve to disinhibit catalytic activity much in the same fashion as does cGMP binding to its protein kinase. Studies have not borne out these expectations, however, inasmuch as no significant effects of cGMP occupation of the allosteric site have been observed on the catalytic activity of the purified enzymes. In the rod outer segment, however, cGMP has been reported to regulate the binding of the γ-subunit to the enzyme thus controlling its capacity to be regulated by the G protein, transducin (Arshavsky *et al.*, 1992). Perhaps conformational changes in the Type V phosphodiesterases in response to cGMP binding regulate its interaction with other proteins as well.

Another potential role for the cGMP binding domain has emerged from studies of the purified lung enzyme. The lung Type V_A phosphodiesterase has been shown to be phosphorylated *in vitro* rather selectively by the cGMP kinase (Thomas *et al.*, 1990b). Cyclic AMP kinase is not as effective, while other kinases (protein kinase C, calmodulin kinase II, for instance) are completely ineffective. Phosphorylation, however, is dependent upon occupation of the allosteric site by cGMP. Corbin and Francis and co-workers have proposed that the role of the binding site is to allow the enzyme to be phosphorylated by the GMP-dependent protein kinase (Thomas *et al.*, 1990b). This begs the question as to the role of phosphorylation of the enzyme. Once again, no answers are yet available since phosphorylation also has no apparent effects on the catalytic properties of the enzyme. Because phosphorylation of the lung Type V phosphodiesterase occurs at the amino-terminal end of the molecule (serine-92), it does not seem likely that this would *directly* alter catalytic activity. On the other hand, it is conceivable that phosphorylation could affect one or more regulatory steps, such as the association of another protein or subunit with the phosphodiesterase. Such an association could conceivably inhibit cGMP hydrolysis thereby promoting larger increases in cGMP levels upon cGMP elevation. Also, phosphorylation may cause translocation of the phosphodiesterase to a different cellular compartment. Further studies will be required to define this potentially important signalling mechanism.

4.4 Role of cyclic GMP kinase in the nervous system

It had been known since the earliest days in cyclic nucleotide research that the central nervous system contains high levels of enzymes involved in the synthesis and degradation of cAMP and cGMP. Nevertheless, because of the heterogeneity of nervous tissue and the difficulties encountered in studying the mechanisms of neuronal function at the biochemical level, the precise roles of cAMP and cGMP are obscure. With more recent studies on the role of NO in cellular signalling and the availability of pharmacologic tools with which to study this pathway, new insights into the role of cGMP in neuronal function have been emerging. On the other hand, the role of cGMP-dependent protein phosphorylation in the nervous system is still obscure.

4.4.1 Long-term potentiation and depression

Studies in several laboratories indicate that NO mimics or fits many of the requirements for the retrograde long-term potentiation (LTP) messenger (Bohme *et al.*, 1991; O'Dell *et al.*, 1991; Schuman and Madison, 1991; see Chapter 7). LTP is mimicked by NO-generating drugs and is inhibited by NOS inhibitors or haemoglobin. The potential role of cGMP as a mediator of LTP has been suggested by studies of Haley *et al.* (1992) who have shown that cGMP analogues can partially reproduce the effects of NO on LTP. Several experimental findings support a role for NO in enhanced neurotransmitter release. NO donors (i.e. SIN-1, nitroprusside, etc.) have been shown to stimulate release of acetylcholine from the basal forebrain and dopamine from rat striatal slices, in addition to stimulating glutamate release from hippocampal areas (Prast and Phillipu, 1992). Furthermore, inhibitors of NO production or action attenuate release of neurotransmitters.

Although cGMP levels are consistently elevated in response to NO production in most of the preparations described above, cGMP analogues often do not reproduce the effects of NO. In the study by Haley *et al.* (1992) dibutyryl cGMP only partially mimicked LTP and reversed the effects of NOS inhibitors. Others have found that cGMP either has no effect on synaptic transmission or, in several cases, depresses postsynaptic currents and neuronal firing (Doerner and Alger, 1988; Schuman *et al.*, 1992). Therefore, it has been suggested that NO stimulates presynaptic transmitter release through non-cGMP mediated mechanisms – perhaps, mechanisms involving NO-evoked ADP ribosylation of presynaptic neuronal proteins (Schuman *et al.*, 1992).

In contrast to LTP, a direct role for cGMP in long-term depression (LTD) has been proposed (Crepel and Jaillard, 1990; Shibuki and Okada, 1991; Daniel *et al.*, 1993). Both NO and cGMP depress Purkinje cell electrical activity, consistent with a role for cGMP mediating the effects of NO. It has been suggested that cGMP kinase mediates AMPA receptor depression, although the biochemical mechanism has not been defined. There have been some studies that have not supported a role for cGMP in mediating AMPA receptor depression and hence LTD. In the first instance, Purkinje cells do not appear to express NOS, suggesting that elevations in Purkinje cell cGMP occur via a non-NO-mediated mechanism (Bredt *et al.*, 1991). On the other hand, both parallel nerve terminals as well as cerebellar basket cells, which also receive input from parallel fibres, express large amounts of NOS. Thus, it is possible that parallel fibres release NO or stimulate basket cells to produce NO, or both, and NO diffuses into the Purkinje cells and elevates cGMP.

Another potential problem is the lack of a robust stimulation of guanylate cyclase by NO in cerebellum. Garthwaite and Garthwaite (1987) were unable to demonstrate elevations in cGMP levels in Purkinje neurons in response to nitroprusside. Other studies, however, have shown that NOS inhibitors do indeed prevent LTD in whole tissue slices, but not when the inhibitors are added to Purkinje cells configured for whole cell recordings (Daniel *et al.*, 1993). The fact that Purkinje cells are not necessarily rich in either NOS or in soluble guanylate cyclase may not be supportive of a role for NO in elevating cGMP levels. It should be kept in mind, however, that small elevations in cGMP in response to NO may be all that is necessary for activation of cGMP kinase in the Purkinje cells since the enzyme is so highly expressed here (Lohmann *et al.*, 1981). Taken together, these results tend to support a role for cGMP in mediating cerebellar LTD even though Purkinje cells may not be particularly rich in NOS and guanylate cyclase.

4.4.2 Cyclic GMP kinase in neurons

Some of the earliest studies in signal transduction research suggested that cGMP may mediate the actions of neurotransmitters in the nervous system. Because of the heterogeneity of the cell population present, more mechanistic approaches to study the role of cGMP in the nervous system have been limited. For these and other reasons, investigators initially approached the role of cyclic nucleotides in the nervous system by studying the distribution of enzymes involved in cyclic nucleotide metabolism and action. Studies in the laboratory of Greengard provided the first clues on the role of cGMP in the central nervous system (see Walter and Greengard, 1981, for a review). Investigators in this laboratory initially used an isolated ganglionic system, the rabbit superior cervical ganglion, to demonstrate that dopamine via the adenylate cyclase–cAMP pathway stimulates postganglionic potentials known as excitatory postsynaptic potentials (EPSP). On the other hand, cGMP was found to mediate cholinergic-evoked inhibitory postsynaptic potentials (IPSP). These initial observations have been extended to studies in the central nervous system, and especially the mammalian cerebellum where the role of cGMP may be to mediate LTD. Greengard and colleagues have determined that the cerebellum contains the highest concentration of cGMP kinase in brain. Indeed, cerebral cortex, hippocampus, hypothalamus, and other major areas contain only sparse amounts of kinase activity compared with the cerebellum. Lohmann *et al.* (1981), using specific antibodies toward mammalian cGMP kinase, have localized the cGMP-dependent enzyme to the Purkinje cells in the cerebellum. Other cells found in the cerebellum (granular cells, stellate neurons, astrocytes, etc.) are essentially devoid of cGMP kinase. As a natural extension of these findings, these investigators examined the role of cGMP-dependent protein phosphorylation in cerebellum. A cerebellar-specific protein substrate for cGMP kinase has been identified and termed the G-substrate (Schlichter *et al.*, 1978; Aswad and Greengard, 1981). This protein, M_r=26 kDa, is phosphorylated selectively by cGMP kinase on threonine residues. The amino acid sequence of the G-substrate has been determined and the protein appears to be related to a class of proteins that are inhibitors of protein phosphatase 1 (Aiken *et al.*, 1981). Whether the G-substrate is indeed a protein phosphatase inhibitor must await further characterization. An interesting observation is that the G-substrate and cGMP kinase are co-localized in Purkinje cells suggesting that this protein substrate may be physiologically important in this cell type (Schlichter *et al.*, 1980).

Other protein substrates for cGMP kinase have been observed in the cerebellum. Studies in our laboratory have shown that the cerebellar IP_3 receptor protein is phosphorylated *in vitro* using cGMP kinase, and that cGMP kinase is at least as good a catalyst as the cAMP kinase (Komalavilas and Lincoln, 1994). The physiologic significance of IP_3 receptor phosphorylation is not known, although Supattopone *et al.* (1988) have reported that IP_3 receptor phosphorylation by cAMP kinase inhibits the Ca^{2+} gating function of the protein. Given that cGMP and cAMP kinases catalyse the phosphorylation of the same seryl residue in the protein, perhaps the decrease in AMPA receptor function is related to depressed Ca^{2+} mobilization due to IP_3 receptor phosphorylation.

Recently, a novel example of cross-activation of cyclic nucleotide-dependent protein kinases has been documented in brain slices. It has been known for more than a decade that the dopaminergic neuronal protein, DARPP-32 (which stands for dopamine and cAMP-regulated phosphoprotein of molecular mass of 32 kDa), is phosphorylated in response to elevations in intracellular cAMP. Upon phosphorylation, DARPP-32 becomes a potent inhibitor of protein phosphatase-1, much in the same fashion as the skeletal muscle inhibitor-1 and possibly the cerebellar G-substrate are phosphatase inhibitors upon phosphorylation. Tsou *et al.* (1993) have demonstrated that nitroprusside and cyclic GMP analogues promote the rapid phosphorylation of DARPP-32 on threonine 34, the residue known to be phosphorylated in response to elevations in cAMP. The effects of nitroprusside on DARPP-32 phosphorylation are completely prevented using haemoglobin in the extracellular fluid to bind NO. Interestingly, cGMP analogues (i.e. 8-Br-cGMP and 8-pCPT-cGMP) are more potent than the corresponding cAMP analogues, suggesting that cGMP kinase may be the *preferred* kinase in DARPP-32 phosphorylation in the intact cell. These findings raise the important issue that cGMP kinase may mediate DARPP-32 phosphorylation in the intact cell in response to either cAMP or cGMP elevations, much in the same manner that cGMP kinase may mediate vascular smooth muscle relaxation in response to both cyclic nucleotides. Thus, it should be kept in mind that studies using broken cell or purified protein phosphorylation may not be indicative of events that occur in the intact cell.

4.4.3 Effects of cGMP on neuronal ion fluxes

Glutamate-evoked LTD in the cerebellum has been correlated with alterations in extracellular K^+ concentrations (Shibuki and Okada, 1990). Conjunctive stimulation of parallel and climbing fibres decreases extracellular K^+ levels, as recorded using an ion-sensitive electrode. The decrease in K^+ levels is blocked by NOS inhibitors but overcome with nitroprusside or cGMP analogues (Shibuki and Okada, 1990, 1991). These results suggest that cGMP inhibits K^+ efflux from cerebellar slices. The mechanism is far from elucidated, but given the recent findings that cGMP kinase may regulate K^+ channel function (White *et al.*, 1993), a potential role for cGMP in K^+ channel regulation is feasible. Other examples of cGMP kinase regulation of neuronal ion channels exist as well. For example, cGMP and cGMP kinase activate Ca^{2+} currents in the ventral neurons from the snail, *Helix aspersa* (Paupardin-Tritsch *et al.*, 1986). Other examples of cGMP regulating ion currents in invertebrate (Liu and Takeuchi, 1993) and vertebrate (Chen and Schofield, 1993) neurons have been documented as well. It is not known at this point what the significance is of the cGMP and cGMP kinase-regulated ion channel function. Clearly, more work needs to be devoted to this area.

References

Ahmad, I., Korbmacher, C., Segal, A.S., Cheung, P., Boulpaep E.L. & Barnstable, C.J. (1992) *Proc. Natl. Acad. Sci. USA* **89**, 10262–10266.

Aiken, A., Bilham, T., Cohen, P., Aswad, D. & Greengard, P. (1981) *J. Biol. Chem.* **256**, 3501–3506.

Arshavsky, V.Y., Dumke, C.L. & Bownds, M.D. (1992) *J. Biol. Chem.* **267**, 24501–24507.

Aswad, D.W. & Greengard, P. (1981) *J. Biol. Chem.* **256**, 3494–3500.

Atkinson, R.W., Saudek, V., Huggins, J.P. & Pelton, J.T. (1991) *Biochemistry* **30**, 9387–9395.

Babcock, D.F., Bosma, M.M., Battaglia, D.E. & Darszon, A. (1992) *Proc. Natl. Acad. Sci. USA* **89**, 6001–6005.

Baltensperger, K., Chiesi, M. & Carafoli, E. (1990) *Biochemistry* **29**, 9753–9760.

Beavo, J.A. & Reifsnyder, D.H. (1990) *TIPS* **11**, 150–155.

Biel, M., Altenhofer, W., Hullin R., Ludwig, J., Freichel, M., Flockerzi, V., Dascal, N., Kaupp, U.B. & Hofmann, F. (1993) *FEBS Lett.* **329**, 134–138.

Bohme, G.A., Bon, C., Stutzmann, J.-M., Doble, A. & Blanchard, J.C. (1991) *Eur. J. Pharmacol.* **199**, 379–381.

Bredt, D.S., Glatt, C.E., Hwang, P.M., Fotuhi, M., Dawson, T.M. & Snyder, S.H. (1991) *Neuron* **7**, 615–624.

Buck, L. & Axel, R. (1991) *Cell* **65**, 175–187.

Casnellie, J.E., Ives, H.E., Jamieson, J.D. & Greengard, P. (1980) *J. Biol. Chem.* **255**, 3770–3776.

Charbonneau, H. (1990) In *Cyclic Nucleotide Phosphodiesterases: Structure, Regulation and Drug Action* (Beavo, J. and Houslay, M.D., eds), pp. 265–296. John Wiley and Sons, Ltd., West Sussex, England.

Charbonneau, H., Prusti, R., LeTrong, H., Sonnenburg, W.K., Mullaney, P.J., Walsh, K.A. & Beavo, J.A. (1990) *Proc. Natl. Acad. Sci. USA* **87**, 288–292.

Chen, C. & Schofield, G.G. (1993) *Eur. J. Pharmacol.* **243**, 83–86.

Colbran, J.L., Francis, S.H., Leach, A.B., Thomas, M.K., Jiang, H., McAllister, L.M. & Corbin, J.D. (1992) *J. Biol. Chem.* **267**, 9589–9594.

Cook, N.J., Hanke, W. & Kaupp, U.B. (1987) *Proc. Natl. Acad. Sci. USA* **84**, 585–589.

Cook, N.J., Zeilinger, C., Koch, K.W. & Kaupp, U.B. (1986) *J. Biol. Chem.* **261**, 17033–17039.

Cook, S.P. & Babcock, D.F. (1993) *J. Biol Chem.* **268**, 22402–22407.

Corbin, J.D., Ogreid, D., Miller, J.P., Suva, R.H., Jastorff, B. & Doskeland, S.O. (1986) *J. Biol. Chem.* **261**, 1208–1214.

Corbin, J.D., Sugden, P.H., West, L., Flockhart, D.A., Lincoln, T.M. & McCarthy, D. (1978) *J. Biol. Chem.* **253**, 3997–4003.

Cornwell, T.L., Pryzwansky, K.B., Wyatt, T.A. & Lincoln, T.M. (1991) *Mol. Pharmacol.* **40**, 923–931.

Crepel, F. & Jaillard, D. (1990) *NeuroReport* **1**, 133–136.

Daniel, H., Hemart, N., Jaillard, D. & Crepel, F. (1993) *Eur. J. Neurosci.* **5**, 1079–1082.

Doerner, D. & Alger, B.E. (1988) *Neuron* **1**, 693–699.

Doskeland, S.O., Vintermayr, O.K., Corbin, J.D. & Ogreid, D. (1987) *J. Biol. Chem.* **262**, 3534–3540.

Ecker, T., Gobel, C., Hullin, R., Rettig, R., Seitz, G. & Hofmann, F. (1989) *Circ. Res.* **65**, 1361–1369.
Ferris, C.D., Cameron, A.M., Bredt, D.S., Huganir, R.L. & Snyder, S.H. (1991) *Biochem. Biophys. Res. Commun.* **175**, 192–198.
Fesenko, E.E., Kolenikov, S.S. & Lyubarsky, A.L. (1985) *Nature* **313**, 310–313.
Forte, L.R., Thorne, P.K., Eber, S.L., Krause, W.J., Freeman, R.H., Francis, S.H. & Corbin, J.D. (1992) *Am. J. Physiol.* **263**, C607–C615.
Foster, J.L., Guttman, J. & Rosen, O.M. (1981) *J. Biol. Chem.* **256**, 5029–5036.
Francis, S.H., Lincoln, T.M. & Corbin, J.D. (1980) *J. Biol. Chem.* **255**, 620–626.
Francis, S.H., Noblett, B.D., Todd, B.W., Wells, J.N. & Corbin, J.D. (1988) *Mol. Pharmacol.* **504**, 506–517.
Francis, S.H., Woodford, T.A., Wolfe, L. & Corbin, J.D. (1989) *Second Messengers Phosphoproteins* **12**, 301–310.
Garbers, D.L. & Kopf, G.S. (1980) *Adv. Cyclic Nucleotide Res.* **13**, 251–306.
Garthwaite, J. & Garthwaite, G. (1987) *J. Neurochem.* **48**, 29–39.
Geiger, J., Nolte, C., Butt, E., Sage, S.O. & Walter, U. (1992) *Proc. Natl. Acad. Sci. USA* **89**, 1031–1035.
Gill, G.N. (1977) *J. Cyclic Nucleotide Res.* **3**, 153–162.
Glass, D.B. & Krebs, E.G. (1979) *J. Biol. Chem.* **254**, 9728–9738.
Guy, H.R., Durell, S.R., Warmke, J., Drysdale, R. & Ganetzky, B. (1991) *Science* **254**, 730.
Halbrugge, M., Friedrich, C., Eigenthaler, M., Schanzenbacher, P. & Walter, U. (1990) *J. Biol. Chem.* **265**, 3088–3093.
Haley, J.E., Wilcox, G.L. & Chapman, P.F. (1992) *Neuron* **8**, 211–216.
Hartzell, H.C. & Fischmeister, R. (1986) *Nature* **323**, 273–275.
Hofmann, F., Dostmann, W., Keilbach, A., Landgraf, W. & Ruth, P. (1992) *Biochim. Biophys. Acta* **1135**, 51–60.
Hubbard, M.J. & Cohen, P. (1993) *TIBS* **18**, 172–177.
Hurley, J.B. (1987) *Ann. Rev. Physiol.* **49**, 793–812.
Ives, H.E., Casnellie, J.E., Greengard, P. & Jamieson, J.D. (1980) *J. Biol. Chem.* **255**, 3777–3785.
Jiang, H., Colbran, J.L., Francis, S.H. & Corbin, J.D. (1992) *J. Biol. Chem.* **267**, 1015–1019.
Joyce, N.C., DeCamilli, P., Lohmann, S.M. *et al.*, (1986). *J. Cyclic Nucleotide Prot. Phospho. Res.* **11**, 191–198.
Kalderon, D. & Rubin, G.M. (1989) *J. Biol. Chem.* **264**, 10738–10748.
Karczewski, P., Kelm, M., Hartmann, M. & Schrader, J. (1992) *Life Sci.* **51**, 1205–1210.
Kaupp, U.B., Niidome, T., Tanabe, T., Terada, S., Bonigk, W., Stuhmer, W., Cook, N.J., Kanagawa, K., Matsuo, H., Hirose, T., Miyata, T. & Numa, S. (1989) *Nature* **342**, 762–766.
Knighton, D.R., Zheng, J., Ten Eyck, L.F., Ashford, V.A., Xuong, N.H., Taylor, S.S. & Sowadski, J.M. (1991) *Science* **253**, 407–414.
Knighton, D.R., Zheng, J., Ten Eyck, L.F., Xuong, N.H., Taylor, S.S. & Sowadski, J.M. (1991) *Science* **253**, 414–420.
Komalavilas, P. & Lincoln, T.M. (1994) *J. Biol. Chem.* **269**, 8701–8707.
Koga, T., Yoshida, Y., Cai, J.Q., Islam, M.U. & Imai, S. (1994) *J. Biol. Chem.* **269**, 11640–11647.
Krebs, E.G. & Beavo, J.A. (1979) *Ann. Rev. Biochem.* **48**, 923–959.
Landgraf, W., Hofmann, F., Pelton, J.T. & Huggins, J.P. (1990) *Biochemistry* **29**, 9921–9928.
Landgraf, W., Hullin, R., Gobel, C. & Hofmann, F. (1986) *Eur. J. Biochem.* **154**, 113–117.
Landgraf, W., Regulla, S., Meyer, H.E. & Hofmann, F. (1991) *J. Biol. Chem.* **266**, 16305–16311.
Light, D.B., Corbin, J.D. & Stanton, B.A. (1990) *Nature* **344**, 336–339.
Lincoln, T.M. & Corbin, J.D. (1977) *Proc. Natl. Acad. Sci. USA* **74**, 3229–3234.
Lincoln, T.M. & Corbin, J.D. (1983) *Adv. Cyclic Nucleotide Res.* **15**, 139–192.

Lincoln, T.M. & Cornwell, T.L. (1993) *FASEB J.* **7**, 328–338.
Lincoln, T.M., Cornwell, T.L. & Taylor, A.E. (1990) *Am. J. Physiol.* **258**, C399–C497.
Lincoln, T.M., Flockhart, D.A. & Corbin, J.D. (1978) *J. Biol. Chem.* **253**, 6002–6009.
Lincoln, T.M., Komalavilas, P. & Cornwell, T.L. (1994) *Hypertension* **23** [part 2], 1141–1147.
Lincoln, T.M., Thompson, M. & Cornwell, T.L. (1988) *J. Biol. Chem.* **163**, 17632–17637.
Liu, G.J. & Takeuchi, H. (1993) *Comp. Biochem. Physiol.* **104C**, 199–204.
Lohmann, S.M., Walter, U., Miller, P.E., Greengard, P. & DeCamilli, P. (1981) *Proc. Natl. Acad. Sci. USA* **78**, 653–657.
Ludwig, J., Margalit, T., Eismann, E., Lancet, D. & Kaupp, U.B. (1990) *FEBS Lett.* **270**, 24–29.
McAllister-Lucas, L.M., Sonnenburg, W.K., Kadlecek, A., Seger, D., Trong, H.L., Colbran, J.L., Thomas, M.K., Walsh, K.A., Francis, S.H., Corbin, J.D. & Beavo, J.A. (1993) *J. Biol. Chem.* **268**, 22863–22873.
MacFarland, R.T., Zelus, B.D. & Beavo, J.A. (1991) *J. Biol. Chem.* **266**, 136–142.
MacMillan-Crow, L.A. & Lincoln, T.M. (1994) *Biochemistry* **33**, 8035–8043.
MacMillan-Crow, L.A., Murphy-Ullrich, J.E. & Lincoln, T.M. (1994) *Biochem. Biophys. Res. Commun.* **201**, 531–537.
Manganiello, V.C., Tanaka, T. & Murashima, S. (1990) In *Cyclic Nucleotide Phosphodiesterases: Structure, Regulation, and Drug Action* (Beavo, J.A. and Houslay, M.D., eds), pp. 61–85. John Wiley & Sons Ltd., West Sussex, England.
Mery, P.F., Pavoine, C., Belhassen, L., Pecker, F. & Fischmeister, R. (1993) *J. Biol. Chem.* **268**, 26286–26295.
Miller, W.H. & Nicol, G.D. (1979) *Nature* **280**, 64–66.
Nakamura, T. & Gold, G.H. (1987) *Nature*, **325**, 442–444.
O'Dell, T.J., Hawkins, R.D., Kandel, E.R. & Arancio, O. (1991) *Proc. Natl. Acad. Sci. USA* **88**, 11285–11289.
Parks, T.P., Nairn, A.C., Greengard, P. & Jamieson, J.D. (1987) *Arch. Biochem. Biophys.* **255**, 361–371.
Paupardin-Tritsch, D., Hammond, C., Gerschenfeld, H.M., Nairn, A.C. & Greengard, P. (1986) *Nature* **323**, 812–814.
Prast, H. & Phillipu, A. (1992) *Eur. J. Pharmacol.* **216**, 139–140.
Pryzwansky, K.B., Wyatt, T.A., Nichols, H. & Lincoln, T.M. (1990) *Blood* **76**, 612–618.
Raeymaekers, L., Hofmann, F. & Casteels, R. (1988) *Biochem. J.* **252**, 269–273.
Reinhard, M., Halbrugge, M., Scheer, U., Wiegand, C., Jockusch, B.M. & Walter, U. (1992) *EMBO J.* **11**, 2063–2070.
Ruth, P., Landgraf, W., Keilbach, A., May, B., Egleme, C. & Hofmann, F. (1991) *Eur. J. Biochem.* **202**, 1339–1344.
Sandberg, M., Natarajan, V., Ronander, I., Kalderon, D., Walter, U., Lohmann, S.M. & Jahnsen, T. (1989) *FEBS Lett.* **255**, 321–329.
Sarcevic, B., Brookes, V., Martin, T.J., Kemp, B.E. & Robinson, P.J. (1989) *J. Biol. Chem.* **264**, 20648–20654.
Schlichter, D.J., Casnellie, J.E. & Greengard, P. (1978) *Nature* **273**, 61–62.
Schlichter, D.J., Detre, J.A., Aswad, D.W., Chehrazi, B. & Greengard, P. (1980) *Proc. Natl. Acad. Sci. USA* **77**, 5537–5541.
Shabb, J.B., Ng, L. & Corbin, J.D. (1990) *J. Biol. Chem.* **265**, 16031–16034.
Shabb, J.B., Buzzeo, B.D., Ng, L. & Corbin, J.D. (1991) *J. Biol. Chem.* **266**, 24320–24326.
Shabb, J.B. & Corbin, J.D. (1992) *J. Biol. Chem.* **267**, 5723–5726.
Shibuki, K. & Okada, D. (1990) *Neurosci. Lett.* **113**, 34–39.
Shibuki, K. & Okada, D. (1991) *Nature* **349**, 326–328.
Schuman, E.M. & Madison, D.V. (1991) *Science* **254**, 1503–1506.
Schuman, E.M., Meffert, M.K., Schulman, H. & Madison, D.V. (1992) *Soc. Neurosci.* **18**, 761.
Sonnenburg, W.K., Mullaney, P.J. & Beavo, J.A. (1991) *J. Biol. Chem.* **266**, 17655–17661.

Supattopone, S., Danoff, S.K., Theibert, A., Joseph, S.K., Steiner, J. & Snyder, S.H. (1988) *Proc. Natl. Acad. Sci. USA* **85**, 8747–8750.
Thomas, M.K., Francis, S.H. & Corbin, J.D. (1990a) *J. Biol. Chem.* **265**, 14964–14970.
Thomas, M.K., Francis, S.H. & Corbin, J.D. (1990b) *J. Biol. Chem.* **265**, 14971–14978.
Tsou, K., Snyder, G.L. & Greengard, P. (1993) *Proc. Natl. Acad. Sci. USA* **90**, 3462–3465.
Uhler, M. (1993) *J. Biol. Chem.* **268**, 13586–13591.
Waldmann, R., Bauer, S., Gobel, C., Hofmann, F., Jakobs, K.H. & Walter, U. (1986) *Eur. J. Biochem.* **158**, 203–210.
Waldmann, R., Nieberding, M. & Walter, U. (1987) *Eur. J. Biochem.* **167**, 441–448.
Walter, U. (1981) *Eur. J. Biochem.* **118**, 339–346.
Walter, U. & Greengard, P. (1981) *Curr. Topics Cell Regul.* **19**, 219–256.
Weber, I.T., Steitz, T.A., Bubis, J. & Taylor, S.S. (1987) *Biochemistry* **26**, 343–351.
Weber, I.T., Shabb, J.B. & Corbin, J.D. (1989) *Biochemistry* **28**, 6122–6127.
Wernet, W., Flockerzi, V. & Hofmann, F. (1989) *FEBS Lett.* **251**, 191–196.
Whalin, M.E., Strada, S.J. & Thompson, W.J. (1988) *Biochim. Biophys. Acta* **972**, 79–94.
Whalin, M.E., Scammell, J.G., Strada, S.J. & Thompson, W.J. (1991) *Mol. Pharmacol.* **39**, 711–717.
White, R.E., Lee, A.B., Shcherbatko, A.D., Lincoln, T.M., Schonbrunn, A. & Armstrong, D.L. (1993) *Nature* **361**, 263–266.
Wolfe, L., Corbin, J.D. & Francis, S.H. (1989) *J. Biol. Chem.* **264**, 7734–7741.
Wyatt, T.A., Lincoln, T.M. & Pryzwansky, K.B. (1991) *J. Biol. Chem.* **266**, 21274–21280.
Yoshida, Y., Sun, H.-T., Cai, J.-Q. & Imai, S. (1991) *J. Biol. Chem.* **266**, 19819–19825.

CHAPTER 5

LOCALIZATION OF NITRIC OXIDE NEURONS IN THE CENTRAL NERVOUS SYSTEM

Steven R. Vincent

Department of Psychiatry, The University of British Columbia, Vancouver, B.C., V6T 1Z3, Canada

Table of Contents

5.1 Introduction 83
5.2 Caveats and cautions 84
5.3 Localization of NOS in the mammalian central nervous system 86
5.3.1. Telencephalon 86
5.3.2 Diencephalon 89
5.3.3 Midbrain, pons, medulla 90
5.3.4 Cerebellum 92
5.3.5 Spinal cord 92
5.4 Non-mammalian species 93
5.5 NOS expression during development and following injury 94
5.6 Localization of soluble guanylyl cyclase, the NO receptor 95
5.7 Conclusions 96

References 97

5.1 Introduction

Nitric oxide has only very recently been identified as a neural messenger molecule. However, we already have a rather detailed understanding of the distribution of the neurons using this signalling molecule throughout the nervous system (Vincent, 1994). This is largely due to the fortuitous fact that the synthetic enzyme nitric oxide synthase (NOS) is able to catalyse selectively a histochemically detectable NADPH diaphorase reaction in aldehyde fixed tissue. Hope *et al.* (1991) initially provided biochemical evidence indicating that NOS accounted for the histochemical NADPH diaphorase reaction. This hypothesis was supported by others who found a one-to-one correspondence between neurons expressing NOS mRNA or NOS immunoreactivity, and those expressing NADPH diaphorase (Dawson *et al.*, 1991;

Nitric Oxide in the Nervous System
ISBN 0-12-721985-4

Bredt *et al.*, 1991a). Recently, the demonstration that in transgenic mice lacking a functional neuronal NOS gene there is a complete loss of NADPH diaphorase staining in the nervous system, provided definitive evidence for the molecular correspondence between these two activities (Huang *et al.*, 1993).

The regional distribution of NOS in the brain could be determined using a biochemical assay (Knowles *et al.*, 1989; Förstermann *et al.*, 1990). Subsequently, the purification of neuronal NOS made possible the development of specific antibodies to this enzyme for use in immunohistochemical studies (Bredt *et al.*, 1990). Antibodies to the NO precursor L-arginine (Aoki *et al.*, 1991), the NO co-product citrulline (Pasqualotto *et al.*, 1991) or the citrulline metabolizing enzyme argininosuccinate synthetase (Nakamura *et al.*, 1990, 1991; Arnt-Ramos *et al.*, 1992) have also been used in immunohistochemical studies. Finally, the molecular cloning of neuronal NOS (Bredt *et al.*, 1991a,b) has allowed *in situ* hybridization studies to localize neurons expressing the NOS gene (Figure 1). Together these varied approaches have provided much insight into the localization of NOS in the central nervous system.

5.2 Caveats and cautions

'I see nobody on the road,' said Alice.

'I only wish *I* had such eyes,' the King remarked in a fretful tone.

'To be able to see Nobody! And at that distance too! Why, it's as much as *I* can do to see real people, by this light!'

Through the Looking-Glass, Lewis Carroll

Although NADPH diaphorase histochemistry provides a robust and reliable method with which to localize NOS, care must be taken in the interpretation of such staining under certain conditions. Hope *et al.* (1991) demonstrated that while NOS accounts for only a small fraction of the NADPH diaphorase activity in a brain homogenate, this activity has the uniquely fortuitous property of surviving formaldehyde fixation. This has also been noted by others (Matsumoto *et al.*, 1993; Tracey *et al.*, 1993). When formaldehyde fixation is omitted, or incomplete, artefacts will occur. Thus much of the early literature on NADPH diaphorase histochemistry which has recently been cited (Sestan and Kostovic, 1994) was undertaken prior to the introduction of aldehyde fixation to this method (Scherer-Singler *et al.*, 1983) and would appear to be unrelated to NOS. Similarly, a recent study which reported staining of CA1 pyramidal neurons used only very short fixation periods, leaving the relationship of this staining to NOS unclear (Wallace and Fredens, 1992). Other studies have employed acetone fixation, which results in NADPH diaphorase staining unrelated to NOS (Schmidt *et al.*, 1992). This fixation also appears to adversely affect the immunohistochemical localization of NOS, resulting in staining of many

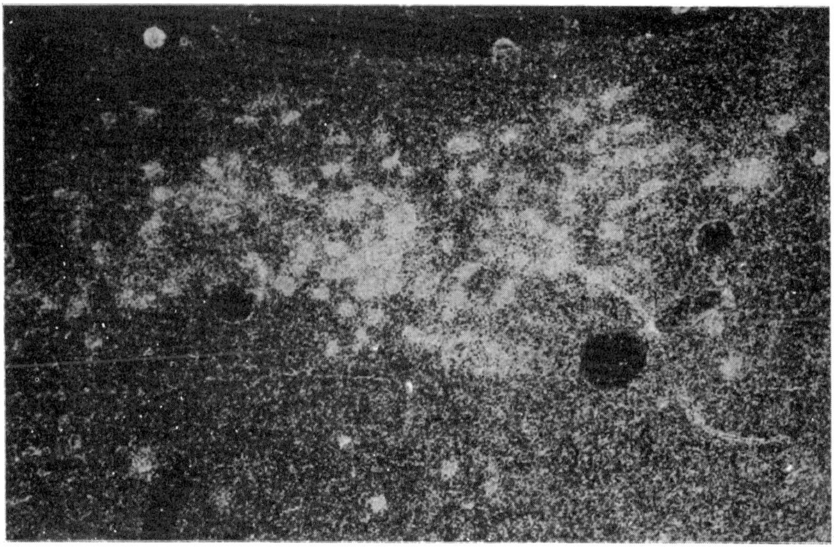

Figure 1 *In situ* hybridization using ^{35}S-labelled oligonucleotide probe to neuronal NO synthase showing high levels of expression in neurons of the laterodorsal tegmental nucleus of the rat.

cells not expressing neuronal NOS mRNA, NADPH diaphorase or NOS immunoreactivity when examined using the same antibodies and conventional methods (Schmidt *et al.*, 1992; Wendland *et al.*, 1994). Others have reported weak NADPH diaphorase staining of Purkinje cells during early postnatal development (Brüning, 1993a; Yan *et al.*, 1993). However, as noted by Brüning (1993a) this weak staining may not necessarily indicate the presence of NOS. Differences in fixation between rats of different ages might explain these results. It is thus key that these studies be confirmed using NOS immunohistochemistry or *in situ* hybridization.

Various dehydrogenase histochemical reactions rely on the reduction of tetrazolium salts, and in situations in which NADP is used as a substrate for the dehydrogenase, NADPH diaphorase can be detected (Scherer-Singler *et al.*, 1983). Thus reports on enzyme histochemistry using this procedure often are detecting NOS, and not the dehydrogenases themselves (i.e. Martinez-Murillo and Martinez-Rodriguez, 1984).

Although the nitro blue tetrazolium-based NADPH diaphorase technique has been used to examine the ultrastructure of some neurons (Vincent and Johansson, 1983; Scott *et al.*, 1987; Mizukawa *et al.*, 1988b; Mizukawa, 1990); it is not very useful for this purpose. Hope and Vincent (1989) used 2-(2′benzothiazolyl)-5-styryl-3-(4′-phthalhydrazidyl) tetrazolium chloride (BSPT), which forms an osmiophilic formazan, to localize NADPH diaphorase staining ultrastructurally to the endoplasmic reticulum of stained neurons. These observations were later confirmed by others (Wolf *et al.*, 1992). However, the electron microscopic detection of low levels of BSPT formazan in pyramidal and granule cells as well as astrocytes in the hippocampus, which were not visibly stained at the light microscopic level (Wolf *et al.*, 1993), indicates that this method may not only be detecting NADPH diaphorase activity associated with NOS. Furthermore, the localization of this formazan to the endoplasmic reticulum may be artefactual, since the staining in these studies is performed after fixation which could crosslink otherwise soluble proteins. Similar considerations apply in studying NOS immunohistochemically at the ultrastructural level (Valtschanoff *et al.*, 1993b). Ultrastructural studies using NOS immunohistochemistry in the enteric nervous system do not indicate a preferential association of the enzyme with any organelle (Llewellyn-Smith *et al.*, 1992).

5.3 Localization of NOS in the mammalian central nervous system

5.3.1 Telencephalon

NOS is found in cortical neurons scattered in layers II–VI and in the subcortical white matter (Mizukawa *et al.*, 1989; Hedlich *et al.*, 1990; Leigh *et al.*, 1990; Vincent and Kimura, 1992) (Figure 2A,B). These neurons are the original 'solitary active cells' described by Thomas and Pearse (1961, 1964) in their pioneering work on dehydrogenase histochemistry. The cortical NOS neurons contain somatostatin and neuropeptide Y (Vincent *et al.*, 1983a; Kowall and Beal, 1988) and express glutamate decarboxylase (GAD) mRNA (Chesselet and Robbins, 1989) and GABA immunoreactivity (Hedlich *et al.*, 1990) but lack parvalbumin (Hedlich *et al.*, 1990). Similar cell and neuropil staining has been reported in human cortex (DeFelipe, 1993; Springall *et al.*, 1992). The staining of cortical neuropil with NADPH diaphorase and NOS immunohistochemistry in primates appears to match the functional organization seen with cytochrome oxidase histochemistry (Sandell, 1986; Aoki *et al.*, 1993). Likewise in the lizard cortex, the laminar organization of these two enzyme markers is similar (Regidor and Poch, 1987).

In the rat, most of the neurons in the subcortical white matter contain NADPH diaphorase (Meyer *et al.*, 1991), while in humans only a few such neurons are reported to display this enzyme (Meyer *et al.*, 1992). In the rat, these neurons have been shown to give rise to projections to the overlying cortex, as well as adjacent cortical areas (Meyer *et al.*, 1991). In embryonic rat cortex, similar scattered NADPH diaphorase-positive neurons have been reported on E18 (Sharp *et al.*, 1986). These interstitial cells appear to be a remnant of the embryonic subplate neuronal population, and are the oldest neurons in the cortex (Chun and Shatz, 1989). Indeed, a recent study has demonstrated that NADPH diaphorase expression occurs in the subplate but is absent from the cortical plate during embryonic development (Derer and Derer, 1993). In human brain Cajal–Retzius cells are

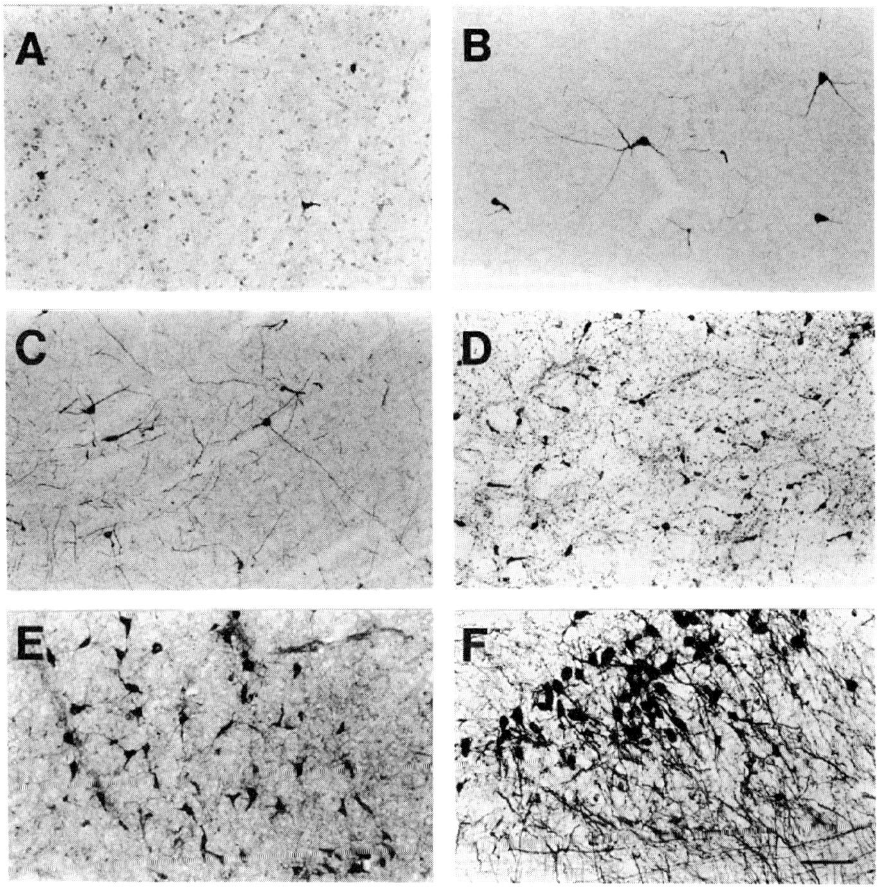

Figure 2 NADPH diaphorase histochemical localization of NO synthase in the brain. (A) Scattered positive neurons in *post-mortem* human cortex. Positive, aspiny neurons in adult rat cortex (B) and striatum (C) and in rat striatum at postnatal day 1 (D). Large, intensely labelled neurons are present in the mesopontine tegmentum of chick (E) and frog (F) corresponding to the laterodorsal and pedunculopontine tegmental cell group of mammals. Bar=100 μm for all figures.

reported to express NADPH diaphorase transiently at 20 weeks gestation (Meyer and Gonzalez-Hernandez, 1993). A very exciting recent observation in schizophrenic brains indicates a loss of NADPH diaphorase-positive neurons in the cortex, together with an increase in the density of these neurons in the subcortical white matter, suggesting a developmental disturbance of these subplate neurons may occur in this disease (Akbarian *et al.*, 1993a,b).

Populations of NADPH diaphorase-stained or NOS-immunoreactive neurons similar to those in the neocortex are also seen in allocortical areas. These have been well described in the olfactory bulb (Scott *et al.*, 1987; Croul-Ottman and Brunjes, 1988; Villalba *et al.*, 1989; Davis, 1991; Vincent and Kimura, 1992).

Like the pyramidal cells of the neocortex, the CA1 pyramidal cells of the hippocampus do not express the mRNA for the neuronal form of NOS, or display NOS immunoreactivity (Bredt *et al.*, 1991a; Bredt and Snyder, 1992). A few reports have appeared suggesting NADPH diaphorase staining is present in these cells (Endoh *et al.*, 1993; Southam and Garthwaite, 1993), but most studies indicate an absence of this activity from the hippocampal pyramidal neurons (Mizukawa *et al.*, 1989; Leigh *et al.*, 1990; Mufson *et al.*, 1990; Seidel *et al.*, 1991; Vincent and Hope, 1992; Vincent and Kimura, 1992). Instead, NADPH diaphorase and NOS are contained in a population of hippocampal interneurons, many of which also express GABA immunoreactivity (Valtschanoff *et al.*, 1993b). The exception to this rule is in the subiculum, where most of the neurons in the inner part of the pyramidal layer are NADPH diaphorase and NOS positive (Vincent and Kimura, 1992; Valtschanoff *et al.*, 1993b).

NOS mRNA is expressed in the dentate gyrus (Bredt and Snyder, 1992), where NADPH diaphorase and NOS immunoreactivity are confined to basket cells and other interneurons in rat, cat and primates (Mizukawa *et al.*, 1989; Leigh *et al.*, 1990; Mufson *et al.*, 1990; Vincent and Hope, 1992; Vincent and Kimura, 1992; Hong *et al.*, 1993; Rebeck *et al.*, 1993). It is interesting that while NADPH diaphorase neurons have often been suggested to be resistant to various insults, those in the dentate gyrus appear to be selectively damaged by transient ischaemia (Hong *et al.*, 1993). There is also a marked loss of NADPH diaphorase staining from the neuropil in the molecular layer of the dentate in Alzheimer's disease (Rebeck *et al.*, 1993).

A further source of hippocampal NO might be the septohippocampal cholinergic projection. NADPH diaphorase has been demonstrated in magnocellular neurons in the medial septum and nucleus of the diagonal band of Broca which project to the hippocampus (Kinjo *et al.*, 1989). These have been shown by double labelling to be cholinergic (Schober *et al.*, 1989; Brauer *et al.*, 1991; Pasqualotto and Vincent, 1991; Kitchener and Diamond, 1993), and many also contain the neuropeptide galanin (Pasqualotto and Vincent, 1991) and express the low affinity p75 NGF receptor (Peng *et al.*, 1994). These cells are distinct from the parvalbumin-immunoreactive septohippocampal GABA neurons (Brauer *et al.*, 1991), and also lack calbindin (Geula *et al.*, 1993). It is interesting that while about 70% of the cholinergic neurons in the rat medial septum and up to a third of those in the entire rat basal forebrain express NADPH diaphorase activity (Geula *et al.*, 1993; Kitchener and Diamond, 1993), essentially no cholinergic neurons in the monkey or human basal forebrain express this enzyme (Ellison *et al.*, 1987; Mesulam *et al.*, 1989; Geula *et al.*, 1993).

Within the amygdala, a scattered population of NADPH diaphorase-positive neurons similar to that in cortical regions has been described in rat, cat and primates (Ellison *et al.*, 1987; Mizukawa *et al.*, 1988a, 1989; Leigh *et al.*, 1990; Sims and Williams, 1990; Pitkänen and Amaral, 1991; Brady *et al.*, 1992; Vincent and Kimura, 1992; McDonald *et al.*, 1993).

The striatal NOS neurons are also similar to those present in the cortex (Figure 2C,D). In the rat, they have been shown to be aspiny interneurons, which contain somatostatin, neuropeptide Y (NPY) (Vincent *et al.*, 1983a,d) and the NPY-related peptide C-PON (Villalba *et al.*, 1988). Similar neurons are seen in cat and human striatum, where the neuropil staining of NADPH diaphorase fibres appears to correspond to the striatal patch-matrix organization (Sandell *et al.*, 1986; Kowall *et al.*, 1987; Sajin *et al.*, 1992; Sadikot *et al.*, 1992; Morton *et al.*, 1993).

Although the striatal NOS-immunoreactive neurons do not appear to express significant GAD mRNA (Chesselet and Robbins, 1989), they are reported to display GAD_{67} immunoreactivity (Kubota *et al.*, 1993). These cells lack the calcium binding proteins parvalbumin (Cowan *et al.*, 1990; Kita *et al.*, 1990; Kubota *et al.*, 1993) and calretinin (Bennett and Bolam, 1993a; Kubota *et al.*, 1993), although some do display calbindin D28k immunoreactivity (Bennet and Bolam, 1993b). Calcineurin, the calmodulin-dependent protein phosphatase, has also been reported to be absent from these cells (Goto *et al.*, 1987).

These striatal neurons express substance P receptor mRNA (Kaneko *et al.*, 1993), and most also express m_1 muscarinic receptor mRNA, while only about 15% express m_4 mRNA (Bernard *et al.*, 1992). A similar small percentage of these neurons express high levels of NMDA receptor mRNA (Price *et al.*, 1993; Augood *et al.*, 1994). The weak or undetectable levels of NMDA receptors expressed in many of these cells may account for their relative lack of sensitivity to NMDA-induced neurotoxicity. Furthermore, these cells also express high levels of Mn-superoxide dismutase (Inagaki *et al.*, 1991), which may also protect them from damage by oxygen radicals (see Chapter 10).

The striatal neurons expressing NADPH diaphorase are spared in Huntington's disease (Ferrante *et al.*, 1985; Kowall *et al.*, 1987; Morton *et al.*, 1993), although a dramatic decrease in the striatal neuropil NADPH diaphorase staining occurs in this disease (Morton *et al.*, 1993).

5.3.2 Diencephalon

NOS is present in various populations of hypothalamic neurons, and NO appears to play a major role in regulating hypothalamic functions (see Chapter 8). The magnocellular neurons of the supraoptic and paraventricular nuclei of the rat show NOS immunoreactivity (Bredt *et al.*, 1990). NADPH diaphorase has also been described in these hypothalamic neurons in rat, cat and humans (Mizukawa *et al.*, 1989; Leigh *et al.*, 1990; Sangruchi and Kowall, 1991; Alonso *et al.*, 1992a; Arevalo *et al.*, 1992; Vincent and Kimura, 1992). The magnocellular neurosecretory NOS neurons in the rostral hypothalamus contain vasopressin, however co-localization appears more frequent with oxytocin (Calka and Block, 1993b; Torres *et al.*, 1993; Sanchez *et al.*, 1994). The NOS-immunoreactive magnocellular neurosecretory neurons display calbindin D28k immunoreactivity (Alonso *et al.*, 1992a) and a few also contain calretinin (Arevalo *et al.*, 1993). A small number of supraoptic and paraventicular neurons contain NADPH diaphorase and angiotensin-(1–7) immunoreactivity (Calka and Block, 1993a), and a small percentage of the NOS neurons in the periventricular parvicellular division of the paraventricular nucleus contains somatostatin (Alonso *et al.*, 1992b). Some parvocellular CRF-immunoreactive neurons also displayed NADPH diaphorase staining (Torres *et al.*, 1993).

Dehydration increased the staining of neurons in the supraoptic nucleus (Pow, 1992), although adrenalectomy did not (Torres *et al.*, 1993). Salt loading also increased NADPH diaphorase activity in the posterior pituitary (Sagar and Ferriero, 1987). An increase in NOS mRNA can be seen in the paraventricular nucleus following immobilization stress (Calza *et al.*, 1993), and in lactating rats (Ceccatelli and Eriksson, 1993).

In addition to being present in hypothalamic neurosecretory neurons, there is also recent evidence for NOS immunoreactivity and NOS mRNA expression in gonadotrophs and folliculo-stellate cells of the anterior pituitary itself (Ceccatelli *et al.*, 1993). This expression is increased following gonadectomy. Furthermore, GH3 cells express a Ca^{2+}/calmodulin-dependent NOS activity (Wolff and Datto, 1992).

NADPH diaphorase has been detected in interneurons in the retina of fish, amphibians, birds and mammals (Kuwabara and Cogan, 1960; Bhattacharjee, 1977; Sandell, 1985; Sagar, 1986, 1990; Mitrofanis, 1989; Vaccaro *et al.*, 1991; Mitrofanis *et al.*, 1992; Vincent and Hope, 1992; Koistinaho *et al.*, 1993; Osborne *et al.*, 1993; Yamamoto *et al.*, 1993). Some of these correspond to GABAergic amacrine cells in the inner margin of the inner nuclear layer (Vaney and Young, 1988). In addition, cells in the ganglion cell layer have been reported in guinea-pig, cat and primate (Sandell, 1985; Cobcroft *et al.*, 1989; Vaccaro *et al.*, 1991). Both NOS activity (Venturini *et al.*, 1991) and an NO-activated guanylyl cyclase have been identified in bovine rod outer segments (Horio and Murad, 1991). In the human eye, cone photoreceptors appear to display NADPH diaphorase activity (Kuwabara and Cogan, 1960; Provis and Mitrofanis, 1990).

NADPH diaphorase-positive cells are rare in the thalamus. Patches of NADPH diaphorase-positive fibres are seen in medial thalamic nuclei, often matching acetylcholinesterase-positive zones (Vincent and Kimura, 1992; Mengual *et al.*, 1993). This likely reflects the origin of these fibres in the mesopontine tegmental cholinergic cell group. These neurons also innervate the lateral geniculate nucleus (Bickford *et al.*, 1993). Intensely stained cells are present in the ventral lateral geniculate nucleus of various mammals (Agarwala *et al.*, 1992a,b; Vincent and Kimura, 1992; González-Hernández *et al.*, 1993). NADPH diaphorase-positive neurons are also present in the dorsal part of the suprageniculate nucleus and the caudal, basal, medial portion of the medial geniculate body (Vincent and Kimura, 1992; Druga and Syka, 1993).

5.3.3 Midbrain, pons, medulla

An extensive pattern of NADPH diaphorase-positive neurons is present in the adult as well as the developing tectum (González-Hernández *et al.*, 1992, 1993; Vincent and Kimura, 1992). Positive neurons are present in the external cortices of the inferior colliculus and the intercollicular commissure (Vincent and Kimura, 1992; Druga and Syka, 1993). The neurons of the magnocellular nucleus of the posterior commissure are very heavily stained (Vincent and Kimura, 1992). A lattice of NADPH diaphorase-positive fibres is present in the intermediate grey layer of the superior colliculus which appears to arise in the cholinergic nuclei of the mesopontine tegmentum (Wallace, 1986; Wallace and Fredens, 1989). Of particular note is the presence of a column of stained cells in the dorsolateral periaqueductal grey, which extend out into the deeper layers of the superior colliculus (Vincent and Kimura, 1992; Onstott *et al.*, 1993).

The most striking NOS cell group in the brainstem is in the laterodorsal and pedunculopontine tegmental nuclei (Figure 2E,F). This is a well-known cholinergic cell group, and double staining experiments in the rat indicate that essentially all of these neurons express NOS (Vincent *et al.*, 1983b). Similar staining is seen in the cat (Reiner and Vincent, 1987).

Many neuropeptides are co-expressed together with NOS in these cholinergic neurons. Substance P, corticotropin-releasing factor (CRF) and gastrin-releasing peptide (GRP) appear to be present in about a third of these cells (Vincent *et al.*, 1983c, 1986). The observations that these neurons all express atrial and brain natriuritic peptides is especially interesting (Standaert *et al.*, 1986; Saper *et al.*, 1989). These peptides act on cell surface receptors which are particulate guanylyl cyclases (see Chapter 3). The release of these peptides from these neurons could activate particulate guanylyl cyclase while NO produced in these same neurons would activate soluble guanylyl cyclase in target cells.

These neurons also appear to express NOS in the human brain (Kowall and Mueller, 1988; Nakamura *et al.*, 1988; Mesulam *et al.*, 1989). There are reports that these neurons show neurofibrillary tangles in Alzheimer's disease (Mufson *et al.*, 1988) and degenerate in progressive supranuclear palsy and idiopathic Parkinson's disease (Hirsch *et al.*, 1987; Zweig *et al.*, 1989). In contrast, there is a report that there is an increase in the number of these neurons in *post-mortem* schizophrenic brains (Karson *et al.*, 1991). In the rat, the development of these neurons has been studied in detail (Skinner *et al.*, 1989), and these cells appear to decrease in size with ageing (Kawamata *et al.*, 1990).

The distribution of NADPH diaphorase-positive neurons in relation to aminergic neurons containing tyrosine hydroxylase and serotonin in the rat brainstem has been reported (Johnson and Ma, 1993). A similar study has also been undertaken using antibodies to NOS (Dun *et al.*, 1994). These reports indicate that NOS and serotonin extensively co-exist in the dorsal and medial raphe nuclei of the pons, but NOS is not present in the serotonin neurons of the ventral medulla.

NOS neurons are widespread in the medulla. Large, multipolar NADPH diaphorase-positive neurons are scattered in the medullary reticular formation in rat and cat (Mizukawa *et al.*, 1989; Vincent and Kimura, 1992). Many neurons in the nucleus of the solitary tract and the ventrolateral medulla are well stained, and these are distinct from the tyrosine hydroxylase-immunoreactive cells in these areas

(Ohta *et al.*, 1993; Dun *et al.*, 1994), although a few tyrosine hydroxylase-positive neurons in the lateral reticular nucleus are reported to contain NOS (Dun *et al.*, 1994). These observations are consistent with the lack of NADPH diaphorase in the C1 adrenaline cells (Iadecola *et al.*, 1993). A small number of the tyrosine hydroxylase-positive neurons in the rostrodorsal ventral tegmental area, rostral linear nucleus and periaqueductal grey contain NADPH diaphorase, but other catecholamine neurons in the brain lack NOS (Johnson and Ma, 1993). The nucleus ambiguus shows weak NADPH diaphorase staining (Vincent and Kimura, 1992), and a few cells and fibres show NOS immunoreactivity (Dun *et al.*, 1994). Axotomy increases NADPH diaphorase staining in rat vagal motoneurons (Gonzalez *et al.*, 1987). NADPH diaphorase-positive neurons have been described in the nucleus of the solitary tract, the dorsal motor nucleus of the vagus, the hypoglossal nucleus and the reticular formation in the human brain (Kowall and Mueller, 1988). Many small neurons in the substantia gelatinosa of the spinal trigeminal nucleus show NADPH diaphorase activity (Vincent and Kimura, 1992).

5.3.4 Cerebellum

The granule cells of the cerebellar cortex are stained with NADPH diaphorase activity (Bredt *et al.*, 1991a; Southam *et al.*, 1992; Vincent and Kimura, 1992; Yan *et al.*, 1993). These neurons also show NOS immunoreactivity and *in situ* hybridization indicates that they express the neuronal NOS gene (Bredt *et al.*, 1991a; Schilling *et al.*, 1994). More recent studies indicate that there are clusters of granule cells which express high levels of NADPH diaphorase and other patches of granule cells that do not appear to express NOS in the rodent cerebellum (Yan *et al.*, 1993; Schilling *et al.*, 1994).

Basket cells in the molecular layer also stain for both NOS immunoreactivity and NADPH diaphorase activity (Southam *et al.*, 1992; Vincent and Kimura, 1992). Stellate cells have also been reported to be stained (Southam *et al.*, 1992). Neurons in the molecular layer are also NOS immunoreactive in the human brain (Springall *et al.*, 1992).

5.3.5 Spinal cord

The localization of NOS in the spinal cord has been described by many groups in rat, mouse, cat and primate. Positive neurons are found scattered in laminae I–IV and X throughout the cord (Mizukawa *et al.*, 1989; Leigh *et al.*, 1990; Aimi *et al.*, 1991; Brüning, 1992; Dun *et al.*, 1992, 1993; Valtschanoff *et al.*, 1992; Terenghi *et al.*, 1993; Zhang *et al.*, 1993; Saito *et al.*, 1994; Vizzard *et al.*, 1994). There is a dense plexus of NOS fibres in lamina II of the dorsal horn, in which GABA co-exists (Valtschanoff *et al.*, 1993a). Indeed, many of the NADPH diaphorase-positive neurons in the dorsal horn contain both GABA and glycine immunoreactivities (Spike *et al.*, 1993).

Much of the neuropil staining in the dorsal horn may arise from primary afferents at certain spinal levels. Aimi *et al.* (1991) have found that about 20% of the dorsal root ganglia neurons at lower thoracic levels display NADPH diaphorase. These appear to also contain CGRP and substance P, and to be visceral afferents (Aimi *et al.*, 1991). Lumbar dorsal root ganglion neurons displaying NOS immunoreactivity have also been described in rat and monkey, co-existing with CGRP and sometimes substance P and galanin (Zhang *et al.*, 1993). Visceral afferent neurons in the L6–S1 ganglia, including sensory neurons innervating the cervix (Papka and McNeill, 1992), penis (McNeill *et al.*, 1992a) and bladder (MacNeil *et al.*, 1992b; Vizzard *et al.*, 1993a,b,c), also contain NADPH diaphorase. In the cat, many sensory cells in the L7–S2 dorsal root ganglia show strong NADPH diaphorase staining (Vizzard *et al.*, 1994). NOS immunoreactivity has been seen in human dorsal root ganglia neurons (Springall *et al.*, 1992; Terenghi *et al.*, 1993).

The cholinergic preganglionic autonomic neurons in the intermediolateral cell columns of the thoracic cord display high levels of NOS immunoreactivity or NADPH diaphorase activity (Anderson, 1992; Blöttner and Baumgarten, 1992; Brüning, 1992; Valtschanoff *et al.*, 1992; Dun *et al.*, 1992, 1993; Blöttner *et al.*, 1993; Spike *et al.*, 1993; Terenghi *et al.*, 1993; Saito *et al.*, 1994; Vizzard *et al.*, 1994). Most sacral parasympathetic preganglionic neurons show NADPH diaphorase activity as well (Anderson, 1992). Many positive neurons are also seen in lamina X throughout the cord. At least 10% of these NOS cells appear to be spinothalamic neurons (Lee *et al.*, 1993). A similar distribution is seen in human spinal cord, although some ventral horn motoneurons also appear to contain NOS (Springall *et al.*, 1992; Terenghi *et al.*, 1993).

5.4 Non-mammalian species

Although the structures of NOS from non-mammalian species have yet to be determined, this enzyme appears to be highly conserved. Neurons expressing NADPH diaphorase activity have been noted in avian (Brüning, 1993b), reptilian (Regidor and Poch, 1987; Luebke *et al.*, 1992), amphibian and fish (Schober *et al.*, 1993) brains (Figure 2E,F). As in mammals, NADPH diaphorase is present in retinal interneurons of fish (Weiler and Kewitz, 1993) and amphibians (Sato, 1990a). The frog pineal also shows NADPH diaphorase activity (Sato, 1990b). In the toad, Li *et al.* (1992) showed that an antiserum against rat brain NOS stained a population of enteric neurons that also displayed NADPH diaphorase activity.

Even before the discovery of NO as a physiological messenger, there was already evidence for a role for arginine metabolism in memory consolidation in insects (D'Alessio *et al.*, 1982). This may reflect a role for NO, since NADPH diaphorase histochemistry detects discrete populations of neurons in invertebrates. NOS activity has also been described in the locust brain (Elphick *et al.*, 1993). Specific NADPH diaphorase staining of neurons in annelids, molluscs and arthropods, but not in coelenterates, turbellarians, nematodes and urochordates, has been described (Elofsson *et al.*, 1993). The buccal ganglia of pulmonate molluscs, where there is electrophysiological evidence for an action of NO on neurons (Moroz *et al.*, 1993), is particularly strongly stained, and displays NOS activity (Elofsson *et al.*, 1993). NOS activity and the reversible binding of NO by an endogenous haem protein have been described in the salivary gland of the bloodsucking insect *Rhodnius prolixus* (Ribeiro and Nussenzveig, 1993; Ribeiro *et al.*, 1993). Here NO would appear to be injected into prey, causing vasodilatation and inhibiting platelet aggregation, to enable the insect to feed on blood.

5.5 NOS expression during development and following injury

The neuronal and endothelial forms of NOS are often referred to as constitutive, in contrast to the inducible form expressed by cells of the immune system. However, the expression of neuronal NOS also appears to be actively regulated. In the early foetal rat, most dorsal root ganglion cells display NADPH diaphorase activity (Wetts and Vaughn, 1993). However, most cells lose this activity during late foetal development and only in mid-thoracic ganglia are a substantial number of positive cells seen in adults (Aimi *et al.*, 1991; Wetts and Vaughn, 1993). In contrast, very little staining is seen in the dorsal horn at birth, but it increases steadily to reach the adult pattern by the third postnatal week (Liuzzi *et al.*, 1993). A group of NOS-positive neurons distinct from but adjacent to the ventral horn motoneurons is present at postnatal day 1, and then slowly disappears over the next 2 weeks (Kalb and Agostini, 1993). These may play a role in shaping motoneuron development, since NOS antagonists appear to inhibit this process (Kalb and Agostini, 1993).

Many studies indicate that glial cells can respond to local injury with an induction of NOS, and glial cells in culture express the calcium-independent, inducible form of this enzyme (see Chapter 9). When cortical or brainstem neurons are cultured, the same populations of neurons that express NOS *in vivo* appear to express the enzyme (Uehara-Kunugi *et al.*, 1991). However, NOS expression in neurons may be induced by injury and other stimuli. Unilateral cortical or hippocampal lesions induced the bilateral appearance of NADPH diaphorase in some pyramidal cells (Kitchener *et al.*, 1993; Regidor *et al.*, 1993). Following hypophysectomy, NOS activity is increased in the perikarya and regenerating axons of supraoptic and paraventricular hypothalamic neurons (Wu and Scott, 1993). Lesions of the medial forebrain bundle and mammillothalamic tract increased NOS and NADPH diaphorase in the mammillary nucleus (Herdegen *et al.*, 1993). Unilateral hindpaw inflammation produces a bilateral increase in NADPH diaphorase-positive fibre staining in the lumbar spinal cord (Solodkin *et al.*, 1992). Wu (1993) found that NOS could be induced in adult rat spinal cord after spinal root avulsion or spinal hemisection. This increase in NOS staining could be prevented by chronic treatment with nitroarginine which also reduced the subsequent death of motoneurons (Wu and Li, 1993). When the sciatic nerve is cut in the mid-thigh of neonatal, but not adult rats, NADPH diaphorase is induced in sciatic motoneurons, during the time of extensive motoneuron death (Clowry, 1993). These lesions increase NOS synthesis in dorsal root ganglia of the rat, but apparently not in the monkey (Verge *et al.*, 1992; Fiallos-Estrada *et al.*, 1993; Zhang *et al.*, 1993). Rhizotomy at L6–S1 doubled the number of NADPH diaphorase-positive neurons in the sacral parasympathetic nucleus (Vizzard *et al.*, 1993c).

5.6 Localization of soluble guanylyl cyclase, the NO receptor

Soluble guanylyl cyclase has been well characterized as the physiological receptor for NO (see Chapter 3). Initial attempts to localize this enzyme histochemically used an antiserum against the purified rat brain enzyme (Zwiller *et al.*, 1981). This stained all neurons and glia in the cerebellum. Another study used monoclonal antibodies raised against the enzyme purified from rat lung, and found staining was primarily confined to Purkinje cells, with some staining also in basket and stellate cells, while glial elements were unstained (Ariano *et al.*, 1982). In cortex, pyramidal neurons showed guanylyl cyclase immunoreactivity, while in the striatum medium spiny neurons were stained (Ariano *et al.*, 1982). Another group prepared a monoclonal antibody to soluble guanylyl cyclase from rat brain and found a similar pattern of staining, and also noted staining in the interpeduncular nucleus, medial habenula and dentate gyrus (Nakane *et al.*, 1983).

Recent studies have begun to use *in situ* hybridization to examine the localization of the various isoforms of soluble guanylyl cyclase that have now been cloned (see Chapter 3). The β_1 subunit was expressed in the striatum, olfactory tubercle, cortical and hippocampal pyramidal cells, dentate granule cells and the medial habenula (Matsuoka *et al.*, 1992). In the cerebellar cortex, Purkinje, basket, stellate and Golgi cells were moderately labelled, while only a weak signal was observed in granule cells, and glial elements were unlabelled (Matsuoka *et al.*, 1992). The distribution of the α_1 subunit was similar to that of the β_1 subunit in most areas, although the CA1 pyramidal cells and dentate granule cells showed only weak expression of this larger subunit (Furuyama *et al.*, 1993). In addition, the α_1 subunit was strongly expressed in the pineal gland, while the β_1 was weak. The particularly strong labelling of the medial habenula with probes to both subunits is of interest (Matsuoka *et al.*, 1992; Furuyama *et al.*, 1993), since this brain area is essentially devoid of NOS-positive fibres or cell bodies (Vincent and Kimura, 1992). Thus in this area, factors other than NO may regulate soluble guanylyl cyclase. Also of particular note is the very high expression of both subunits in the laterodorsal tegmental nucleus (Furuyama *et al.*, 1993). As discussed above, these cells express very high levels of NOS as well, indicating that in this cell group soluble guanylyl cyclase may act as a NO autoreceptor.

5.7 Conclusions

As with other neurotransmitters, NO is produced by a wide variety of neurons. It is common in both the central and peripheral nervous systems. It is found in local interneurons, as in the striatum, and in long projection neurons, such as those in the mesopontine tegmentum. Neurosecretory neurons in the hypothalamus may release NO into the portal system, together with peptide messengers. NOS is frequently found co-existing with amine or peptide neurotransmitters, but can also be expressed in excitatory glutamate neurons or inhibitory GABA neurons (Table 1). Indeed, NO would appear to rank with these latter two as one of the most common of all neural messenger molecules.

The differential staining of nerve cell bodies, dendrites and axon terminals in various brain regions may well indicate different modes of NO transmission. In some situations, NO would be expected to be released from the cell bodies and dendrites in response to calcium influx following synaptic activation. In other areas, where NOS is concentrated in nerve terminals, one might predict NO synthesis to be triggered by calcium influx following action potential invasion of the nerve ending. Given that the NO receptor appears to be soluble guanylyl cyclase, the varied actions of NO in these different situations may well be mediated by the multiple actions of cGMP and its targets in the neurons responding to NO. Thus now

Table 1 Nitric oxide synthase co-exists with various neurotransmitters

Glutamate neurons	Cerebellar granule cells	(Bredt *et al.*, 1991a)
GABA neurons	Cerebellar basket cells	(Bredt *et al.*, 1991a)
	Cortical interneurons	(Chesselet and Robbins, 1989)
	Hippocampal interneurons	(Valtschoff *et al.*, 1993b)
	Spinal interneurons	(Spike *et al.*, 1993)
	Retinal interneurons	(Vaney and Young, 1988)
Cholinergic neurons	Septohippocampal cells	(Pasqualotto and Vincent, 1991)
	Mesopontine tegmental neurons	(Vincent *et al.*, 1983b)
	Preganglionic sympathetic neurons	(Anderson, 1992)
Catecholamine neurons	Rostral ventral tegmentum	(Johnson and Ma, 1993)
Serotonergic neurons	Dorsal raphe	(Johnson and Ma, 1993)
Peptidergic neurons	Forebrain interneurons	(Vincent *et al.*, 1983a)
	Hypothalamic neurosecretory neurons	(Bredt *et al.*, 1990)
	Sensory neurons	(Aimi *et al.*, 1991)
	Enteric neurons	(Aimi *et al.*, 1993)

that we know which neurons are saying NO, attention can focus on what the cells hearing this message are doing in response.

References

Agarwala, S., Günlük, A.E., May, J.G. III & Petry, H.M. (1992a) *J. Comp. Neurol.* **318**, 267–276.

Agarwala, S., May, J.G. III, Moore, J.K. & Petry, H.M. (1992b) *J. Comp. Neurol.* **318**, 255–266.

Aimi, Y., Fujimura, M., Vincent, S.R. & Kimura, H. (1991) *J. Comp. Neurol.* **382**, 382–392.

Aimi, Y., Kimura, H., Kinoshita, T., Minani, Y., Fujimura, M. & Vincent, S.R. (1993) *Neuroscience* **53**, 553–560.

Akbarian, S., Bunney, W.E.Jr., Potkin, S.G., Wigal, S.B., Hagman, J.O., Sandman, C.A. & Jones, E.G. (1993a) *Arch. Gen. Psychiat.* **50**, 169–177.

Akbarian, S., Vinuela, A., Kim, J.J., Potkin, S.G., Bunney, W.E.Jr. & Jones, E.G. (1933b) *Arch. Gen. Psychiat*, **50**, 178–187.

Alonso, J.R., Sánchez, F., Arévalo, R., Carretero, J., Aijón, J. & Vázquez, R. (1992) *NeuroReport* **3**, 249–252.

Alonso, J.R., Sánchez, F., Arévalo, R., Carretero, J., Vázquez, R. & Aijón, J. (1992b) *Neurosci. Lett.* **148**, 101–104.

Anderson, C.R. (1992) *Neurosci. Lett.* **139**, 280–284.

Aoki, E., Semba, R., Mikoshiba, K. & Kashiwamata, S. (1991) *Brain Res.* **547**, 190–192.

Aoki, C., Fenstemaker, S., Lubin, M. & Go. C.-G. (1993) *Brain Res.* **620**, 97–113.

Arevalo, R. Sanchez, F., Alonso, J.R., Carretero, J., Vaquez, R. & Aijon, J. (1992) *Brain Res. Bull.* **28**, 599–603.

Arevalo, R., Sanchez, F., Alonso, J.R., Rubio, M., Aijon, J. & Vazquez, R. (1993) *J. Chem. Neuroanat.* **6**, 335–341.

Ariano, M.A., Lewicki, J.A., Brandwein, H.J. & Murad, F. (1982) *Proc. Natl. Acad. Sci. USA* **79**, 1316–1320.

Arnt-Ramos, L.R., O'Brien, W.E. & Vincent, S.R. (1992) *Neuroscience* **51**, 773–789.

Augood, S.J., McGowan, E.M. & Emson, P.C. (1994) *Neuroscience* **59**, 7–12.

Bennett, B.D. & Bolam, J.P. (1993a) *Brain Res.* **609**, 137–148.

Bennett, B.D. & Bolam, J.P. (1993b) *Brain Res.* **610**, 305–310.

Bernard, V., Normand, E. & Bloch, B. (1992) *J. Neurosci.* **12**, 3591–3600.

Bhattacharjee, J. (1977) *J. Anat.* **123**, 273–282.

Bickford, M.E., Gunluk, A.E., Guido, W. & Sherman, S.M. (1993) *J. Comp. Neurol.* **334**, 410–430.

Blöttner, D. & Baumgarten, H.-G. (1992) *J. Comp. Neurol.* **316**, 45–55.

Blöttner, D., Schmidt, H.H.H.W. & Baumgarten, H.G. (1993) *NeuroReport* **4**, 923–926.

Brady, D.R., Carey, R.G. & Mufson, E.J. (1992) *Brain Res.* **577**, 236–248.

Brauer, K., Schober, A., Wolff, J.R., Winkelmann, E., Luppa, H., Lüth, H.-J. & Böttcher, H. (1991) *J. Hirnforsch.* **32**, 1–17.

Bredt, D.S. & Snyder, S.H. (1992) *Neuron* **8**, 3–11.

Bredt, D.S., Hwang, P.M. & Snyder, S.H. (1990) *Nature* **347**, 768–770.

Bredt, D.S., Glatt, C.E., Hwang, P.M., Fotuhi, M., Dawson, T.M. & Snyder, S.H. (1991a) *Neuron* **7**, 615–624.

Bredt, D.S., Hwang, P.M., Glatt, C.E., Lowenstein, C., Reed, R.R. & Snyder, S.H. (1991b) *Nature* **351**, 714–718.

Brüning, G. (1992) *Acta Histochem.* **93**, 397–401.

Brüning, G. (1993a) *J. Neurosci. Res.* **36**, 580–587.
Brüning, G. (1993b) *J. Comp. Neurol.* **334**, 192–208.
Calka, J. & Block, C.H. (1993a) *Brain Res. Bull.* **30**, 677–685.
Calka, J. & Block, C.H. (1993b) *Brain Res. Bull.* **32**, 207–210.
Calza, L., Giardino, L. & Ceccatelli, S. (1993) *NeuroReport* **4**, 627–630.
Ceccatelli, S. & Eriksson, M. (1993) *Brain Res.* **625**, 177–179.
Ceccatelli, S., Hulting, A.-L., Zhang, X., Gustafsson, L., Villar, M. & Hökfelt, T. (1993) *Proc. Natl. Acad. Sci. USA* **90**, 11292–11296.
Chesselet, M.-F. & Robbins, E. (1989) *Brain Res.* **492**, 237–244.
Chun, J.J.M. & Shatz, C.J. (1989) *J. Comp. Neurol.* **282**, 555–569.
Clowry, G.J. (1993) *NeuroReport* **5**, 361–364.
Cobcroft, M., Vaccaro, T. & Mitrofanis, J. (1989) *Neurosci. Lett.* **103**, 1–7.
Cowan, R.L., Wilson, C.J., Emson, P.C. & Heizmann, C.W. (1990) *J. Comp. Neurol.* **302**, 197–205.
Croul-Ottman, C.E. & Brunjes, P.C. (1988). *Brain Res.* **460**, 323–328.
D'Alessio, G., Di Donato, A., Jaffe, K., Maldonado, H. & Zabala, N.A. (1982) **147**, 231–235.
Davis, B.J. (1991) *J. Comp. Neurol.* **314**, 493–511.
Dawson, T.D., Bredt, D.S., Fotuhi, M., Hwang, P.M. & Snyder, S.H. (1991) *Proc. Natl. Acad. Sci. USA* **88**, 7797–7001.
DeFelipe, J. (1993) *Brain Res.* **615**, 342–346.
Derer, P. & Derer, M. (1993) *Neurosci. Lett.* **152**, 21–24.
Druga, R. & Syka, J. (1993) *NeuroReport* **4**, 999–1002.
Dun, N.J., Dun, S.L., Förstermann, U. & Tseng, L.F. (1992) *Neurosci. Lett.* **147**, 217–220.
Dun, N.J., Dun, S.L., Wu, S.Y., Förstermann, U., Schmidt, H.H.H.W. and Tseng, L.F. (1993) *Neuroscience* **54**, 845–857.
Dun, N.J., Dun, S.L. & Förstermann, U. (1994) *Neuroscience* **59**, 429–445.
Ellison, D.W., Kowall, N.W. & Martin, J.B. (1987) *J. Comp. Neurol.* **260**, 233–245.
Elofsson, R., Carlberg, M., Moroz, L., Nezlin, L. & Sakharov, D. (1993) *NeuroReport* **4**, 279–282.
Elphick, M.R., Green, I.C. & O'Shea, M. (1993) *Brain Res.* **619**, 344–346.
Endoh, M., Maiese, K., Pulsinelli, W.A. & Wagner, J.A. (1993) *Neurosci. Lett.* **154**, 125–128.
Ferrante, R.J., Kowall, N.W., Beal, M.F., Richarson, E.P.Jr., Bird, E.D. & Martin, J.B. (1985) *Science* **230**, 561–563.
Fiallos-Estrada, C.E., Kummer, W., Mayer, B., Bravo, R., Zimmermann, M. & Herdegen, T. (1993) *Neurosci. Lett.* **150**, 169–173.
Förstermann,U., Gorsky, L.E., Pollock, J.S., Schmidt, H.H.H.W., Heller, M. & Murad, F. (1990) *Biochem. Biophys. Res. Commun.* **168**, 727–732.
Furuyama, T., Inagaki, S. & Takagi, H. (1993) *Molec. Brain Res.* **20**, 335–344.
Geula, C., Schatz, C.R. & Mesulam, M.-M. (1993) *Neuroscience* **54**, 461–476.
Gonzalez, M.F., Sharp, F.R. & Sagar, S.M. (1987) *Brain Res. Bull.* **18**, 417–427.
González-Hernández, T., Conde-Serdín, M. & Meyer, G. (1992) *Anat. Embryol.* **186**, 245–250.
González-Hernández, T., Conde-Serdín, M., González-González, B., Mantolán-Sarmiento, B., Pérez-González, H. & Meyer, G. (1993) *Dev. Brain Res.* **76**, 141–145.
Goto, S., Matsukado, Y., Miyamoto, E. & Yamada, M. (1987) *Neuroscience* **22**, 189–201.
Hedlich, A., Lüth, H.-J., Werner, L., Bär, B., Hanisch, U. & Winkelmann, E. (1990) *J. Hirnforsch.* **31**, 681–687.
Herdegen, T., Brecht, S., Mayer, B., Leah, J., Kummer, W., Bravo, T. & Zimmermann, M. (1993) *J. Neurosci.* **13**, 4130–4145.
Hirsch, E.C., Graybiel, A.M., Duyckaerts, C. & Javoy-Agid, F. (1987) *Proc. Natl. Acad. Sci. USA* **84**, 5976–5980.
Hong, S.-C. Lanzino, G., Collins, J., Kassell, N.F. & Lee, K.S. (1993) *NeuroReport* **5**, 84–86.
Hope, B.T. & Vincent, S.R. (1989) *J. Histochem. Cytochem.* **37**, 653–661.

Hope, B.T., Michael, G.J., Knigge, K.M. & Vincent, S.R. (1991) *Proc. Natl. Acad. Sci. USA* **88**, 2811–2814.
Horio, Y. & Murad, F. (1991) *J. Biol. Chem.* **266**, 3411–3415.
Huang, P.L., Dawson, T.M., Bredt, D.S., Snyder, S.H. & Fishman, M.C. (1993) *Cell* **75**, 1273–1286.
Iadecola, C., Faris, P.L., Hartman, B.K. & Xu, X. (1993) *Brain Res.* **603**, 173–179.
Inagaki, S., Suzuki, K., Taniguchi, N. & Takagi, H. (1991) *Brain Res.* **549**, 174–177.
Johnson, M.D. & Ma, P.M. (1993) *J. Comp. Neurol.* **332**, 391–406.
Kalb, R.G. & Agostini, J. (1993) *Neuroscience* **57**, 1–8.
Kaneko, T., Shigemoto, R., Nakanishi, S. & Mizuno, N. (1993) *Brain Res.* **631**, 297–303.
Karson, C.N., Garcia-Rill, E., Biedermann, J., Mrak, R.E., Husain, M.M. & Skinner, R.D. (1991) *Psychiat. Res.* **40**, 31–48.
Kawamata, T., Nakamura, S., Akiguchi, I., Kimura, J., Kameyama, M., Kimura, H. & Takeda, T. (1990) *Neurobiol. Aging* **11**, 185–192.
Kinjo, N., Skinner, R.D. & Powell, E.W. (1989) *Neurosci. Res.* **7**, 154–158.
Kita, H., Kosaka, T. & Heizmann, C.W. (1990) *Brain Res.* **536**, 1–15.
Kitchener, P.D. & Diamond, J. (1993) *J. Comp. Neurol.* **335**, 1–15.
Kitchener, P.D., Van der Zee, C.E.E.M. & Diamond, J. (1993) *NeuroReport* **4**, 487–490.
Knowles, R.G., Palacios, M., Palmer, R.M.J. & Moncada, S. (1989) *J. Biol. Chem.* **267**, 11374–11378.
Koistinaho, J., Swanson, R.A., de Vente, J. & Sagar, S.M. (1993) *Neuroscience* **57**, 587–597.
Kowall, N.W. & Beal, M.F. (1988) *Ann. Neurol.* **23**, 105–114.
Kowall, N.W. & Mueller, M.P. (1988) *Neuroscience* **26**, 645–654.
Kowall, N.W., Ferrante, R.J., Beal, M.F., Richardson, E.P.Jr., Sofroniew, M.V., Cuello, A.C. & Martin, J.B. (1987) *Neuroscience* **20**, 817–828.
Kubota, Y., Mikawa, S. & Kawaguchi, Y. (1993) *NeuroReport* **5**, 205–208.
Kuwabara, T. & Cogan, D.G. (1960) *J. Histochem. Cytochem.* **8**, 214–224.
Lee, J.-H., Price, R.H., Williams, F.G., Mayer, B. & Beitz, A.J. (1993) *Brain Res.* **608**, 324–333.
Leigh, P.N., Connick, J.H. & Stone, T.W. (1990) *Comp. Biochem. Physiol.* **97C**, 259–264.
Li, Z.S., Furness, J.B., Young, H.M. & Campbell, G. (1992) *Arch. Histol. Cytol.* **55**, 333–350.
Liuzzi, F.J., Wu, W., Scoville, S.A. & Schinco, F.P. (1993) *Exp. Neurol.* **121**, 275–278.
Llewellyn-Smith, I.J., Song, Z.-M., Costa, M., Bredt, D.S. & Snyder, S.H. (1992) *Brain Res.* **577**, 337–342.
Luebke, J.I., Weider, J.M., McCarley, R.W. & Greene, R.W. (1992) *Neurosci. Lett.* **148**, 129–132.
Martinez-Murillo, R. & Martinez-Rodriguez, R. (1984) *Cell. Molec. Biol.* **30**, 217–222.
Matsumoto, T., Nakane, M., Pollock, J.S., Kuk, J.E. & Förstermann, U. (1993) *Neurosci. Lett.* **155**, 61–64.
Matsuoka, I., Giuili, G., Poyard, M., Stengel, D., Parma, J., Guellaen, G. & Hanoune, J. (1992) *J. Neurosci.* **12**, 3350–3360.
McDonald, A.J., Payne, D.R. & Mascagni, F. (1993) *Neuroscience* **52**, 97–106.
McNeill, D.L., Papka, R.E. & Harris, C.H. (1992a) *Peptides* **13**, 1239–1246.
McNeill, D.L., Traugh, Jr., N.E., Vaidya, A.M., Hua, H.T. & Papka, R.E. (1992b) *Neurosci. Lett.* **147**, 33–36.
Mengual, E., Velayos, J.L. & Reinoso-Suarez, F. (1993) *Brain Res.* **625**, 165–168.
Mesulam, M.-M., Geula, C., Bothwell, M.A. & Hersh, L.B. (1989) *J. Comp. Neurol.* **281**, 611–633.
Meyer, G. & Gonzalez-Hernandez, T. (1993) *J. Comp. Neurol.* **338**, 317–336.
Meyer, G., Gonzalez-Hernandez, T., Galindo-Mireles, D., Castañeyra-Perdomo, A. & Ferres-Torres, R. (1991) *Anat. Embryol.* **184**, 99–102.
Meyer, G., Wahle, P., Castaneyra-Perdomo, A. & Ferres-Torres, R. (1992) *Exp. Brain Res.* **88**, 204–212.

Mitrofanis, J. (1989) *Neurosci. Lett.* **102**, 165–172.
Mitrofanis, J., Robinson, S.R. & Ashwell, K. (1992) *J. Comp. Neurol.* **319**, 560–585.
Mizukawa, K. (1990) *Meth. Neurosci.* **3**, 457–472.
Mizukawa, K., McGeer, P.L., Vincent, S.R. & McGeer, E.G. (1988a) *Brain Res.* **452**, 286–292.
Mizukawa, K., Vincent, S.R., McGeer, P.L. & McGeer, E.G. (1988b) *Brain Res.* **461**, 274–281.
Mizukawa, K., McGeer, P.L., Vincent, S.R. & McGeer, E.G. (1989) *J. Comp. Neurol.* **279**, 281–311.
Moroz, L.L., Park, J.-H. & Winlow, W. (1993) *NeuroReport* **4**, 643–646.
Morton, A.J., Nicholson, L.F.B. & Faull, R.L.M. (1993) *Neuroscience* **53**, 159–168.
Mufson, E.J., Mash, D.C. & Hersh, L.B. (1988) *Ann. Neurol.* **24**, 623–629.
Mufson, E.J., Brady, D.R. & Carey, R.G. (1990) *Brain Res.* **516**, 237–247.
Nakamura, H., Saheki, T. & Nakagawa, S. (1990) *Brain Res.* **539**, 108–112.
Nakamura, H., Saheki, T., Ichiki, H., Nakata, K. & Nakagawa, S. (1991) *J. Comp. Neurol.* **312**, 652–679.
Nakamura, S., Kawamata, T., Kimura, T., Akiguchi, I., Kameyama, M., Nakamura, N., Wakata, Y. & Kimura, H. (1988) *Brain Res.* **455**, 144–147.
Nakane, M., Ichikawa, M. & Deguchi, T. (1983) *Brain Res.* **273**, 9–15.
Ohta, A., Takagi, H., Matsui, T., Hamai, Y., Iida, S. & Esumi, H. (1993) *Neurosci. Lett.* **158**, 33–35.
Onstott, D., Mayer, B. & Beitz, A.J. (1993) *Brain Res.* **610**, 317–324.
Osborne, N.N., Barnett, N.L. & Herrera, A.J. (1993) *Brain Res.* **610**, 194–198.
Papka, R.E. & McNeill, D.L. (1992) *Neurosci. Lett.* **147**, 224–228.
Pasqualotto, B.A. & Vincent, S.R. (1991) *Brain Res.* **551**, 78–86.
Pasqualotto, B.A., Hope, B.T. & Vincent, S.R. (1991) *Neurosci. Lett.* **128**, 155–160.
Peng, Z.-C., Chen, S., Bertini, G., Schmidt, H.H.H.W. & Bentivoglio, M. (1994) *Neurosci. Lett.* **166**, 153–156.
Pitkänen, A. & Amaral, D.G. (1991) *J. Comp. Neurol.* **313**, 326–348.
Pow, D.V. (1992) *J. Neuroendocrinol.* **4**, 377–380.
Price, Jr., R.H., Mayer, B. & Beitz, A.J. (1993) *NeuroReport* **4**, 807–810.
Provis, J.M. & Mitrofanis, J. (1990) *Visual Neurosci.* **4**, 619–623.
Rebeck, G.W., Marzloff, K. & Hyman, B.T. (1993) *Neurosci. Lett.* **152**, 165–168.
Regidor, J. & Poch, L. (1987) In *The Forebrain of Reptiles* (ed. Schwerdtfeger, W.K. & Smeets, W.J.A.J.), pp. 77–84. Basel, S. Karger.
Regidor, J., Montesdeoca, J., Ramirez-Gonzalez, J.A., Hernanez-Urquia, C.M. & Divac, I. (1993) *Brain Res.* **631**, 171–174.
Reiner, P.B. & Vincent, S.R. (1987) *Brain Res. Bull.* **19**, 705–714.
Ribeiro, J.M.C., & Nusseuzveig, R.H. (1993). *FEBS Lett.* 330, 165–168.
Ribeiro, J.M.C., Hazzard, J.M.H., Nussenzveig, R.H., Champagne, D.E. & Walker, F.A. (1993) *Science* **260**, 539–541.
Sadikot, A.F., Parent, A., Smith, Y. & Bolam, J.P. (1992) *J. Comp. Neurol.* **320**, 228–242.
Sagar, S.M. (1986) *Brain Res.* **373**, 153–158.
Sagar, S.M. (1990) *J. Comp. Neurol.* **300**, 309–319.
Sagar, S.M. & Ferriero, D.M. (1987) *Brain Res.* **400**, 348–352.
Saito, S., Kidd, G.J., Trapp, B.D., Dawson, T.M., Bredt, D.S., Wilson, D.A., Traystman, R.J., Snyder, S.H. & Hanley, D.F. (1994) *Neuroscience* **59**, 447–456.
Sajin, B., Sestan, N. & Dmitrovic, B. (1992) *Neurosci. Lett.* **140**, 117–120.
Sanchez, F., Alonso, J.R., Arevalo, R., Blanco, E., Aijon, J. & Vazquez, R. (1994) *Cell Tiss. Res.* **276**, 31–34.
Sandell, J.H. (1985) *J. Comp. Neurol.* **238**, 466–472.
Sandell, J.H. (1986) *J. Comp. Neurol.* **251**, 388–397.
Sandell, J.H., Graybiel, A.M. & Chesselet, M.-F. (1986) *J. Comp. Neurol.* **243**, 326–334.

Sangruchi, T. & Kowall, N.W. (1991) *Neuroscience* **40**, 713–724.
Saper, C.B., Hurley, K.M., Moga, M.M., Holmes, H.R., Adams, S.A., Leahy, K.M. & Nedleman, P. (1989) *Neurosci. Lett.* **96**, 29–34.
Sato, T. (1990a) *Arch. Histol. Cytol.* **53**, 63–69.
Sato, T. (1990b) *Arch. Histol. Cytol.* **53**, 141–146.
Scherer-Singler, U., Vincent, S.R., Kimura, H. & McGeer, E. (1983) *J. Neurosci. Meth.* **9**, 229–234.
Schilling, K., Schmidt, H.H.H.W. & Baader, S.L. (1994) *Neuroscience* **59**, 893–903.
Schmidt, H.H.H.W., Gagne, G.D., Nakane, M., Pollock, J.S., Miller, M.F. & Murad, F. (1992) *J. Histochem. Cytochem.* **40**, 1439–1456.
Schober, A., Brauer, K. & Luppa, H. (1989) *Acta Histochem. Cytochem.* **22**, 669–674.
Schober, A., Brauer, K., Johansson, O., Luppa, H. & Schober, W. (1990) *Acta Histochem. Cytochem.* **23**, 663–670.
Schober, A., Malz, C.R. & Meyer, D.L. (1993) *Neurosci. Lett.* **151**, 67–70.
Scott, J.W., McDonald, J.K. & Pemberton, J.L. (1987) *J. Comp. Neurol.* **260**, 378–391.
Seidel, I., Bernstein, H.-G., Becker, A., Grecksch, G., Luppa, H. & Muller, M. (1991) *Acta Histochem.* **91**, 157–164.
Sestan, N. & Kostovic, I. (1994) *TINS* **17**, 105–106.
Sharp, F.R., Gonzalez, M.F., Ferriero, D.M. & Sagar, S.M. (1986) *Neurosci. Lett.* **65**, 204–208.
Sims, K.S. & Williams, R.S. (1990) *Neuroscience* **36**, 449–472.
Skinner, R.D., Conrad, C., Henderson, V., Gilmore, S.A. & Garcia-Rill, E. (1989) *Exp. Neurol.* **104**, 15–21.
Solodkin, A., Traub, R.J. & Gebhart, G.F. (1992) *Neuroscience* **51**, 495–499.
Southam, E. & Garthwaite, J. (1993) *Neuropharmacology* **32**, 1267–1278.
Southam, E., Morris, R. & Garthwaite, J. (1992) *Neurosci. Lett.* **137**, 241–244.
Spike, R.C., Todd, A.J. & Johnston, H.M. (1993) *J. Comp. Neurol.* **335**, 320–333.
Springall, D.R., Riveros-Moreno, V., Buttery, L., Suburo, A., Bishop, A.E., Merrett, M., Moncada, S. & Polak, J.M. (1992) *Histochemistry* **98**, 259–266.
Standaert, D.G., Saper, C.B., Rye, D.B. & Wainer, B.H. (1986) *Brain Res.* **382**, 163–168.
Terenghi, G., Riveros-Moreno, V., Hudson, L.D., Ibrahim, N.B.N. & Polak, J.M. (1993) *J. Neurol. Sci.* **118**, 34–37.
Thomas, E. & Pearse, A.G.E. (1961) *Histochemie* **2**, 266–282.
Thomas, E. & Pearse, A.G.E. (1964) *Acta Neuropath.* **3**, 238–249.
Torres, G., Lee, S. & Rivier, C. (1993) *Molec. Cell. Neurosci.* **4**, 155–163.
Tracey, W.R., Nakane, M., Pollock, J.S. & Förstermann, U. (1993) *Biochem. Biophys. Res. Commun.* **195**, 1035–1040.
Uehara-Kunugi, Y., Terai, K., Taniguchi, T., Tooyama, I. & Kimura, H. (1991) *Dev. Brain Res.* **59**, 157–162.
Vaccaro, T.M., Cobcroft, M.D., Provis, J.M. & Mitrofanis, J. (1991) *Cell Tissue Res.* **265**, 371–379.
Valtschanoff, J.G., Weinberg, R.J., Rustioni, A. & Schmidt, H.H.H.W. (1992) *Neurosci. Lett.* **148**, 6–10.
Valtschanoff, J.G., Weinberg, R.J., Kharazia, V.N., Schmidt, H.H.H.W. & Nakane, M. (1993a) *Neurosci. Lett.* **157**, 157–161.
Valtschanoff, J.G., Weinberg, R.J., Kharazia, V.N., Nakane, M. & Schmidt, H.H.H.W. (1993b) *J. Comp. Neurol.* **331**, 111–121.
Vaney, D.I. & Young, H.M. (1988) *Brain Res.* **474**, 380–385.
Venturini, C.M., Knowles, R.G., Palmer, R.M.J. & Moncada, S. (1991) *Biochem. Biophys. Res. Commun.* **180**, 920–925.
Verge, V., Xu, Z., Xu, X.-J., Wiesenfeld-Hallin, Z. & Hökfelt, T. (1992) *Proc. Natl. Acad. Sci. USA* **89**, 11617–11621.
Villalba, R.M., Martínez-Murillo, R., Blasco, I., Alvarez, F.J. & Rodrigo, J. (1988) *Brain Res.* **462**, 359–362.

Villalba, R.M., Rodrigo, J., Alvarezy, F.J., Achaval, M. & Martinez-Murillo, R. (1989) *Neuroscience* **33**, 373–382.
Vincent, S.R. (1994) *Prog. Neurobiol.* **42**, 129–160.
Vincent, S.R. & Hope, B.T. (1992) *Trends Neurosci.* **15**, 108–113.
Vincent, S.R. & Johansson, O. (1983) *J. Comp. Neurol.* **217**, 264–270.
Vincent, S.R. & Kimura, H. (1992) *Neuroscience* **46**, 755–784.
Vincent, S.R., Johansson, O., Hökfelt, T., Skirboll, L., Elde, R.P., Terenius, L., Kimmel, J. & Goldstein, M. (1983a) *J. Comp. Neurol.* **217**, 252–263.
Vincent, S.R., Satoh, K., Armstrong, D.M. & Fibiger, H.C. (1983b) *Neurosci. Lett.* **43**, 31–36.
Vincent, S.R., Satoh, K., Armstrong, D.M. & Fibiger, H.C. (1983c) *Nature* **306**, 688–691.
Vincent, S.R., Staines, W.A. & Fibiger, H.C. (1983d). *Neurosci. Lett.* **35**, 111–114.
Vincent, S.R., Satoh, K., Armstrong, D.M., Panula, P., Vale, W. & Fibiger, H.C. (1986) *Neuroscience* **17**, 167–182.
Vizzard, M.A., Erdman, S.L. & de Groat, W.C. (1993a) *J. Auton. Nerv. Syst.* **44**, 85–90.
Vizzard, M.A., Erdman, S.L. & de Groat, W.C. (1993b) *Neurosci. Lett.* **152**, 72–76.
Vizzard, M.A., Erdman, S.L. & de Groat, W.C. (1993c). *Brain Res.* **607**, 349–353.
Vizzard, M.A., Erdman, S.L., Erickson, V.L., Stewart, R.J., Roppolo, J.R. & de Groat, W.C. (1994) *J. Comp. Neurol.* **339**, 62–75.
Wallace, M.N. (1986) *Neuroscience* **19**, 381–391.
Wallace, M.N. & Fredens, K. (1989) *Exp. Brain Res.* **78**, 435–445.
Wallace, M.N. & Fredens, K. (1992) *NeuroReport* **3**, 953–956.
Weiler, R. & Kewitz, B. (1993) *Neurosci. Lett.* **158**, 151–154.
Wendland, R., Schweizer, F.E., Ryan, T.A., Nakane, M., Murad, F., Scheller, R.H. & Tsien, R.W. (1994) *Proc. Natl. Acad. Sci. USA* **91**, 2151–2155.
Wetts, R. & Vaughn, J.E. (1993) *Dev. Brain Res.* **76**, 278–282.
Wolf, G., Würdig, S. & Schünzel, G. (1992) *Neurosci. Lett.* **147**, 63–66.
Wolf, G., Henschke, G. & Würdig, S. (1993) *Neurosci. Lett.* **161**, 49–52.
Wolff, D.J. & Datto, G.A. (1992) *Biochem. J.* **285**, 201–206.
Wu, W. (1993) *Exp. Neurol.* **120**, 1–7.
Wu, W. & Li, L. (1993) *Neurosci. Lett.* **153**, 121–124.
Wu, W. & Scott, D.E. (1993) *Exp. Neurol.* **121**, 279–283.
Yamamoto, R., Bredt, D.S. Snyder, S.H. & Stone, R.A. (1993) *Neuroscience* **54**, 189–200.
Yan, X.X., Jen, L.S. & Garey, L.J. (1993) *NeuroReport* **4**, 1227–1230.
Zhang, X., Verge, V., Wiesenfeld-Hallin, Z., Ju, G., Bredt, D., Snyder, D. & Hökfelt, T. (1993) *J. Comp. Neurol.* **335**, 563–575.
Zweig, R.M., Jankel, W.R., Hedreen, J.C., Mayeux, R. & Price, D.L. (1989) *Ann. Neurol.* **26**, 41–46.
Zwiller, J., Ghandour, M.S., Revel, M.O. & Basset, P. (1981) *Neurosci. Lett.* **23**, 31–36.

CHAPTER 6

NITRIC OXIDE AND EXCITATORY AMINO ACID-COUPLED SIGNAL TRANSDUCTION IN THE CEREBELLUM AND HIPPOCAMPUS

Paul L. Wood

Department of Pharmacology, Cocensys, Inc., 213 Technology Drive, Irvine, CA 92718, USA

Table of Contents

6.1 Historical background 104
6.2 Nitric oxide synthase (1.14.23) 105
  6.2.1 Glial NOS 106
  6.2.2 NO actions 107
6.3 Cerebellum 107
  6.3.1 EAA receptor populations 107
  6.3.2 Endogenous EAA receptor ligands 110
  6.3.3 NOS/cGMP compartments 112
  6.3.4 Noradrenergic afferents 114
  6.3.5 Citrulline recycling 115
  6.3.6 NO and cerebellar LTD 115
  6.3.7 NO and cerebellar neurotoxicity 115
6.4 Hippocampus 115
  6.4.1 EAA receptor populations 115
  6.4.2 Endogenous EAA receptor ligands 116
  6.4.3 NOS/cGMP compartments 116
  6.4.4 Noradrenergic afferents 117
  6.4.5 Citrulline recycling 117
  6.4.6 NO and hippocampal LTP 117
  6.4.7 NO and hippocampal neurotoxicity 118
6.5 Conclusions 118

References 118

Nitric Oxide in the Nervous System
ISBN 0-12-721985-4

6.1 Historical background

Clearly, our rapidly evolving concepts of nitric oxide (NO) function in the CNS have been preceded by prior intensive studies of endothelium derived relaxing factor (EDRF) in the peripheral vasculature and many years of research and debate that led to the conclusion that EDRF was indeed NO. Additionally, the pioneers in this area generated the first NOS inhibitors which have been used extensively to define roles for NO in the brain. However, prior to the extensive use of these selective nitric oxide synthase (NOS) inhibitors, several key publications appeared in the literature which were landmarks along the path to our current flood of information in this area. The hallmark citations include the following.

(1) The independent demonstrations of activation of soluble guanylate cyclase, by NO donors, in mouse cortex (Miki *et al.*, 1977) and rat cerebellum and cortex (Katsuki *et al.*, 1977).
(2) The demonstration that L-arginine could act as an endogenous activator of soluble guanylate cyclase in N1E 115 neuroblastoma cells (Deguchi and Yoshioka, 1982).
(3) The first demonstration that the excitatory amino acid (EAA) agonist glutamate and the NO donor, *N*-methyl-*N*′-nitro-*N*-nitroguanidine, could evoke increased cGMP efflux from rat cerebellar slices (Tjörnhammer *et al.*, 1986).
(4) The first functional studies demonstrating that *N*-methyl-D-aspartate (NMDA) receptor activation in cerebellar slices resulted in NO generation which in turn activated guanylate cyclase to generate cGMP as an intracellular messenger (Garthwaite *et al.*, 1988).
(5) These *in vitro* observations were confirmed and further extended to demonstrate kainate- and AMPA receptor-mediated activation of cerebellar NOS (Garthwaite *et al.*, 1989a,b; Bredt and Snyder, 1989).
(6) The first *in vivo* demonstrations of EAA modulation of cerebellar NOS activity (Wood *et al.*, 1990c; Rao *et al.*, 1990a; Bansinath *et al.*, 1993).
(7) The cloning of neuronal NOS (Bredt *et al.*, 1990) and the demonstration that NADPH diaphorase histochemistry was in fact NOS activity (Hope *et al.*, 1991; Bredt *et al.*, 1991; Dawson *et al.*, 1991).
(8) *In vivo* brain dialysis studies demonstrating NO release (Vallebuona and Raiteri, 1993; Luo *et al.*, 1993, 1994).

These studies formed the basis of many subsequent *in vitro* and *in vivo* approaches to study the roles of NO in signal transduction pathways within the CNS. A number of receptor systems are coupled to the generation of NO but it is the focus of this review to concentrate on EAA pathways within the cerebellum and hippocampus which modulate neuronal function via NO signalling.

6.2 Nitric oxide synthase (1.14.23)

After the development of a purification strategy (Bredt and Snyder, 1989), using ADP agrose columns to isolate neuronal NOS (nNOS) from rat cerebellum, a number of other laboratories utilized this approach to isolate NOS from rat cerebellum (Giovanelli *et al.*, 1991), from porcine cerebellum (Mayer *et al.*, 1991), and from human cerebellum (Klatt *et al.*, 1992). These studies of purified nNOS, as well as studies of the cloned enzyme (McMillan *et al.*, 1992), have clearly established that nNOS is a calcium-dependent cytosolic enzyme of 150–160 kDa and is an arginine- and NADPH-dependent flavoprotein containing FAD, FMN, terahydrobiopterin, iron, a concensus sequence for calmodulin binding and a protein kinase A consensus sequence for a phosphorylation site (see Chapters 1 and 2). Phosphorylation of this site has been demonstrated, but no change in nNOS activity was noted (Brüne and Lapetina, 1991). Another interesting feature of nNOS is that it possesses NADPH diaphorase activity and upon purification both activities co-purify to homogeneity (Hope *et al.*, 1991; Bredt *et al.*, 1991). This NADPH diaphorase activity can be monitored by a simple histochemical staining procedure (Hope *et al.*, 1991) which has been used extensively to map nNOS-positive cells in the CNS (Vincent and Kimura, 1992; see Chapter 5).

With NOS, one of the atoms of molecular oxygen is incorporated into the guanidinium nitrogen atoms of arginine, resulting in the generation of equimolar quantities of NO and citrulline (Figure 1; Leonne *et al.*, 1991; Bush *et al.*, 1992). Citrulline is subsequently metabolized by argininosuccinate synthetase (6.3.4.5) to form argininosuccinate, which in turn can be hydrolysed by argininosuccinate lyase (4.3.2.1) to replenish arginine supplies (Figure 1; Ratner *et al.*, 1960; Jones *et al.*, 1961; Kemp and Woodbury, 1965; Sadasivudu and Rao, 1976). The cellular distributions of these enzymes will be presented in further discussions of specific hippocampal and cerebellar circuitry. Another feature which should not be ignored is the apparent high capacity of NOS-containing cells to accumulate extracellular arginine via specific uptake carriers which are up-regulated during enhanced NOS activity (Bogle *et al.*, 1992; Westergaard *et al.*, 1993).

While the initial studies of nNOS defined the potential roles for this enzyme in the generation of NO, it rapidly became apparent that there are multiple NOS isoforms (Förstermann *et al.*, 1991); however, the compartmentation of these different NOS isoforms is still not completely defined. The isoforms to consider include: (1) neuronal NOS of which two variants have been described, with the most recent isoform resulting from an inframe deletion in the middle of the nNOS cDNA (Ogura *et al.*, 1993); this isoform may represent the membrane-bound form of nNOS that has been described in the endoplasmic reticulum (Wolf *et al.*, 1992; Hiki *et al.*, 1991; Matsumoto *et al.*, 1993); (2) inducible NOS which appears to involve a cytosolic form in astrocytes (Simmons and Murphy, 1992) and both cytosolic and membrane-bound isoforms in microglia, the resident CNS macrophages (Corradin *et al.*, 1993; Wood *et al.*, 1994b).

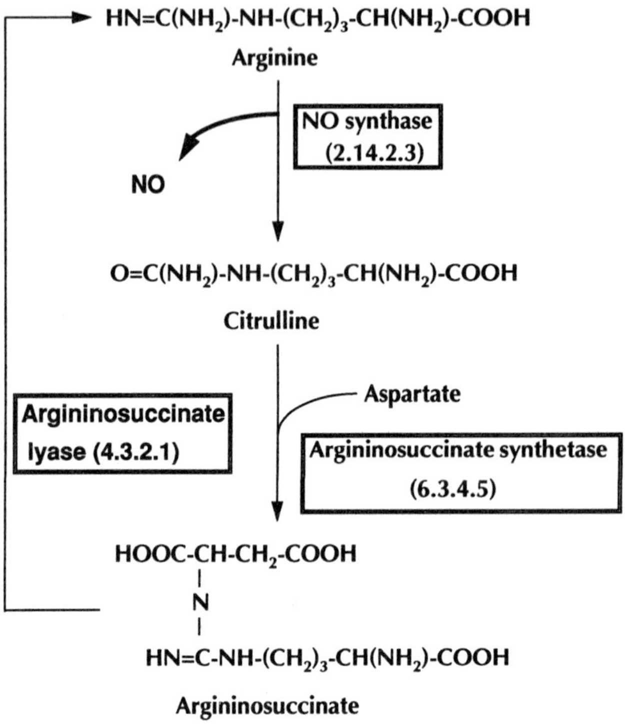

Figure 1 Generation of NO from arginine and recycling of metabolic products back to arginine.

6.2.1 Glial NOS

Glia possess a constitutive level of NOS activity (Murphy *et al.*, 1990; see Chapter 9) which can be augmented by α_1-adrenergic and metabotropic EAA receptor activation (Murphy *et al.*, 1991; Agullo and García, 1991, 1992a,b). Subsequent studies using lipopolysaccharide (LPS)- or cytokine-activated primary glial and C6 glioma cultures have demonstrated that these cells also possess inducible NOS (iNOS) which is a cytosolic, Ca^{2+}-independent form of the enzyme which can produce sufficient NO levels to activate guanylate cyclase and increase glial cGMP levels (Galea *et al.*, 1992; Simmons and Murphy, 1992; Demerlé-Pallardy *et al.*, 1993; Vigne *et al.*, 1993; Feinstein *et al.*, 1994). These observations are consistent with the pioneering studies which demonstrated glial cGMP levels which could be augmented by phosphodiesterase inhibition (Bloch-Tardy *et al.*, 1980) and immunocytochemical studies demonstrating the highest CNS levels of arginine-like immunoreactivity in astrocytes (Aoki *et al.*, 1991).

In contrast to studies with rodent astrocytes, the literature is not consistent in the case of human glial NOS. While LPS activation of a human astrocytoma augmented NOS activity (Salvemini *et al.*, 1992), studies with primary cultures of human astrocytes were unable to demonstrate cytokine activation of NOS (Lees, 1993). Further studies are clearly needed with human astrocytes.

6.2.2 NO actions

While NO has a short half-life (≈5 s), this *diffusible* intra- and intercellular messenger molecule possesses tremendous potential to modulate neurotransmission (Gally *et al.*, 1990), particularly in brain regions with laminar cell populations, where diffusional barriers might be less. Within this context, NO will have many actions which can act to alter both short-term and long-term synaptic events. With regard to potential cellular sites of action, these include a number of enzymes involved in cellular metabolism and DNA synthesis (Table 1).

These actions of NO clearly play roles in the modulation of cellular activity and cellular proliferation. In the case of cerebellar glial cells, the NO donors 5-nitro-*N*-acetylpenicillamine (SNAP) and sodium nitroprusside (SNP) both decrease DNA synthesis and inhibit mitrochondrial activity, actions observed at concentrations of the NO donors which are not cytotoxic (Garg *et al.*, 1992).

6.3 Cerebellum

6.3.1 EAA receptor populations

Within the cerebellum all three major ionotropic EAA receptor populations (NMDA, AMPA/quisqualate, kainate) are present and activation of each subtype results in increased cerebellar cGMP levels, each with its own demonstrated pharmacological specificity (Figure 2; Lehmann *et al.*, 1988, 1989; Wood *et al.*, 1982, 1989, 1990a,b; Wood and Rao, 1991). Agonists of the NMDA-associated glycine receptor, a positive allosteric site on the NMDA receptor complex, also increase *in vivo* levels of cerebellar cGMP. These include studies both with glycine (Danysz *et al.*, 1989) and the more selective agonist D-serine (Figure 2; Wood *et al.*, 1989; Rao *et al.*, 1990b). These data indicate that the basal tone of EAA transmission in the cerebellum can be augmented by pharmacological activation of the NMDA-associated glycine receptor and argue against the saturation of this receptor site by endogenous glycine. Cerebellar granule cell cultures have also been used to demonstrate direct modulation of NMDA receptors by glycine and D-serine on these neurons (Wroblewski *et al.*, 1989).

In situ studies have demonstrated that cerebellar granule cells possess a constitutive message for the NMDA receptor subunits NR1, NR2A and NR2C (Seeburg, 1993). These cells also possess mRNA for AMPA (GluR2 and GluR4 subunits) and kainate (KA2 and GluR6 subunits) receptors. Purkinje cells only possess a message

Table 1 Cellular sites of action and molecular targets for NO as an intra- and intercellular messenger

Cellular compartments
(1) Intracellular at the site of generation
(2) Other cellular compartments after diffusion from the site of generation:
  (i) local nerve endings (homologous neuronal types as well as modulatory afferents)
  (ii) local neuronal cell bodies
  (iii) glia
  (iv) microglia

Molecular targets
(1) Activation of guanylate cyclase via binding to the haem moiety (Miki *et al.*, 1977; Katsuki *et al.*, 1977)
(2) Activation of cyclo-oxygenase (Salvemini *et al.*, 1993)
(3) Activation of cytosolic ADP ribosyltransferase (Brüne and LaPetina, 1989; Dimmeler and Brüne, 1991) resulting in the ADP ribosylation of glyceraldehyde-3-phosphate dehydrogenase (Dimmeler *et al.*, 1992, 1993; Zhang and Snyder, 1992; Brüne *et al.*, 1994) and inhibition of glycolysis
(4) Destruction of iron-containing enzymes, including:
  (i) aconitase leading to decreases in TCA cycle activity (Welsh and Sandler, 1992)
  (ii) ribonucleotide reductase, resulting in decreased deoxyribonucleotide synthesis (Lepoivre *et al.*, 1992)
  (iii) mitochondrial electron transfer proteins (Garg *et al.*, 1992)
(5) Potentiation Ca^{2+}-dependent protein kinase A activation of the transcriptional factor CREB (Peunova and Enikolopov, 1993)
(6) Possible S-nitrosylation of proteins (Stamler *et al.*, 1992) and deamination of N-terminal amino acid residues in proteins (Moriguchi *et al.*, 1992)
(7) Alterations in neurotransmitter release, including potentiation of glutamate release (Dickie *et al.*, 1992) and acetylcholine release (Prast and Philippu, 1992)
(8) Feedback inhibition to regulate NOS (Assreuy *et al.*, 1993)
(9) Blockade of EAA receptors including NMDA (East *et al.*, 1991b; Manzoni and Blockaert, 1993) and kainate (Tanaka *et al.*, 1993)

for the NR1 subunit of MNDA receptor but have high levels of mRNA for AMPA (GluR1-3 subunits) and kainate (KA1 and GluR5 subunits) receptors (Seeburg, 1993).

NMDA, kainate and AMPA/quisqualate receptor agonists also have more

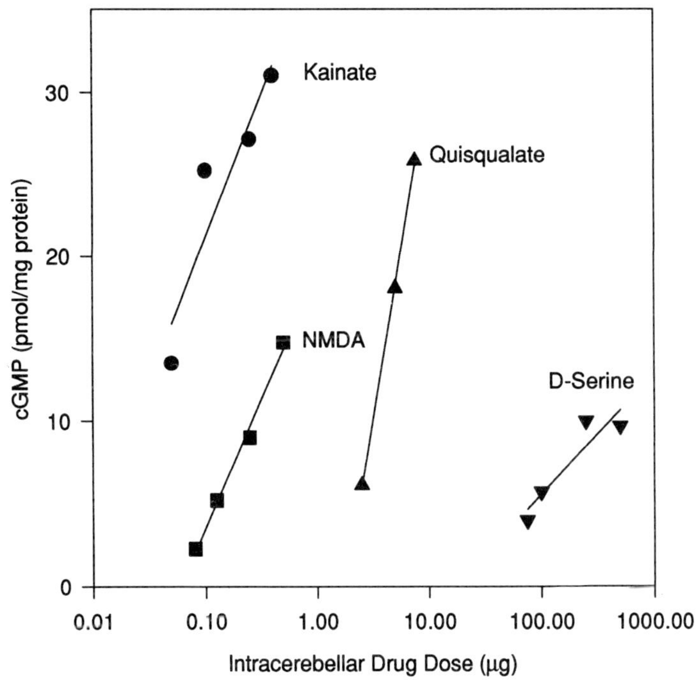

Figure 2 Comparison of the dose–response curves for intracerebellar EAA agonists, with respect to increases (10 min) in mouse cerebellar cGMP levels.

recently been demonstrated to evoke cGMP efflux is monitored by *in vivo* cerebellar microdialysis (Luo *et al.*, 1994; Vallebuona and Raiteri, 1993), again each with its own pharmacological specificity. In the case of AMPA/quisqualate receptors, both AMPA and quisqualate increase cerebellar cGMP via a CNQX-reversible mechanism, indicating that the actions of quisqualate do not involve its affinity for the metabotropic EAA receptor. The cellular locations of the ionotropic EAA receptors include AMPA/quisqualate receptors on Purkinje cells, kainate receptors on granule cells, and NMDA receptors on granule cells and possibly noradrenergic afferent nerve endings. AMPA- and kainate-dependent augmentation of cerebellar cGMP levels involves direct cellular activation since these actions are not blocked by tetrodotoxin (TTX) in cerebellar slices (Southam *et al.*, 1991). In contrast, NMDA-dependent increases in cGMP were significantly antagonized by TTX, indicating trans-synaptic mechanisms in the generation of NO and cGMP after NMDA receptor activation (Southam *et al.*, 1991).

A further technical advantage in studies of cerebellar EAA-mediated neurotransmission is the ability to pharmacologically invoke the release of EAA ligands within the cerebellum. The tools utilized are (Wood *et al.*, 1988, 1990a,b): (1) harmaline, which specifically activates inferior olive neurons and

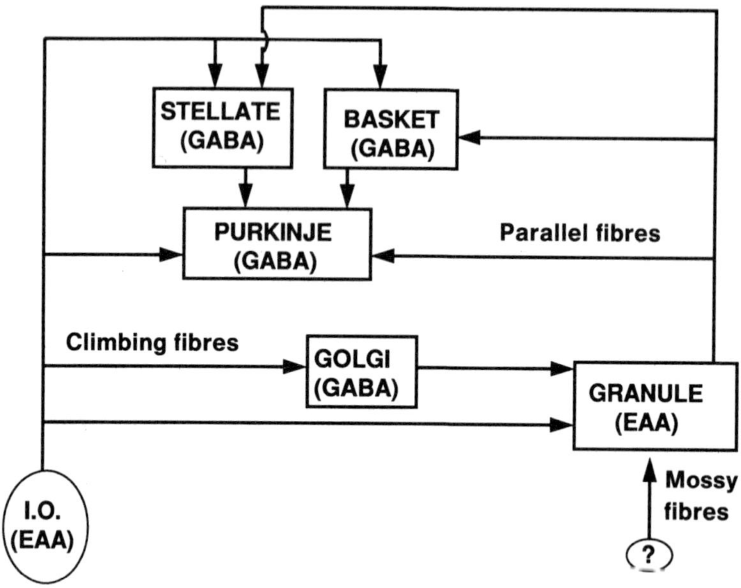

Figure 3 Major neuronal elements and interconnections of the cerebellum, including the climbing fibre and mossy fibre afferent inputs to the cerebellum (Palay and Chan-Palay, 1974). I.O., inferior olive.

evokes EAA release from the climbing fibre pathway (Figure 3); and (2) pentylenetetrazol (PTZ) which reduces GABAergic tone and increases cerebellar granule cell activity, thereby increasing EAA release from parallel fibres (Figure 3). This activation of intrinsic EAA release in the cerebellum results in elevated cerebellar cGMP levels (Wood, 1991a,b) and cGMP monitored by *in vivo* microdialysis (Luo *et al.*, 1994). Additionally, a key point which has been noted in both measurements of tissue cGMP levels (Wood *et al.*, 1982) and measurements of cGMP in cerebellar dialysates (Luo *et al.*, 1993), is that cGMP increases evoked by harmaline augmentation of endogenous EAA release are almost completely abolished by competitive and non-competitive NMDA antagonists and glycine antagonists (Rao *et al.*, 1990b, 1993), indicating that a large portion of observed changes in cerebellar cGMP levels or efflux are mediated by NMDA receptors. Similarly, modulation of cerebellar cGMP levels and neuronal activity has been demonstrated to involve mainly NMDA receptor activation, both *in vitro* (Garthwaite and Beaumont, 1989) and *in vivo* (Wood *et al.*, 1982).

6.3.2 Endogenous EAA receptor ligands

The endogenous EAA ligand(s) for these receptors is thought to be glutamate and/or aspartate (reviewed by Wood, 1991a) while the co-transmitter at the NMDA receptor is glycine (Johnson and Ascher, 1987; Kleckner and Dingeldine, 1988; Thomson *et al.*, 1988; Thomson, 1990) and/or D-serine (Hashimoto *et al.*, 1992, 1993a,b.c; Nagata, 1992; Chouinard *et al.*, 1993; Nagata *et al.*, 1994). The cellular sources and metabolic pathways which supply synaptic glycine and/or D-serine involved in NMDA-type EAA neurotransmission remain to be accurately defined. However, experiments in cerebellar slices (Southam *et al.*, 1991) have demonstrated that TTX antagonism of NMDA-dependent increases in cGMP levels can be reversed by D-serine, suggesting that NMDA trans-synaptically releases an endogenous ligand for the NMDA-associated glycine receptor. Such a complex mechanism involved in the supply of synaptic glycine requires further validation but the high glycine levels in Golgi cell terminals may be involved (Ottersen *et al.*, 1988).

As with other neurotransmitter systems, the concentrations of these ligands availabe to the synaptic cleft are tightly regulated. In the case of glutamate or aspartate, there are both glial and neuronal glutamate transporter sites (Pines *et al.*, 1992; Kanai and Hediger, 1992; Danbolt *et al.*, 1992; Raghavendra Rao and Murthy, 1993). In the cerebellum, the glial transporter is highly expressed in Bergmann glia in the Purkinje cell layer, while the neuronal transport site is most concentrated in the granular cell layer. These data indicate that the removal of glutamate from EAA synapses, prior to the metabolism of this amino acid, is probably the prime mechanism for neurotransmitter inactivation. In the case of glycine, several glycine transporter sites have also recently been cloned and localized with one subtype highly localized in the granular cell layer and correlating with NMDA receptor densities (Smith *et al.*, 1992). A second subtype of the glycine transporter is expressed heavily by the Bergmann glia cells of the Purkinje cell layer (Guastella *et al.*, 1992).

The enzymatic machinery responsible for the subsequent inactivation of the potential co-transmitter glycine is the glycine cleavage system (Figure 4), which is a multienzyme complex located in cerebellar astrocytes (Sato *et al.*, 1991). D-Serine is more likely to be metabolized by D-amino acid oxidase (Figure 4), which is localized in microperoxisomes of astrocytes around cerebellar Purkinje cells and

a. <u>Glycine Cleavage System</u>

$$\text{Glycine} + \text{THF} + \text{NAD}^+ \rightarrow \text{CH}_2\text{-THF} + \text{CO}_2 + \text{NH}_3 + \text{NADH} + \text{H}^+$$

$$\text{Glycine} + \text{CH}_2\text{-THF} + \text{H}_2\text{O} \rightarrow \text{Serine} + \text{THF}$$

$$2\ \text{Glycine} + \text{H}_2\text{O} + \text{NAD}^+ \rightarrow \text{Serine} + \text{CO}_2 + \text{NH}_2 + \text{NADH} + \text{H}^+$$

b. <u>D-Amino Acid Oxidase (1.4.3.3)</u>

$$\text{D-Serine} + \text{H}_2\text{O} + \text{O}_2 \rightarrow \text{Hydroxypyruvate} + \text{H}_2\text{O}_2 + \text{NH}_3$$

Figure 4 Metabolism of glycine by the glycine cleavage system and D-serine by D-amino acid oxidase.

cerebellar glomeruli (Arnold *et al.*, 1979). Indeed, in mutant mice lacking D-amino acid oxidase (ddY/DAO$^-$) cerebellar D-serine levels are increased 4.2-fold relative to normal mice (Nagata *et al.*, 1994; Hashimoto *et al.*, 1993b). These data support a key role for cerebellar D-amino acid oxidase in the metabolism of D-serine; however, they do not negate the possibility that other inactivation pathways are also present.

Two other potential endogenous EAA receptor ligands include homocysteic acid which, however, appears to be confined to Bergmann glia of the Purkinje layer of the cerebellum (Grandes *et al.*, 1991; Zhang and Ottersen, 1992; Grieve and Griffiths, 1992), and quinolinic acid which appears to be mainly present in microglia, the CNS resident macrophages (Moffett *et al.*, 1993; Wood, 1994). The function(s) of glial homocysteic acid remains to be defined, but in the case of microglial quinolinate, this tryptophan metabolite may play a role in the cytotoxic actions of these immune cells (Wood, 1994).

6.3.3 NOS/cGMP compartments

Within the rat neuroaxis, the cerebellum contains the highest levels of cGMP and has therefore been used extensively for studies of drug effects on this second messenger molecule (reviewed in Wood, 1991a). Based on the immunocytochemical demonstration of high levels of cGMP-dependent protein kinases in cerebellar Purkinje cells, it was initially suggested that these cells were the main site of cGMP synthesis in the cerebellum. However, as specific antibodies were developed for cGMP, it became apparent that the main sites of cGMP synthesis in the cerebellum were not Purkinje cells but were astroglia, granule cells and GABAergic interneurons (Chan-Palay and Palay, 1979; Chan-Palay, 1982; De Vente *et al.*, 1989, 1990; De Vente and Steinbusch, 1992; Southam *et al.*, 1992). In agreement with these observations, studies of rat cerebellar brain slices combined with chemical lesion procedures, concluded that granule cells and glia were sites of cGMP synthesis but that Purkinje and Golgi cells were not sites of cGMP generation (Garthwaite and Garthwaite, 1987). *In vitro* studies of cerebellar co-cultures of granule cells and astroglia have demonstrated that NO released from granule cells can activate astroglial guanylate cyclase and augment cGMP levels (Kiedrowski *et al.*, 1992a).

Interpretation of these data became clearer once it was established that cGMP was not the second messenger after activation of EAA receptors in the cerebellum, but that NO was the second messenger, which could activate guanylate cyclase at the site of its generation or in other neurons or glia after its diffusion across cellular compartments. Subsequently, the cGMP generated can act as an intracellular messenger in the subsequent signal transduction cascade. The cerebellar neuronal compartments, along with their associated neurochemistry, are briefly outlined in Table 2. As presented the Purkinje cell layer is devoid of nNOS activity while the GABAergic interneurons (basket and stellate) of the molecular layer demonstrate high nNOS levels. In the granular layer, the GABAergic interneurons (Golgi) lack nNOS while the EAA-utilizing neurons, the granule cells and their afferent inputs, the mossy fibres, possess high levels of nNOS. The other major afferent fibre system,

Table 2 The major neuronal components of each cerebellar layer and the associated enzymes of NO and arginine metabolism. Also listed are the cell lesions in genetic and lesion models

Parameter	Purkinje layer	Molecular layer		Granular layer		
		Basket	Stellate	Granule	Golgi	Mossy fibres
Transmitter (1)	GABA	GABA	GABA	EAA	GABA	EAA
NO synthase (2)	No	Yes	Yes	Yes	No	Yes
ASS (3)	No	Yes	Yes	No	Yes	Yes
ASL (3)	No	No	No	No	No	No
Nervous mice (4)	↓	↔	↔	↔	↔	↔
MAM-1 mice (5)	↔	↓	↓	↓	↔	↓
MAM-5 mice (5)	↔	↔	↔	↓	↔	↓
Kainate (6)	↓	↓	↓	↔	↓	↔

(1) Palay and Chan Palay (1974).
(2) Bredt *et al.* (1990), Aoki *et al.* (1991), Vincent and Kimura (1992), Southam *et al.* (1992).
(3) Nakamura *et al.* (1990, 1991), Arnt-Ramos *et al.* (1992).
(4) McBride *et al.* (1976).
(5) Slevin *et al.* (1982), Debarry *et al.* (1987).
(6) Herndon and Coyle (1977), Biggio *et al.* (1978), Tran and Snyder (1979).

the climbing fibres, originating in the inferior olive, is devoid of NOS. These data argue in favour of afferent fibres activating the cerebellar granule, basket and stellate cells, resulting in the activation of nNOS and the subsequent diffusion of NO to activate guanylate cyclase in diverse cell types. Additionally, the mossy fibres will activate basket and stellate cells to evoke NO production. Consistent with these conclusions are the observations that: (1) EAA agonists increase NOS activity in cerebellar granule cell cultures (Kiedrowski *et al.*, 1992a,b); (2) EAA agonists activate NOS in cerebellar brain slices and result in augmented cGMP levels (immunocytochemistry) in granule, basket, stellate and glial cells as well as cerebellar glomeruli, the synaptic complexes of mossy fibre, Golgi fibre and climbing fibre terminals with granule cell dendrites (De Vente and Steinbusch, 1992; Southam *et al.*, 1992); (3) EAA-dependent increases in cerebellar cGMP are independent of Purkinje cell loss ('nervous' mice; see Figure 4; Wood *et al.*, 1994a); (4) granule cell losses (MAM lesioned mice; see Figure 4) drastically reduce cGMP responses to EAA receptor activation (Wood *et al.*, 1994a) and cerebellar NOS activity (Staggerer mutant mouse; Ieda *et al.*, 1994); (5) NO donors dramatically elevate cGMP levels in cerebellar slices, up to 400-fold (Southam and Garthwaite, 1991a), and *in vivo* up to 38-fold, after direct intracerebellar administration (Wood, 1991b); NO donors also augment cGMP efflux in cerebellar dialysates (Vallebuona and Raiteri, 1993).

Action-dependent formation of NO in cerebellar slices has been demonstrated to induce dramatic increases in NO and cGMP generation, with these actions being dependent upon intact climbing fibre inputs (Dickie *et al.*, 1990; Southam and Garthwaite, 1991b). These data support the widespread effects of climbing fibre inputs on cerebellar neuronal activity (Figure 3). Using NO-sensitive electrodes and electrical stimulation of the climbing fibre pathway, NO increases have been demonstrated in cerebellar slices with highest levels in the molecular layer (Shibuki and Okada, 1991). However, NO increases in other layers were also evident. The neuronal activity within the climbing and mossy fibre systems may play an important role in maintaining the high basal tone of cGMP in the cerebellum, since NOS inhibitors produce a dose-dependent decrease in the basal levels of cGMP *in vivo* (Wood *et al.*, 1990c).

The possible roles of microglia in modulating neuronal function remain to be defined but these CNS macrophages produce large quantities of NO which can be neurotoxic (Boje and Arora, 1992; Chao *et al.*, 1992) and do participate in cell death during development in the cerebellum (Ashwell, 1990).

6.3.4 Noradrenergic afferents

The cerebellum receives widespread noradrenergic inputs, mainly from the locus coeruleus and innervating Purkinje cells, granule cells and cerebellar glomeruli (reviewed in Wood *et al.*, 1992a). *In vivo* electrochemical monitoring has demonstrated the release of noradrenaline (NA) in both the molecular and Purkinje/granular cell layers of the cerebellum (Gerhardt *et al.*, 1987; Bickford-Wimer *et al.*, 1991). Additionally, experiments with rat cerebellar slices have demonstrated that these noradrenergic nerve terminals possess NMDA receptors which augment NA release (Yi *et al.*, 1988). The potential role for NMDA-stimulated NA release to alter cerebellar NO and cGMP has therefore been studied. Prior studies of intraventricular NA and α_1-adrenergic agonists had demonstrated increased tissue levels of cGMP via these treatments and decreased levels after α_1-adrenergic antagonists (Haidamous *et al.*, 1980). Subsequent experiments have demonstrated that α_1-adrenergic antagonists (Wood and Rao, 1991; Rao *et al.*, 1991) and prior monoaminergic depletion with reserpine (Wood *et al.*, 1992a) block NMDA- but not kainate- or quisqualate-dependent increases in cerebellar cGMP *in vivo*. Additionally, these treatments blocked harmaline-dependent increases in cerebellar cGMP, suggesting that the climbing fibre pathway modulates NMDA receptors on noradrenergic fibres which in turn activate NOS.

The sites of NOS activation by NA remain to be defined but NO generation within the noradrenergic fibres is unlikely since the locus coeruleus is devoid of nNOS activity (Johnson and Ma, 1993). Within the cerebellum, α_1-adrenergic receptors are present in both the molecular and granular cell layers with about a 3.5-fold enrichment in the molecular layer (Jones *et al.*, 1985). These data would suggest that the noradrenergic activation of NOS might occur in basket and stellate cells in the molecular layer, and in granule cells and cerebellar glomeruli in the granular layer of the cerebellum (Table 2). Additionally, since α_1-adrenergic receptors are also present on astrocytes (Ebersolt *et al.*, 1981) and α_1-dependent cGMP increases in astroglia, which are also NO dependent, have been demonstrated (Agullo and García, 1991, 1992a,b), glial contributions may also be significant in the *in vivo* effects of NA on cerebellar cGMP.

6.3.5 Citrulline recycling

At the sites of NOS activity, citrulline will be generated (granule, basket and stellate cells and mossy fibres; Figure 4). This citrulline can be converted to argininosuccinate (AS) by the basket and granule cells, but in the case of granule cells and mossy fibres, this intermediate may have to be exported to other cells for conversion to AS (Nakamura *et al.*, 1990, 1991). Other compartments containing AS synthetase include glia, climbing fibres and noradrenergic afferents (Nakamura *et al.*, 1991). The climbing fibre and noradrenergic afferent pools are speculations based on the observations that the cell bodies of these pathways, the inferior olive and locus coeruleus possess high levels of AS synthase (Nakamura *et al.*, 1991).

The distribution of AS lyase appears to be more restricted, with high levels present in Golgi cells in the Purkinje cell layer (Nakamura *et al.*, 1991).

6.3.6 NO and cerebellar LTD

Conjunctive high frequency stimulation of parallel and climbing fibre input to Purkinje cells results in depression of neurotransmission at parallel fibre–Purkinje cell synapses. This phenomenon has been termed long-term depression (LTD) and has been invoked to involve NO as a mediator (Shibuki and Okada, 1991; Daniel *et al.*, 1993). However, this conclusion has been challenged (Linden and Connor, 1992) indicating that the role of NO in cerebellar LTD needs further investigation (see Chapter 7).

6.3.7 NO and cerebellar neurotoxicity

The role of NO in EAA-dependent cell death is a controversial issue and appears to be model dependent (see Chapter 10). In the case of the cerebellum, studies of both cerebellar slices (Garthwaite and Garthwaite, 1987) and granule cell cultures (Kiedrowski *et al.*, 1991; Puttfarcken *et al.*, 1992) indicate that EAA-induced neuronal losses are independent of NO and cGMP.

6.4 Hippocampus

6.4.1 EAA receptor populations

In situ studies have demonstrated that dentate granule cells and pyramidal cells in CA1 and CA3 possess a constitutive message for the NMDA receptor subunits

NR1, NR2A and NR2B (Seeburg, 1993). These cells also possess mRNA for AMPA and kainate receptors, with CA1 pyramidal cells expressing the KA2 and GluR6 subunits, CA3 cells expressing KA1, KA2 and GluR6 subunits, and dentate granular cells possessing message for KA1, KA2, GluR6 and GluR7 subunits (Seeburg, 1993).

Studies with hippocampal slices were the first to demonstrate that NMDA receptor activation induces dramatic increases in hippocampal cGMP levels (East and Garthwaite, 1991a). Subsequent *in vivo* studies verified these findings and extended the initial data to also demonstrate that kainate and AMPA/quisqualate receptor activation also augmented hippocampal cGMP levels (Wood *et al.*, 1992b). As in the cerebellum, D-serine also increases *in vivo* hippocampal cGMP levels, as did the convulsant PTZ (Wood *et al.*, 1992b).

6.4.2 Endogenous EAA receptor ligands

As in other brain regions, the endogenous EAA transmitter candidate in the hippocampus is thought to be glutamate with high densities of mRNA for the high affinity transporter for this amino acid being present in pyramidal cells from CA1 to CA4 and in the granular layer of the dentate gyrus (Kanai and Hediger, 1992). Similarly, mRNA for the high affinity glycine transporter is evident in pyramidal cells from CA1 to CA4 and in the granular layer of the dentate gyrus (Smith *et al.*, 1992). In contrast low message levels for the glial glycine transporter have been reported for the hippocampus (Guastella *et al.*, 1992).

Within the hippocampus, D-serine, which may also serve as an agonist for the NMDA-associated glycine receptor, represents 33% of the total serine pool (Chouinard *et al.*, 1993). These levels of D-serine would be sufficient to supply synaptic levels of co-agonist in NMDA synapses and indicate that our understanding of the synthesis and compartmentation of this D-amino acid needs to be increased.

6.4.3 NOS/cGMP compartments

As in the cerebellum, NOS is contained extensively within GABAergic interneurons throughout the hippocampus (Valtschanoff *et al.*, 1993). Additionally, the enzyme has been noted in CA1 pyramidal neurons (Wendland *et al.*, 1994). Activation of NOS has been reported with NMDA both *in vitro* (East and Garthwaite, 1991) and *in vivo* (Wood *et al.*, 1992b) and *in vivo* with quisqualate, kainate and PTZ (Wood *et al.*, 1992b). NO-dependent cGMP efflux into hippocampal brain dialysates has also been recently reported (Vallebuona and Raiteri, 1994).

Since microglia produce high levels of NO and glutamate (Piani *et al.*, 1991, 1992; Wood, 1994), the contributions of microglial NOS in injured or ischaemic hippocampus remains to be defined since microglia are activated within hours of such injuries (Gehrmann *et al.*, 1992; Jorgensen *et al.*, 1993; Koprowski *et al.*, 1993; Lees, 1993; Mitchell *et al.*, 1993).

6.4.4 Noradrenergic afferents

Noradrenergic projections from the locus coeruleus to the hippocampus also possess presynaptic NMDA receptors which stimulate NA release (Keith *et al.*, 1989; Monnet *et al.*, 1992; Pittaluga *et al.*, 1993; Brown, 1993). This NMDA receptor also possesses a glycine binding domain since glycine potentiates NA release in hippocampal slices (Hu *et al.*, 1992); the partial glycine agonist, D-cycloserine, restores decreased NMDA receptor-mediated NA release in hippocampal slices prepared from aged rats (Pittaluga *et al.*, 1993); and D-serine increases hippocampal cGMP *in vivo* after intrahippocampal injections (Wood *et al.*, 1992b). The actions of NMDA on hippocampal cGMP were reversed by NOS inhibitors, both *in vitro* (East and Garthwaite, 1991) and *in vivo* (Wood *et al.*, 1992b). Additionally, NO donors have been shown to evoke NA release in hippocampal brain slices (Lonart *et al.*, 1992; Lauth *et al.*, 1993). The actions of NMDA were also antagonized by α_1-adrenergic antagonists suggesting that presynaptic NMDA receptors on noradrenergic terminals were involved in the observed cGMP changes (Wood *et al.*, 1992b). In contrast, kainate- and quisqualate-dependent increases in hippocampal cGMP were only antagonized by NOS inhibitors but not by α_1-adrenergic antagonists, demonstrating specificity for the presynaptic NMDA-dependent modulation of NA release (Wood *et al.*, 1992b). The presynaptic localization of these NMDA receptors has been confirmed by studies with synaptosomes (Pittaluga and Raiteri, 1992).

6.4.5 Citrulline recycling

Argininosuccinate synthetase (Figure 1) has been mapped in the hippocampus and appears to be concentrated in interneurons in the subiculum and CA1, with pyramidal neurons occasionally demonstrating immunoreactivity (Nakamura *et al.*, 1991; Arnt-Ramos *et al.*, 1992). Slight staining was noted with some interneurons in the granular layer, but not in granule cells.

6.4.6 NO and hippocampal LTP

Hippocampal pyramidal neuronal activity is potentiated by NMDA and glycine ligands (Minota *et al.*, 1989). In addition, LTP can be enhanced by glycine (Shahi *et al.*, 1993) and the glycine partial agonist, D-cycloserine (Ramakers *et al.*, 1993). However, the role(s) of NO in hippocampal LTP is/are more controversial (see Chapter 7). While many studies supporting a key role for NO in LTP have been published (Böhme *et al.*, 1991, Schuman and Madison, 1991; Bon *et al.*, 1992; Iga *et al.*, 1993; Zhuo *et al.*, 1993), others have suggested that NO actually may mediate a retrograde inhibitory action and that in some cases prior studies may have recorded non-specific drug effects of the pharmacological tools used (Izumi *et al.*, 1992; Musleh *et al.*, 1993; Tanaka *et al.*, 1993).

6.4.7 NO and hippocampal neurotoxicity

The role of NO in neuronal cell loss in the hippocampus has been studied with quinolinate lesions, in which case lesions were potentiated by inhibition of NOS (Haberny *et al.*, 1992). In contrast, NOS inhibitors have no effect on glutamate toxicity (Pauwells and Leysen, 1992).

6.5 Conclusions

In conclusion, the role of NO as the true second messenger after activation of ionotropic EAA receptors is now well established. The physiological implications of the varied actions of this diffusible messenger require more in-depth studies since both feed-forward and feed-back actions can be activated by NO. In addition, this messenger molecule has the unique property of diversification of a cellular signal to many cell types in the local area. The full implications of such activities still remain to be unravelled.

References

Agulló, L. & García, A. (1991) *Eur. J. Pharmacol.* **206**, 343–346.
Agulló, L. & García, A. (1992a) *Biochem. Biophys. Res. Commun.* **182**, 1362–1368.
Agulló, L. & García, A. (1992b) *Biochem. J.* **288**, 619–624.
Aoki, E., Semba, R., Mikoshiba, K. & Kashiwamata, S. (1991) *Brain Res.* **547**, 190–192.
Arnold, G., Liscum, L. & Holtzman, E. (1979) *J. Histochem. Cytochem.* **27**, 735–745.
Arnt-Ramos, L.R., O'Brien, W.E. & Vincent, S.R. (1992) *Neuroscience* **51**, 773–789.
Ashwell, K. (1990) *Dev. Brain Res.* **55**, 219–230.
Assreuy, J., Cunha, F.Q., Liew, F.Y. & Moncada, S. (1993) *Br. J. Pharmacol.* **108**, 833–837.
Bansinath, M., Arbabha, B., Turndorf, H. & Garg, U.C. (1993) *Neurochem. Res.* **18**, 1063–1066.
Bickford-Wimer, P., Pang, K., Rose, G.M. & Gerhardt, G.A. (1991) *Brain Res.* **558**, 305–311.
Biggio, G., Corda, M.G., Casu, M., Salis, M. & Gessa, G.L. (1978) *Brain Res.* **154**, 203–208.
Bloch-Tardy, M., Fages, C. & Gonnard, P. (1980) *J. Neurochem.* **35**, 612–615.
Bogle, R.G., Baydoun, A.R., Pearson, J.D., Moncada, S. & Mann, G.E. (1992) *Biochem. J.* **284**, 15–18.
Böhme, G.A., Bon, C., Stutzmann, J.M., Doble, A. & Blanchard, J.C. (1991) *Eur. J. Pharmacol.* **199**, 379–381.
Boje, K.M. & Arora, P.K. (1992) *Brain Res.* **587**, 250–256.
Bon, C., Böhme, G.A., Doble, A., Stutzmann, J.-M. & Blanchard, J.-C. (1992) *Eur. J. Neurosci.* **4**, 420–424.
Bredt, D.S. & Snyder, S.H. (1989) *Proc. Natl. Acad. Sci. USA* **86**, 9030–9033.
Bredt, D.S., Hwang, P.M. & Synder, S.H. (1990) *Nature* **347**, 768–770.
Bredt, D.S., Glatt, C.E., Hwang, P.M., Fotuhi, M., Dawson, T.M. & Snyder, S.H. (1991) *Neuron* **7**, 615–624.
Brown, L.M. (1993) *Neurosci. Lett.* **154**, 43–46.

Brüne, B.B. & Lapetina, E.G. (1989) *J. Biol. Chem.* **264**, 8455–8458.
Brüne, B.B. & Lapetina, E.G. (1991) *Biochem. Biophys. Res. Commun.* **181**, 921–926.
Brüne, B., Dimmeler, S., Molina y Vedia, L. & Lapetina, E.G. (1994) *Life Sci.* **54**, 61–70.
Bush, P.A., Gonzalez, N.E., Griscavage, J.M. & Ignarro, L.J. (1992) *Biochem. Biophys. Res. Commun.* **185**, 960–966.
Chan-Palay, V. (1982) In *The Cerebellum – New Vistas* (eds Palay, S.L. & Chan-Palay, V.), pp. 552–585. Berlin, Springer-Verlag.
Chan-Palay, V. & Palay, S.L. (1979) *Proc. Natl. Acad. Sci. USA* **76**, 1485–1488.
Chao, C.C., Hu, S., Molitor, T.W., Shaskan, E.G. & Peterson, P.K. (1992) *J. Immunol.* **149**, 2736–2741.
Chouinard, M.L., Gaitan, D. & Wood, P.L. (1993) *J. Neurochem.* **61**, 1561–1564.
Corradin, S.B., Mauël, J., Donini, S.D., Quattrocchi, E. & Ricciardi-Castagnoli, P. (1993) *Glia* **7**, 255–262.
Danbolt, N.C., Storm-Mathisen, J. & Kanner, B.I. (1992) *Neuroscience* **51**, 295–310.
Daniel, H., Hemart, N., Jaillard, D. & Crepel, F. (1993) *Eur. J. Neurosci.* **5**, 1070–1082.
Danysz, W., Wroblewski, J.T., Brooker, G. & Costa, E. (1989) *Brain Res.* **479**, 270–276.
Dawson, T.M., Bredt, D.S., Fotuhi, M., Hwang, P.M. & Snyder, S.H. (1991) *Proc. Natl. Acad. Sci. USA* **88**, 7797–7801.
Debarry, J., Gombos, S., Klupp, T. & Hamori, J. (1987) *J. Comp. Neurol.* **261**, 253–265.
Deguchi, T. & Yoshioka, M. (1982) *J. Biol. Chem.* **257**, 10147–10151.
Demerlé-Pallardy, C., Lonchampt, M.-O., Chabrier, P.-E. & Braquet, P. (1993) *Life Sci.* **52**, 1883–1890.
De Vente, J. & Steinbusch, H.W.M. (1992) *Acta Histochem.* **92**, 13–38.
De Vente, J., Bol, J.G.J.M. & Steinbusch, H.W.M. (1989) *Brain Res.* **504**, 332–337.
De Vente, J., Bol, J.G.J.M., Berkelmans, H.S., Schipper, J. & Steinbusch, H.M.W. (1990) *Eur. J. Neurosci.* **2**, 845–862.
Dickie, B.G.M., Lewis, M.J. & Davies, J.A. (1990) *Br. J. Pharmacol.* **101**, 8–9.
Dickie, B.G.M., Lewis, M.J. & Davies, J.A. (1992) *Neurosci. Lett.* **138**, 145–148.
Dimmeler, S. & Brüne, B. (1991) *Biochem. Biophys. Res. Commun.* **178**, 848–855.
Dimmeler, S., Lottspeich, F. & Brüne, B. (1992) *J. Biol. Chem.* **267**, 16771–16774.
Dimmeler, S., Ankarcrona, M., Nicotera, P. & Brüne, B. (1993) *J. Immunol.* **150**, 2964–2971.
East, S.J. & Garthwaite, J. (1991) *Neurosci. Lett.* **123**, 17–19.
East, S.J., Batchelor, A.M. & Garthwaite, J. (1991) *Eur. J. Pharmacol.* **209**, 119–121.
Ebersolt, C., Perez, M. & Bockaert, J. (1981) *J. Neurosci. Res.* **6**, 643–652.
Feinstein, D.L., Galea, E., Roberts, S., Berquist, H., Wang, H. & Reis, D.J. (1994) *J. Neurochem.* **62**, 315–321.
Förstermann,U., Schmidt, H.H.H.W., Pollock, J.S., Sheng, H., Mitchell, J.A., Warner, T.D., Nakane, M. & Murad, F. (1991) *Biochem. Pharmacol.* **42**, 1849–1857.
Galea, E., Feinstein, D.L. & Reis, D.J. (1992) *Proc. Natl. Acad. Sci. USA* **89**, 10945–10949.
Gally, J.A., Montague, P.R., Reeke, Jr., G.N. & Edelman, G.M. (1990) *Proc. Natl. Acad. Sci. USA* **87**, 3547–3551.
Garg, U.C., Devi, L., Turndorf, H., Goldfrank, L.R. & Bansinath, M. (1992) *Brain Res.* **592**, 208–212.
Garthwaite, J. & Beaumont, P.S. (1989) *Neurosci. Lett.* **107**, 151–156.
Garthwaite, J. & Garthwaite, G. (1987) *J. Neurochem.* **48**, 29–39.
Garthwaite, J., Charles, S.L. & Chess-Williams, R. (1988) *Nature* **336**, 385–388.
Garthwaite, J., Garthwaite, G., Palmer, R.M.J. & Moncada, S. (1989a) *Eur. J. Pharmacol.* **172**, 413–416.
Garthwaite, J., Southam, E. & Anderton, M. (1989b) *J. Neurochem.* **53**, 1952–1954.
Gehrmann, J., Bonnekoh, P., Miyazawa, T., Oschlies, U., Dux, E., Hossmann, K.-A. & Kreutzberg, G.W. (1992) *Acta Neuropathol. (Berl.)* **84**, 588–595.
Gerhardt, G.A., Rose, G.M. & Hoffer, B.J. (1987) *Brain Res.* **413**, 327–335.
Giovanelli, J., Campos, K.L. & Kaufman, S. (1991) *Proc. Natl. Acad. Sci. USA* **88**, 7091–7095.

Grandes, P., Do, K.Q., Morino, P., Cuénod, M. & Streit, P. (1991) *Eur. J. Neurosci.* **3**, 1370–1373.
Grieve, A. & Griffiths, R. (1992) *Neurosci. Lett.* **145**, 1–5.
Guastella, J., Brecha, N., Weigmann, C., Lester, H.A. & Davidson, N. (1992) *Proc. Natl. Acad. Sci. USA* **89**, 7189–7193.
Haberny, K.A., Pou, S. & Eccles, C.U. (1992) *Neurosci. Lett.* **146**, 187–190.
Haidamous, M., Kouyoumdjian, T.C., Briley, P.A. & Gonnard, P. (1980) *Eur. J. Pharmacol.* **63**, 287–294.
Hashimoto, A., Nishikawa, T., Hayashi, T., Fujii, N., Harada, K., Oka, T. & Takahashi, K. (1992) *FEBS Lett.* **296**, 33–36.
Hashimoto, A., Kumashiro, S., Nishikawa, T., Oka, T., Takahashi, K., Mito, T., Takashima, S., Doi, N., Mizutani, Y., Yamazaki, T., Kaneko, T. & Ootomo, E. (1993a) *J. Neurochem.* **61**, 348–351.
Hashimoto, A., Nishikawa, T., Konno, R., Niwa, A., Yasumura, Y., Oka, T. & Takahashi, K. (1993b) *Neurosci. Lett.* **152**, 33–36.
Hashimoto, A., Nishikawa, T., Oka, T. & Takahashi, K. (1993c) *J. Neurochem.* **60**, 783–786.
Herndon, R.M. & Coyle, J.T. (1977) *Science* **198**, 71–72.
Hiki, K., Yui, Y., Hattori, R., Eizawa, H., Kosuga, K. & Kawai, C. (1991) *Jpn. J. Pharmacol.* **56**, 217–220.
Hope, B.T., Micheal, G.J., Knigge, K.M. & Vincent, S.R. (1991) *Proc. Natl. Acad. Sci. USA* **88**, 2811–2814.
Hu, P.-S., Jin, S. & Fredholm, B.B. (1992) *Acta Physiol. Scand.* **145**, 77–78.
Iga, Y., Yoshioka, M., Togashi, H. & Saito, H. (1993) *Eur. J. Pharmacol.* **238**, 395–398.
Ikeda, M., Matsui, K., Ishihara, Y., Morita, I., Murota, S., Yuasa, T. & Miyatake, T. (1994) *Neurosci. Lett.* **168**, 65–68.
Izumi, Y., Clifford, D.B. & Zorumski, C.F. (1992) *Science* **257**, 1273–1276.
Johnson, J.W. & Ascher, P. (1987) *Nature* **325**, 529–531.
Johnson, M.D. & Ma, P.M. (1993) *J. Comp. Neurol.* **332**, 391–406.
Jones, L.S., Gauger, L.L. & Davis, J.N. (1985) *J. Comp. Neurol.* **231**, 190–208.
Jones, M.E., Anderson, A.D., Anderson, C.A. & Hodes, S. (1961) *Arch. Biochem. Biophys.* **95**, 499–507.
Jorgensen, M.B., Finsen, B.R., Jensen, M.B., Castellano, B., Diemer, N.H. & Zimmer, J. (1993) *Exp. Neurol.* **120**, 70–88.
Kanai, Y. & Hediger, M.A. (1992) *Nature* **360**, 467–471.
Katsuki, S., Arnold, W., Mittal, C. & Murad, F. (1977) *J. Cyclic Nucleotide Res.* **3**, 23–35.
Keith, R.A., Mangano, T.J., Meiners, B.A., Stumpo, R.J., Klika, A.B., Patel, J. & Salama, A.I. (1989) *Eur. J. Pharmacol.* **166**, 393–400.
Kemp, J.W. & Woodbury, D.M. (1965) *Biochim. Biophys. Acta* **111**, 23–31.
Kiedrowski, L., Manev, H., Costa, E. & Wroblewski, J.T. (1991) *Neuropharmacology* **30**, 1241–1243.
Kiedrowski, L., Costa, E. & Wroblewski, J.T. (1992a) *J. Neurochem.* **58**, 335–341.
Kiedrowski, L., Costa, E. & Wroblewski, J.T. (1992b) *Neurosci. Lett.* **135**, 59–61.
Klatt, P., Heinzel, B., Mayer, B., Ambach, E., Werner-Felmayer, G., Wachter, H. & Werner, E.R. (1992) *FEBS Lett.* **305**, 160–162.
Kleckner, N.W. & Dingeldine, R. (1988) *Science* **241**, 835–837.
Koprowski, H., Zheng, Y.M., Heber-Katz, E., Fraser, N., Rorke, L., Fu, Z.F., Hanlon, C. & Dietzschold, B. (1993) *Proc. Natl. Acad. Sci. USA* **90**, 3024–3027.
Lauth, D., Hertting, G. & Jackisch, R. (1993) *Eur. J. Pharmacol.* **236**, 165–166.
Lees, G.J. (1993) *J. Neurol. Sci.* **114**, 119–122.
Lehmann, J., Schneider, J., McPherson, S., Murphy, D.E., Bernard, P., Tsai, C., Bennett, D.A., Pastor, G., Steel, D., Boehm, C., Cheney, D.L., Liebman, J.M., Williams, M. & Wood, P.L. (1987) *J. Pharmacol. Exp. Ther.* **240**, 737–746.

Lehmann, J., Hutchison, A.J., McPherson, S.E., Mondadori, C., Schmutz, M., Sinton, C.M., Tsai, C., Murphy, D.E., Steel, D.J., Williams, M., Cheney, D.L. & Wood, P.L. (1988) *J. Pharmacol. Exp. Ther.* **246**, 65–75.
Leone, A.M., Palmer, R.M.J., Knowles, R.G., Francis, P.L., Ashton, D.S. & Moncada, S. (1991) *J. Biol. Chem.* **266**, 23790–23795.
Lepoivre, M., Flaman, J.-M. & Henry, Y. (1992) *J. Biol. Chem.* **267**, 22994–23000.
Linden, D.J. & Connor, J.A. (1992) *Eur. J. Neurosci.* **4**, 10–15.
Lonart, G., Wang, J. & Johnson, K.M. (1992) *Eur. J. Pharmacol.* **220**, 271–272.
Luo, D., Knezevich, S. & Vincent, S.R. (1993) *Neuroscience* **57**, 897–900.
Luo, D., Leung, E. & Vincent, S.R. (1994) *J. Neurosci.* **14**, 263–271.
Manzoni, O. & Bockaert, J. (1993) *J. Neurochem.* **61**, 368–370.
Matsumoto, T., Pollock, J.S., Nakane, M. & Förstermann, U. (1993) *Dev. Brain Res.* **73**, 199–203.
Mayer, B., John, M., Heinzel, B., Werner, E.R., Wachter, H., Schultz, G. & Böhme, E. (1991) *FEBS Lett.* **288**, 187–191.
McBride, W.J., Aprison, M.H. & Kusano, K. (1976) *J. Neurochem.* **26**, 867–870.
McMillan, K., Bredt, D.S., Hirsch, D.J., Snyder, S.H., Clark, J.E. & Masters, B.S.S. (1992) *Proc. Natl. Acad. Sci. USA* **89**, 11141–11145.
Miki, N., Kawabe, Y. & Kuriyama, K. (1977) *Biochem. Biophys. Res. Commun.* **75**, 851–856.
Minota, S., Miyazaki, T., Wang, M.Y., Read, H.L. & Dunn, N.J. (1989) *Neurosci. Lett.* **100**, 237–242.
Mitchell, J., Sundstrom, L.E. & Wheal, H.V. (1993) *Exp. Neurol.* **121**, 224–230.
Moffett, J.R., Espey, M.G., Gaudet, S.J. & Namboodiri, M.A.A. (1993) *Brain Res.* **623**, 337–340.
Monnet, F.P., Blier, P., Debonnel, G. & De Montigny, C. (1992) *Naunyn-Schmiedebergs Arch. Pharmacol.* **346**, 32–39.
Moriguchi, M., Manning, L.R. & Manning, J.M. (1992) *Biochem. Biophys. Res. Commun.* **183**, 598–604.
Murphy, S., Minor, Jr., R.L., Welk, G. & Harrison, D.G. (1990) *J. Neurochem.* **55**, 349–351.
Murphy, S., Minor, Jr., R.L., Welk, G. & Harrison, D.G. (1991) *J. Cardiovasc. Pharmacol.* **17** Suppl. 3, S265–S268.
Musleh, W.Y., Shahi, K. & Baudry, M. (1993) *Synapse* **13**, 370–375.
Nagata, Y. (1992) *Experientia* **48**, 753–755.
Nagata, Y., Horiike, K. & Maeda, T. (1994) *Brain Res.* **634**, 291–295.
Nakamura, H., Saheki, T. & Nakagawa, S. (1990) *Brain Res.* **530**, 108–112.
Nakamura, H., Saheki, T., Ichiki, H., Nakata, K. & Nakagawa, S. (1991) *J. Comp. Neurol.* **312**, 652–679.
Ogura, T., Yokoyama, T., Fujisawa, H., Kurashima, Y. & Esumi, H. (1993) *Biochem. Biophys. Res. Commun.* **193**, 1014–1022.
Ottersen, O.P., Storm-Mathisen, J. & Somogyi, P. (1988) *Brain Res.* **450**, 342–353.
Palay, S.L. & Chan-Palay, V. (1974) *Cerebellar Cortex Cytology and Organization.* New York, Springer-Verlag.
Pauwels, P.J. & Leysen, J.E. (1992) *Neurosci. Lett.* 27–30.
Peunova, N. & Enikolopov, G. (1993) *Nature* **364**, 450–453.
Piani, D., Frei, K., Do, K.Q., Cuénod, M. & Fontana, A. (1991) *Neurosci. Lett.* **133**, 159–162.
Piani, D., Spranger, M., Frei, K., Schaffner, A. & Fontana, A. (1992) *Eur. J. Immunol.* **22**, 2429–2436.
Pines, G., Danbolt, N.C., Bjoras, M., Zhang, Y., Bendahan, A., Eidie, L., Koepsell, H., Storm-Mathisen, J., Seeburg, E. & Kanner, B.I. (1992) *Nature* **360**, 464–467.
Pittaluga, A. & Raiteri, M. (1992) *J. Pharmacol. Exp. Ther.* **260**, 232–237.
Pittaluga, A., Fedele, E., Risiglione, C. & Raiteri, M. (1993) *Eur. J. Pharmacol.* **231**, 129–134.
Prast, H. & Philippu, A. (1992) *Eur. J. Pharmacol.* **216**, 139–140.
Puttfarcken, P.S., Lyons, W.E. & Coyle, J.T. (1992) *Neuropharmacology* **31**, 565–575.

Raghavendra Rao, V.L. & Murthy, C.R.K. (1993) *Neurochem. Res.* **18**, 647–654.
Ramakers, G.M.J., Urban, I.J.A., De Graan, P.N.E., Di Luca, M., Cattabeni, F. & Gispen, W.H. (1993) *Neuroscience* **54**, 49–60.
Rao, T.S., Cler, J.A., Emmett, M.R., Mick, S.J., Iyengar, S. & Wood, P.L. (1990a) *Neuropharmacology* **29**, 1075–1080.
Rao, T.S., Cler, J.A., Mick, S.J., Emmett, M.R., Iyengar, S. & Wood, P.L. (1990b) *Neuropharmacology* **29**, 1031–1035.
Rao, T.S., Contreras, P.C., Cler, J.A., Emmett, M.R., Mick, S.J., Iyengar, S. & Wood, P.L. (1991) *Neuropharmacology* **30**, 557–565.
Rao, T.S., Gray, N.M., Dappen, M.S., Cler, J.A., Mick, S.J., Emmett, M.R., Iyengar, S., Monahan, J.B., Cordi, A.A. & Wood, P.L. (1993) *Neuropharmacology* **32**, 139–147.
Ratner, S., Morell, H. & Carvalho, E. (1960) *Arch. Biochem. Biophys.* **91**, 280–289.
Sadasivudu, B. & Rao, T.J. (1976) *J. Neurochem.* **27**, 785–794.
Salvemini, D., Mollace, V., Pistelli, A., Änggard, E. & Vane, J. (1992) *Br. J. Pharmacol.* **106**, 931–936.
Salvemini, D., Misko, T.P., Masferrer, J.L., Seibert, K., Currie, M.G. & Needleman, P. (1993) *Proc. Natl. Acad. Sci. USA* **90**, 7240–7244.
Sato, K., Yoshida, S., Fujiwara, K., Tada, K. & Tohyama, M. (1991) *Brain Res.* **567**, 64–70.
Schuman, E.M. & Madison, D.V. (1991) *Science* **254**, 1503–1506.
Seeburg, P.H. (1993) *TIPS* **14**, 297–303.
Shahi, K., Marvizon, J.-C. & Baudry, M. (1993) *Neurosci. Lett.* **149**, 185–188.
Shibuki, K. & Okada, D. (1991) *Nature* **349**, 326–328.
Simmons, M.L. & Murphy, S. (1992) *J. Neurochem.* **59**, 897–905.
Slevin, J.T., Johnston, M.V., Biziere, K. & Coyle, J.T. (1982) *Dev. Neurosci.* **5**, 3–12.
Smith, K.E., Borden, L.A., Hartig, P.R., Branchek, T. & Weinshank, R.L. (1992) *Neuron* **8**, 927–935.
Southam, E. & Garthwaite, J. (1991a) *Neurosci. Lett.* **130**, 107–111.
Southam, E. & Garthwaite, J. (1991b) *Eur. J. Neurosci.* **3**, 279–382.
Southam, E., East, S.J. & Garthwaite, J. (1991) *J. Neurochem.* **56**, 2072–2081.
Southam, E., Morris, R. & Garthwaite, J. (1992) *Neurosci. Lett.* **137**, 241–244.
Stamler, J.S., Simon, D.I., Osborne, J.A., Mullins, M.E., Jarki, O., Michel, T., Singel, D.J. & Loscazo, J. (1992) *Proc. Natl. Acad. Sci. USA* **89**, 444–448.
Tanaka, T., Saito, H. & Matsuki, N. (1993) *Brain Res.* **631**, 72–76.
Thomson, A.M. (1990) *Prog. Neurobiol.* **35**, 53–74.
Thomson, A.M., Walker, V.E. & Flynn, D.M. (1988) *Nature* **338**, 422–424.
Tjörnhammar, M.-L., Lazaridis, G. & Bartfai, T. (1986) *Neurosci. Lett.* **68**, 95–99.
Tran, V.T. & Snyder, S.H. (1979). *Brain Res.* **167**, 345–353.
Vallebuona, F. & Raiteri, M. (1993) *Neuroscience* **57**, 577–585.
Vallebuona, F. & Raiteri, M. (1994) *J. Neurosci.* **14**, 134–139.
Valtschanoff, J.G., Weinberg, R.J., Kharazia, V.N., Nakane, M. & Schmidt, H.H.H.W. (1993) *J. Comp. Neurol.* **331**, 111–121.
Vigne, P., Damais, C. & Frelin, C. (1993). *Brain Res.* **606**, 332–334.
Vincent, S.R. & Kimura, H. (1992) *Neuroscience* **46**, 755–784.
Welsh, N. & Sandler, S. (1992) *Biochem. Biophys. Res. Commun.* **182**, 333–340.
Wendland, B., Schweizer, F.E., Ryan, T.A., Nakane, M., Murad, F., Scheller, R.H. & Tsien, R.W. (1994) *Proc. Natl. Acad. Sci. USA* **91**, 2151–2155.
Westergaard, N., Beart, P.M. & Schousboe, A. (1993) *J. Neurochem.* **61**, 364–367.
Wolf, G., Würdig, S. & Schünzel, G. (1992) *Neurosci. Lett.* **147**, 63–66.
Wood, P.L. (1991a) *Pharmacol. Rev.* **43**, 1–25.
Wood, P.L. (1991b) In *Excitatory Amino Acids* (eds Meldrum, B.S., Moroni, F., Simon, R.P. & Woods, J.H.), pp. 317–321. New York, Raven Press.
Wood, P.L. (1994) *Drug News and Perspectives* **7**, 138–157.
Wood, P.L. and Rao, T.S. (1991) *Prog. Neuro-Psychopharmacol. Biol. Psychiat.* **15**, 229–235.

Wood, P.L., Richard, J.W., Pilapil, C. & Nair, N.P.V. (1982) *Neuropharmacology* **21**, 1235–1238.
Wood, P.L., Steel, D.J., Kim, H.S., Petrack, B. & Altar, C.A. (1988) *J. Pharmacol. Exp. Ther.* **244**, 58–62.
Wood, P.L., Emmett, M.R., Rao, T.S., Mick, S., Cler, J. & Iyengar, S. (1989) *J. Neurochem.* **53**, 979–981.
Wood, P.L., Emmett, M.R., Rao, T.S., Mick, S., Cler, J., Oei, E. & Iyengar, S. (1990a) *Neuropharmacology* **29**, 675–679.
Wood, P.L., Rao, T.S., Iyengar, S., Lanthorn, T., Monahan, J., Cordi, A., Sun, E., Vazquez, M., Gray, N. & Contreras, P. (1990b) *Neurochem. Res.* **15**, 217–230.
Wood, P.L., Emmett, M.R., Rao, T.S., Cler, J., Mick, S. & Iyengar, S. (1990c) *J. Neurochem.* **55**, 346–348.
Wood, P.L., Ryan, R. & Li, M. (1992a) *Life Sci.* **51**, PL267–PL270.
Wood, P.L., Ryan, R. & Li, M. (1992b) *Life Sci.* **51**, 601–606.
Wood, P.L., Emmett, M. & Wood, J.A. (1994a) *Life Sci.* **54**, 615–620.
Wood, P.L., Choksi, S. & Bocchini, V. (1994b) *NeuroReport*, **5**, 977–980.
Wroblewski, J.T., Fadda, E., Mazzetta, J., Lazarewicz, J.W. & Costa, E. (1989) *Neuropharmacology* **28**, 447–452.
Yi, S.J., Snell, L.D. & Johnson, K.M. (1988) *Brain Res.* **445**, 147–151.
Zhang, J. & Snyder, S.H. (1992) *Proc. Natl. Acad. Sci. USA* **89**, 9382–9385.
Zhang, N. & Ottersen, O.P. (1992) *Exp. Brain Res.* **90**, 11–20.
Zhuo, M., Small, S.A., Kandel, E.R. & Hawkins, R.D. (1993) *Science* **260**, 1946–1950.

CHAPTER 7

NITRIC OXIDE SIGNALLING, LONG-TERM POTENTIATION AND LONG-TERM DEPRESSION

Erin M. Schuman

California Institute of Technology, Division of Biology, Pasadena, CA 91125, USA

Table of Contents

7.1	Introduction	125
7.2	Long-term potentiation, the hippocampus and memory	126
	7.2.1 Glutamate-activated ion channels and LTP induction	127
	7.2.2 LTP expression and quantal analysis	128
	7.2.3 A requirement for a retrograde signal	129
7.3	Nitric oxide and LTP	130
	7.3.1 Long-term potentiation and diffusible signalling	134
	7.3.2 Functional significance of distributed potentiation	142
7.4	Long-term depression in the hippocampus	143
	7.4.1 The role of nitric oxide in hippocampal LTD	143
7.5	Long-term depression in the cerebellum	144
7.6	Conclusions	147
	References	147

7.1 Introduction

The fidelity of information flow in the nervous system is determined by the strength of synaptic connections between neurons. Synaptic strength is not a static process, but rather it can be regulated on time scales ranging from milliseconds to many days. Many forms of synaptic regulation are use dependent; that is, when the synapse is activated in a particular way the potency of the synaptic transmission is changed. For example, a single presynaptic action potential, which results in the exocytosis of neurotransmitter, can increase the amount of neurotransmitter released following a second action potential, if the second action potential occurs soon after the first. This phenomenon, known as paired pulse facilitation, is a type of short-term plasticity that can last for several hundred milliseconds. In addition

Nitric Oxide in the Nervous System
ISBN 0-12-721985-4

to several forms of short-term plasticity (Zucker, 1989), synaptic strength can also be persistently enhanced or depressed for periods of hours to days (Bliss and Collingridge, 1993; Linden, 1994). Many of these forms of long-lasting plasticity can be elicited in *in vitro* preparations such as brain slices enabling researchers to probe the cellular mechanisms responsible for these changes.

7.2 Long-term potentiation, the hippocampus and memory

Long-term potentiation (LTP) is the long-lasting increase in synaptic transmission that occurs when postsynaptic neurons experience depolarization coupled with presynaptic activity. LTP has been observed in many brain areas but has been studied most extensively in the hippocampus, a brain region associated with learning and memory in humans and animals (Squire and Zola-Morgan, 1991; Eichenbaum *et al.*, 1992). Damage to the hippocampal formation in humans causes severe amnesia for both events preceding the damage (retrograde amnesia) and events occurring after the damage (anterograde amnesia). Likewise, hippocampal lesion studies in rodents and non-human primates have revealed a variety of selective deficits including the inability to learn and remember relations among environmental cues, the inability to transfer knowledge to a new context, and the inability to make comparisons between temporally dissociated events (Morris *et al.*, 1982; Cohen and Eichenbaum, 1993). These types of deficits have led to the idea that the hippocampus stores information relevant to relations between events, rather than information dedicated to a particular sensory function. Indeed, the notion of relational processing makes sense given the multimodal nature of the inputs to the hippocampus from various association cortices as well as the olfactory bulb and pyriform cortex (Witter *et al.*, 1989).

Is LTP a potential storage mechanism for memories? There are several features of LTP that make it an attractive candidate for a neuronal mnemonic device. First, LTP is observed at all the major excitatory connections in the hippocampus. Second, LTP can be quite long-lived: *in vitro* LTP can last as long as the hippocampal slice remains healthy and viable (4–8 hours); *in vivo* LTP has been reported to last from days to weeks. LTP also exhibits the property of associativity (Barrionuevo and Brown, 1983; Sastry *et al.*, 1986): when a weak stimulus, normally subthreshold for inducing LTP, is paired with a strong stimulus, LTP of the weak pathway will result. A simple analogy can be made with Pavlovian or associative learning: when a stimulus which elicits no response on its own (the conditioned stimulus) is paired with a stimulus that elicits a response (the unconditioned stimulus), the conditioned stimulus comes to elicit a response. There have also been attempts to establish pharmacological correlations between LTP and certain types of learning. For example, blocking certain receptors or enzymes previously implicated in LTP (with *in vivo* administration of inhibitors or targeted gene disruption) can block the acquisition of some kinds of spatial learning (Morris *et al.*, 1986; Morris, 1989;

Chapman *et al.*, 1992). For example, animals with targeted disruption of the Ca^{2+}/calmodulin-dependent protein kinase gene (Silva *et al.*, 1992a,b) or the *fyn* non-receptor tyrosine kinase gene (Grant *et al.*, 1992) exhibit impaired spatial learning as well as attenuated LTP.

7.2.1 Glutamate-activated ion channels and LTP induction

The major excitatory neurotransmitter in the brain is excitatory amino acid L-glutamate or a related molecule (Fonnum, 1984). There are two general classes of glutamate receptor channels responsible for synaptic transmission at most excitatory synapses: the amino-3-hydroxyl-5-methyl-4-isoxazole propionate (AMPA)/kainate (KA) and *N*-methyl-D-aspartate (NMDA) receptor channels (Hollmann and Heinemann, 1994). At normal resting potentials, most of the excitatory postsynaptic potential (EPSP) in hippocampal pyramidal neurons is due to ion flux, primarily Na^+ and K^+, through the AMPA/KA class of glutamate-activated channels. However, at depolarized membrane potentials, the voltage-dependent block of NMDA receptor channels (Mayer *et al.*, 1984; Ascher and Nowak, 1988) by Mg^{2+} is relieved and ions, including Ca^{2+} (MacDermott *et al.*, 1986), move across the membrane and contribute to the EPSP. At the synapses between the Schaffer collaterals and the CA1 pyramidal neurons, LTP is induced when the following two conditions are met: postsynaptic neurons must be sufficiently depolarized to relieve the Mg^{2+} block of the NMDA receptor channel and glutamate must be released from presynaptic terminals. The necessity of these two conditions is demonstrated by studies showing that NMDA receptor antagonists (Collingridge *et al.*, 1983), Ca^{2+} chelators (Lynch *et al.*, 1983; Malenka *et al.*, 1988), or membrane hyperpolarization (Malinow and Miller, 1986) can prevent LTP. In addition, postsynaptic depolarization alone (Kelso *et al.*, 1986; Kullman *et al.*, 1992; Wyllie *et al.*, 1994), without presynaptic stimulation, only produces short-lasting potentiation. LTP can be initiated by several different experimental protocols which satisfy the above requirements, but is most frequently accomplished with high frequency stimulation (tetanus) of presynaptic fibres or by pairing postsynaptic depolarization (from d.c. current injection) with low frequency stimulation of presynaptic fibres.

What happens in postsynaptic spines after Ca^{2+} influx through the NMDA receptor? Many different Ca^{2+}-dependent enzymes have been implicated in LTP. In fact, it is probably fair to say that almost every Ca^{2+}-dependent enzyme that has been investigated has been reported to participate in LTP. This list includes protein kinase C (Lovinger *et al.*, 1987; Malinow *et al.*, 1988, 1989), Ca^{2+}/calmodulin-dependent protein kinase II (Malenka *et al.*, 1989; Malinow *et al.*, 1989; Silva *et al.*, 1992b), calpain (Lynch and Baudry, 1984), phospholipase A_2 (Williams *et al.*, 1989; Kato *et al.*, 1994) and nitric oxide synthase (Böhme *et al.*, 1991; O'Dell *et al.*, 1991; Schuman and Madison, 1991; Haley *et al.*, 1992). That so many different biochemical cascades have been implicated in LTP indicates that the underlying mechanisms are much more complicated than previously appreciated. In addition, the possibility that some of these enzymes may contribute to normal synaptic transmission

(Waxham *et al.*, 1993), rather than playing an exclusive role in synaptic plasticity, needs to be considered.

7.2.2 LTP expression and quantal analysis

Despite the implication of many different biochemical pathways and cascades in LTP, we still do not understand the essential nature of the mechanism(s) responsible for increasing synaptic strength. Several possibilities have been entertained, including an increase in the number and/or sensitivity of postsynaptic glutamate receptors, a structural change in postsynaptic spines leading to increased current flow, an increase in the probability and/or amount of neurotransmitter release or, of course, a combination of these possibilities. Using push–pull cannulae, two early *in vivo* studies (Dolphin *et al.*, 1982; Bliss *et al.*, 1986) observed that the concentration of glutamate in the extracellular fluid increased after tetanic stimulation and remained elevated for the duration of LTP. However, these experiments were not entirely conclusive since it is unclear whether the observed increase in extracellular glutamate was due to an increase in release from synapses undergoing LTP, rather than glutamate released from other compartments, or a decrease in glutamate uptake. If neurotransmitter release is increased following LTP then one might expect that an increased postsynaptic response should be observed for both NMDA- and AMPA-mediated synaptic currents. On the contrary, two early studies (Kauer *et al.*, 1988; Muller and Lynch, 1988) reported that only the AMPA component of the postsynaptic response was increased following LTP, suggesting a postsynaptic change. Subsequently, an enhancement of the NMDA-mediated synaptic current following LTP was detected by several groups (Bashir *et al.*, 1991; Berretta *et al.*, 1991; Xie *et al.*, 1992), although some labs report that the isolated AMPA component of the postsynaptic response is enhanced to a greater extent than isolated NMDA component (Asztely *et al.*, 1992; Muller *et al.*, 1992; Perkel and Nicoll, 1993).

Because the different possible mechanisms for LTP expression fall on different sides of the synaptic cleft, many labs have focused on determining the synaptic locus of LTP by employing quantal analysis techniques. This type of analysis was first utilized by Fatt and colleagues (Fatt and Katz, 1952; del Castillo and Katz, 1954) at the neuromuscular junction where the amplitudes of postsynaptic end-plate potentials are observed as integral multiples of a single small unit, suggesting that the release of neurotransmitter is quantized. By measuring the size of many spontaneous miniature end-plate potentials it is possible to estimate a, the amplitude of a postsynaptic response to release of a single quantum, N, the number of presynaptic release sites, and p, the probability of neurotransmitter release. By estimating each of these parameters before and after a change in synaptic efficacy it is possible to ascertain the essential nature of the synaptic change. Under most circumstances, changes in a would be indicative of a postsynaptic change whereas changes in N and/or p would indicate a presynaptic change. Although these types of measurements are well suited for analysis of synaptic transmission at peripheral synapses, it is less clear whether these techniques can be applied to central synapses (Korn and Faber, 1991).

Accurately measuring *a*, *N* and *p* requires that unitary synaptic events are relatively large and distinguishable from background membrane noise. In addition, it is desirable to study synaptic transmission at monosynaptically connected pairs of neurons and to study miniature synaptic events which occur at a defined subset of synapses. Obviously, these particular circumstances are not easy to come by at central synapses where thousands of synapses impinge on a single, highly branched neuron.

Despite these obstacles many groups have attempted to apply quantal analysis techniques to study LTP. Several labs have attempted to circumvent the problems associated with central synapses by utilizing statistical techniques which may allow one to extract quantal information from central synapse data. Utilizing whole-cell recording to measure evoked synaptic transmission in slices and cultured neurons, several different observations have been made: LTP is associated with changes in *N* and/or *p* accompanied by small occasional changes in *a* (Malinow and Tsien, 1990; Malinow, 1991) or no changes in *a* (Bekkers and Stevens, 1990) or robust changes in *a* accompanied by occasional observations of a decrease in the number of synaptic failures (Kullman and Nicoll, 1992). Synaptic failures occur when a presynaptic action potential is elicited but no neurotransmitter release ensues. A decrease in synaptic failures is usually assumed to reflect a presynaptic change, assuming that the number of functional postsynaptic receptors and the ability to detect very small synaptic events remains constant over the duration of the experiment. Two groups have measured miniature synaptic events (minis) before and after LTP. In cultured hippocampal neurons, a short pulse of glutamate can increase the size of evoked synaptic transmission as well as the frequency of miniature synaptic events (Malgaroli and Tsien, 1992). In contrast, an evaluation of minis in hippocampal slices showed that the amplitude, but not the frequency of the minis was enhanced following LTP (Manabe *et al.*, 1992). One study, utilizing conventional intracellular recordings, reported changes in quantal size, without changes in *N* or *p* (Foster and McNaughton, 1991). Another study obtained different results in individual experiments: some experiments exhibited predominantly presynaptic changes, some exhibited postsynaptic changes, and some exhibited both pre- and postsynaptic changes (Larkman *et al.*, 1992). The authors of the latter study made the interesting suggestion that the initial probability of neurotransmitter release may ultimately influence the type of LTP expression manifest: when initial probabilities are low, presynaptic changes may occur; when initial release probabilities are high, postsynaptic mechanisms will predominate. Taken together, there is no apparent consensus among the studies that would indicate on which side of the synapse LTP expression lies. What is clear, however, is that LTP can be expressed by both pre- and postsynaptic mechanisms; the task of future experiments will be to unravel the cellular circumstances that predispose the plasticity to take a particular form.

7.2.3 A requirement for a retrograde signal

As indicated above, the cellular events responsible for initiating LTP occur in the postsynaptic neuron but the events responsible for making LTP persist may occur, at least in part, in the presynaptic neuron. This indicates that the synapse must devise a way of informing the presynaptic cell that LTP induction has occurred. To account for the communication that must occur between post- and presynaptic neurons the existence of a retrograde messenger has been postulated (Bliss *et al.*, 1986). Although other possibilities exist, most studies have focused on the idea that the retrograde messenger takes the form of a diffusible signal which is generated in the postsynaptic cell and diffuses across the synaptic cleft to interact with the presynaptic release machinery. According to this idea, the retrograde messenger is produced in response to Ca^{2+} influx through postsynaptic NMDA receptor channels. Many different diffusible signals have been put forth as potential retrograde signals including arachidonic acid (Williams and Biss, 1989; Drapeau *et al.*, 1990; O'Dell *et al.*, 1991), carbon monoxide (Stevens and Wang, 1993; Zhuo *et al.*, 1993), platelet activating factor (Del Cerro *et al.*, 1990; Arai and Lynch, 1992; Clark *et al.*, 1992; Wieraszko *et al.*, 1993; Kato *et al.*, 1994) and nitric oxide (Böhme *et al.*, 1991; O'Dell *et al.*, 1991; Schuman and Madison, 1991; Haley *et al.*, 1992). It is quite possible that each of these molecules plays a role in LTP, either by acting in concert with one another or by having particular cellular circumstances or induction procedures which preferentially activate a given messenger pathway. However, the remainder of this chapter will focus on the evidence that nitric oxide participates in LTP.

7.3 Nitric oxide and LTP

The apparent need for a retrograde signal as well as the established Ca^{2+} dependence of LTP has prompted several groups to examine whether NO might serve in this capacity. As indicated above, the Ca^{2+} dependence of the NOS makes it an attractive target for Ca^{2+} rises mediated by the NMDA receptor. Is an NOS present in the appropriate postsynaptic neurons responsible for LTP induction? Early studies utilizing antibodies raised against the cerebellar NOS (Bredt *et al.*, 1990; Valtschanoff *et al.*, 1993) *in situ* hybridization to NOS mRNA (Bredt *et al.*, 1991) or NADPH diaphorase staining did not detect NOS in CA1 pyramidal neurons (Vincent and Kimura, 1992), but detected robust staining in non-pyramidal neurons (presumably interneurons) interspersed in both the dendritic and cellular layers of the hippocampus. The failure to observe staining in CA1 pyramidal neurons was problematic for interpretations of electrophysiological experiments that showed that postsynaptic injections of NOS inhibitors into CA1 cells effectively prevented LTP (O'Dell *et al.*, 1991; Schuman and Madison, 1991; see below). However, several more recent studies indicate that the NOS may actually be present in CA1 cells, albeit at lower levels than those detected in the putative interneurons (Wendland *et al.*, 1994). Utilizing the same monoclonal antibody that had previously been shown not to stain CA1 cells (Valtschanoff *et al.*, 1993), Wendland *et al.* (1993) combined gentle fixation techniques (acetone, rather than paraformaldehyde) and confocal microscopy to detect NOS in pyramidal cell bodies and dendrites.

There is yet another potentially interesting twist on the story of NOS localization in the hippocampus. Quite recently, Fishman and colleagues utilized homologous recombination to generate mice that lack the neuronal NOS gene (Huang *et al.*, 1993). While Northern and Western blot analysis confirm that neuronal NOS is entirely absent from the neuronal tissue, there is a low level of residual NOS catalytic activity in the homozygous mutant mice. The amount of residual NOS activity varies dramatically in different brain areas, with the hippocampus possessing the highest amount of activity. Indeed, hippocampal slices prepared from homozygous mutant mice still exhibit LTP (O'Dell *et al.*, 1994); however, this LTP can be blocked by NOS inhibitors. Antibodies to the endothelial NOS (eNOS) stain CA1 pyramidal cells and dendrites (Dinerman *et al.*, 1994) raising the possibility that this enzyme may participate in LTP. Since, unlike nNOS, eNOS is myristoylated and associated with membranes (Pollock *et al.*, 1992) it may be optimally positioned to detect rises in Ca^{2+} that occur during LTP induction.

What is the electrophysiological evidence that NO is involved in LTP? Initial studies showed that when hippocampal slices are bathed in solutions containing NOS inhibitors, subsequent attempts to induce LTP (100 Hz stimulation) were blocked in an arginine-reversible manner (Böhme *et al.*, 1991; O'Dell *et al.*, 1991; Schuman and Madison, 1991; Haley *et al.*, 1992). Furthermore, LTP can also be blocked when NOS inhibitors are selectively introduced into postsynaptic cells through intracellular recording electrodes (O'Dell *et al.*, 1991; Schuman and Madison, 1991). This latter result strongly suggests that the requirement for NOS activity during LTP induction occurs in the postsynaptic cell. Further evidence supporting a role for NO as a retrograde signal comes from studies showing that extracellularly applied haemoglobin, which binds NO, also attenuates LTP (O'Dell *et al.*, 1991; Schuman and Madison, 1991; Haley *et al.*, 1992), suggesting that NO must travel between neurons. It should be noted that this latter result has also been interpreted to support a role for carbon monoxide (CO; Stevens and Wang, 1993; Zhuo *et al.*, 1993) as a retrograde signal since haemoglobin will also bind CO.

The role of nitric oxide in LTP is not as straightforward as the above results would indicate, however. Following the early reports that NOS inhibitors prevent LTP (Böhme *et al.*, 1991; O'Dell *et al.*, 1991; Schuman and Madison, 1991; Haley *et al.*, 1992), several additional studies have been conducted which suggest that under different experimental conditions NOS-independent LTP can be elicited (Table 1). The variable that has received the most attention is the recording temperature. Although two of the four original NOS inhibitor studies were conducted at higher temperatures (29–35°C vs room temperature; Böhme *et al.*, 1991; O'Dell *et al.*, 1991), several more recent studies conducted at higher temperatures show that with particular stimulation protocols NOS-independent LTP can be observed. Most of these studies have found that at 30–32°C NOS inhibitors only block LTP induced by weaker forms of synaptic stimulation (but see Gribkoff and Lum-Ragan, 1992). Weaker stimulation means either shorter periods of stimulation delivered at a given stimulus intensity (Haley *et al.*, 1993) or less intense stimulation (Chetkovich *et al.*, 1993). Unfortunately, there are even discrepancies in these later studies: one

Table 1 Summary of experimental conditions used in 10 different published studies exploring the effects of NOS inhibitors on LTP. Inhibitor abbreviations are L-nitro-arginine (L-NOArg), N^{G}-monomethyl-arginine (L-NMMA), and L-nitro-arginine methyl ester (L-NAME). Length of application indicates how long the inhibitor was applied before LTP induction was attempted. Temp. refers to the recording temperature of the experiment. Stimulation protocol refers to the manner in which LTP was induced. Age refers to the age of animals used in the experiments. The final column indicates whether LTP was blocked or not. In summary three out of three room temperature studies (including Williams *et al.*, 1993b) found that NOS inhibitors block LTP; of the eight studies conducted at 29–35°C four out of eight observed a block of LTP, two studies observed a block of LTP with weak, but not strong induction stimuli, one observed a block with strong, but not weak induction stimuli, and only one observed no block of LTP at all.

Authors	Inhibitors	Length of application	Temp.	Stimulation protocol	Age	Result
Bohme *et al.* (1991)	L-NOArg (100 nM)	15 min	32°C	100 Hz/s 2× (2× test intensity)	30–60 days	No LTP
Schuman and Madison (1991)	L-NMMA (100 μM) L-NOArg (100 μM)	1.5–2 h	22–25°C	100 Hz/s 4x (test intensity)	30–60 days	No LTP
O'Dell *et al.* (1991)	L-NOArg (50 μM)	30 min	30–32°C	100 Hz/s 2x (test intensity)	ND	No LTP
Haley *et al.* (1992)	L-NAME (100 nM)	15 min	22°C	100 Hz/40 ms 10x (test intensity)	ND	No LTP
Gribkoff *et al.* (1992)	L-NMMA (100 μM) L-NOArg (100 μM)	1 h	32°C	100 Hz/s 1x; (test intensity) *or* 2x (max. current)	30–90 days	1x–LTP 2x–no LTP
Izumi *et al.* (1992)	L-NMMA (100 μM) L-NOArg (50 μM)	10 min 30 min	30°C 30°C	100 Hz/s 1x 100 Hz/s 1x	ND ND	10 min–LTP 30 min–no LTP
Haley *et al.* (1993)	L-NOArg (1 mM)	30 min	30–32°C	100 Hz/250 ms 2x 100 Hz/500 ms 2x	30–60 days	250 ms–no LTP 500 ms–LTP
Chetkovich *et al.* (1993)	L-NOArg (100 μM)	1.5 h	32°C	100 Hz/s 3x (max *or* 1/2 max intensity)	30–60 days	Max–LTP 1/2 max–attenuated LTP
Musleh *et al.* (1993)	L-NMMA (125 μM)	1.8 h	35°C	100 Hz/40 ms 10x	30–60 days	No LTP
Williams et al. (1993)	L-NOArg (100 μM) L-NAME (100 μM)	2.5 h	24°C 29°C	100 Hz/40 ms 10x (test intensity) *or* 100 Hz/200 ms 6x (test intensity)	35–42 days *or* 4–6 months	24°C or young–no LTP 29°C or old–LTP

study found that NOS inhibitors do not block LTP induced by theta burst stimulation at 29°C (Williams *et al.*, 1993b), while another group (Musleh *et al.*, 1993) found that NOS inhibitors do block LTP induced by theta stimulation at 35°C, a temperature even closer to body temperature. In the latter study, the investigators suggested that the block of LTP observed in the presence of NOS inhibitors is due to a reduction of the summed postsynaptic depolarization that occurs during theta burst stimulation. However, a similar analysis conducted by Haley and colleagues found NOS inhibitors did not alter the amount of depolarization produced during tetanic stimulation (Haley *et al.*, 1993). A summary of the studies listed in Table 1 indicates the following: of the three studies conducted at 22–25°C, all found that NOS inhibitors block LTP; of the eight studies conducted at 29–35°C, four observed a block of LTP, two studies observed a block of LTP with weak, but not strong induction stimuli, one observed a block with strong, but not weak induction stimuli, and only one observed no block of LTP at all.

An additional issue that has been raised is whether NOS inhibitors block LTP in adult animals (Williams *et al.*, 1993b). Williams *et al.* (1993b) examined this variable by comparing the block of LTP by NOS inhibitors in two age groups of animals: 5–7 weeks old or 4–6 months old. These investigators found that NOS inhibitors were only effective in preventing LTP in the 5–7-week-old animals. On the basis of this finding, it was argued that the block of LTP by NOS inhibitors occurs only under specialized circumstances. Implicit in this criticism is the assumption that most studies of LTP are conducted on older animals (4–6 months); unfortunately, this is not the case. A thorough survey of animal age in all previous LTP studies is difficult, owing to the fact that some papers do not report animal age at all, while others report weight rather than age. However, as indicated in Figure 1, an informal survey of the LTP references cited in a recent major review (Bliss and Collingridge, 1993) indicates that an overwhelming majority of LTP studies are conducted in animals under 2 months of age, not 4–6 months. Thus, contrary to what has been suggested, the age of animal used in most LTP studies is the same age that LTP has been reported to be sensitive to NOS inhibitors. Taken together, these data indicate that NO plays a role in LTP, but under certain experimental conditions, namely when stronger stimulation parameters are used, alternative biochemical signalling systems may also participate. In future experiments much more attention should be paid to the various LTP induction procedures and experimental conditions which are used in different labs. The breadth of biochemical pathways implicated in LTP is ever expanding. The task of future experiments will be to understand how various putative signalling pathways (e.g. arachidonic acid, carbon monoxide, and nitric oxide) may interact to ultimately orchestrate the observed increases in synaptic strength. In addition, the possibility that given biochemical pathway(s) may be selectively invoked based on the stimulation parameters used to induce LTP (Gribkoff and Lum-Ragan, 1992; Chetkovich *et al.*, 1993; Haley *et al.*, 1993) or the pre-existing history of the synapse (Larkman *et al.*, 1992) needs to be explored further.

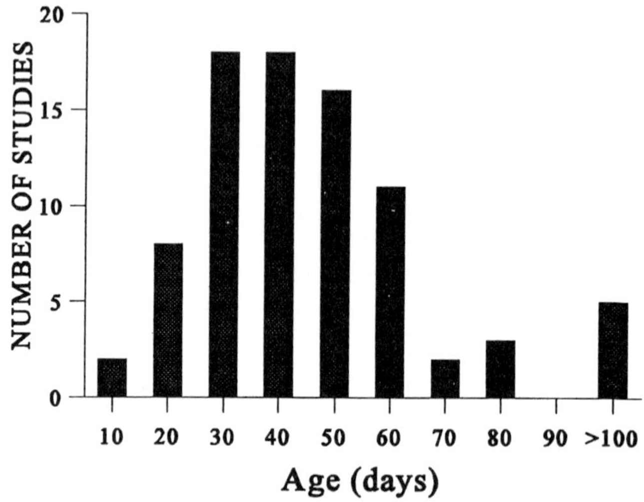

Figure 1 Animal age information for 83 of 129 papers cited in a recent LTP review article (Bliss and Collingridge, 1993). Of the remaining 46 citations, 39 did not list the animal age in the paper (or in papers referenced) and seven studies were conducted on cultured neurons or on rabbit hippocampus. The numbers on the *x* axis indicate the age in days for a 10-day period. For example, the bar above 20 indicates that eight studies were conducted on animals 21–30 days old. Seventy-eight out of 83 studies (94%) were conducted utilizing animals less than 90 days of age, while only five studies (6%) were conducted in animals approaching 4–6 months old, the age suggested to be most relevant in Williams *et al.* (1993b).

7.3.1 Long-term potentiation and diffusible signalling

The involvement of a diffusible signal such as NO in LTP introduces an apparent problem regarding signalling specificity: are individual synapses specifically targeted for messenger action, or rather, does the messenger influence a local ensemble of synapses? Previous studies by Bonhoeffer and colleagues suggested that in cultured hippocampal neurons many synapses near LTP-induced synapses may be enhanced (Bonhoeffer *et al.*, 1989). Schuman and Madison (1994) have recently made a similar observation in acute hippocampal slices. We observed that inducing LTP at the synapses onto a single CA1 pyramidal neuron can enhance synaptic transmission at the synapses onto a neighbouring cell.

Simultaneous intracellular recordings were made from two neighbouring CA1 pyramidal neurons in a hippocampal slice (Figure 2). At the beginning of each experiment, the two pyramidal neurons were tested for synaptic and electrical coupling by injecting current into one cell and observing the response of the second cell, and vice versa. None of the pairs of CA1 pyramidal neurons recorded from showed any evidence of either a synaptic or an electrical connection. [This finding is in good agreement with previous studies by Thomson and colleagues (e.g.

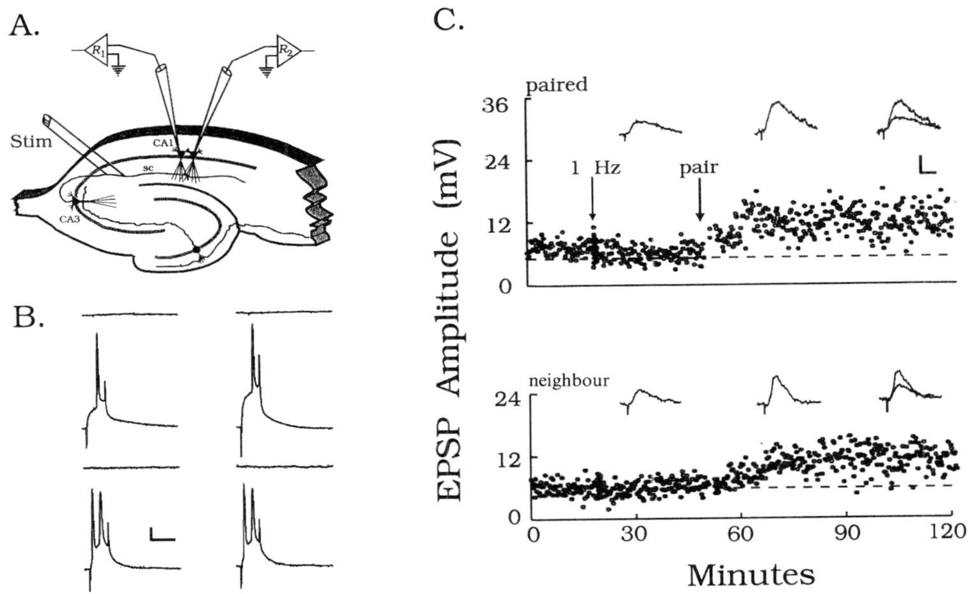

Figure 2 Individual experiment demonstrating neighbouring synapse potentiation. (A) Schematic diagram showing the position of stimulating and recording electrodes in the hippocampal slice preparation. Two nearby CA1 pyramidal cells were impaled and the excitatory responses to test shocks delivered to the Schaffer collaterals by a bipolar stimulating electrode were recorded. (B) Electrophysiological traces testing for electrical and/or synaptic connectivity between the cells shown in (C). Top: shown are action potentials elicited by depolarizing current injection into the 'paired' cell before (left) and after (right) LTP induction; above each action potential is the corresponding response of the neighbouring cell. Bottom: shown are action potentials elicited by depolarizing current injection into the neighbouring cell before (left) and after (right) LTP induction in the 'paired' cell; above each action potential is the corresponding response of the 'paired' cell. Scale bar is 15 mV and 50 ms. (C) Plots of EPSP amplitude obtained with intracellular recording from the 'paired' (top) and neighbouring (bottom) neuron. Low frequency stimulation (1 Hz) and low frequency stimulation paired with depolarization (pair) are indicated with arrows. The first two traces in each plot are representative EPSPs before and after LTP induction of the 'paired' neuron and the third trace is the superimposition of the two. Scale bar is 10 mV and 10 ms. Figures 2–6 are reprinted with permission from *Science*.

Thomson, 1991) who found that only six of 795 pairs of CA1 pyramidal neurons tested were synaptically coupled.] For the remainder of each experiment, a single stimulating electrode was placed in the dendritic region of stratum radiatum to stimulate the Schaffer collaterals and the excitatory postsynaptic potentials (EPSP) were recorded in each pyramidal neuron in response to single test stimuli delivered once every 15 s.

After obtaining a period of baseline responses, LTP was induced in one pyramidal cell by postsynaptic depolarization coupled with 1 Hz stimulation of the afferent fibres. The LTP induced in the paired cell was accompanied by an increase in the synaptic strength at the synapses onto the neighbouring cell, indicating that potentiation can be communicated between synapses (Figures 2 and 3A). When measured 50–60 minutes after LTP induction in the paired cell, the synaptic strength of the neighbouring synapses increased to 135% of baseline values on average (Figure 3A). In individual experiments the enhancement of synaptic transmission ranged from no potentiation to 250%, with 16 of 20 neighbouring cells exhibiting synaptic potentiation that was at least 115% of baseline values. The paired cells exhibited LTP that was on average 180% of baseline values, with individual experiments ranging from no potentiation to 330%.

7.3.1.1 Spatial restrictions

How close are the pyramidal neurons that exhibit this distributed potentiation? In many experiments, biocytin was included in the intracellular recording electrodes and allowed to diffuse into the neurons during the course of the recording. Slices were then fixed, processed for horseradish peroxidase (HRP), and sectioned. Individual pyramidal neurons were reconstructed using camera lucida techniques. Ideally, one would like to know the distance between the potentiated synapses on the paired cell and the potentiated synapses on the neighbouring cell. However, since the specific sites where the stimulated presynaptic fibres contact postsynaptic dendrites cannot be readily visualized, the distance between pyramidal cell somas was measured to estimate the actual intersynaptic distance. Although the microeletrode placement was accomplished with an ordinary dissecting microscope, we were able to obtain recordings from neurons that were very close together; in the example shown in Figure 3B the cell bodies were approximately 35 μm apart. On average, the cell bodies of the neurons were 142.1 ± 36.1 μm apart (range 20.0–285.0 μm). In almost all cases there were detectable regions of overlap in the dendritic arbors of the two neurons.

The diffusion properties of nitric oxide predict that the spatial extent of its influence should be limited. In order to test this idea in the context of distributed potentiation, recordings were obtained from two spatially remote pyramidal neurons. As before, LTP was produced in one neuron by pairing postsynaptic depolarization with low frequency stimulation of the presynaptic fibres; on average the paired cells exhibited potentiation that was 174% of baseline values. However, no potentiation was detected at the spatially remote cell synapses; the mean level of synaptic

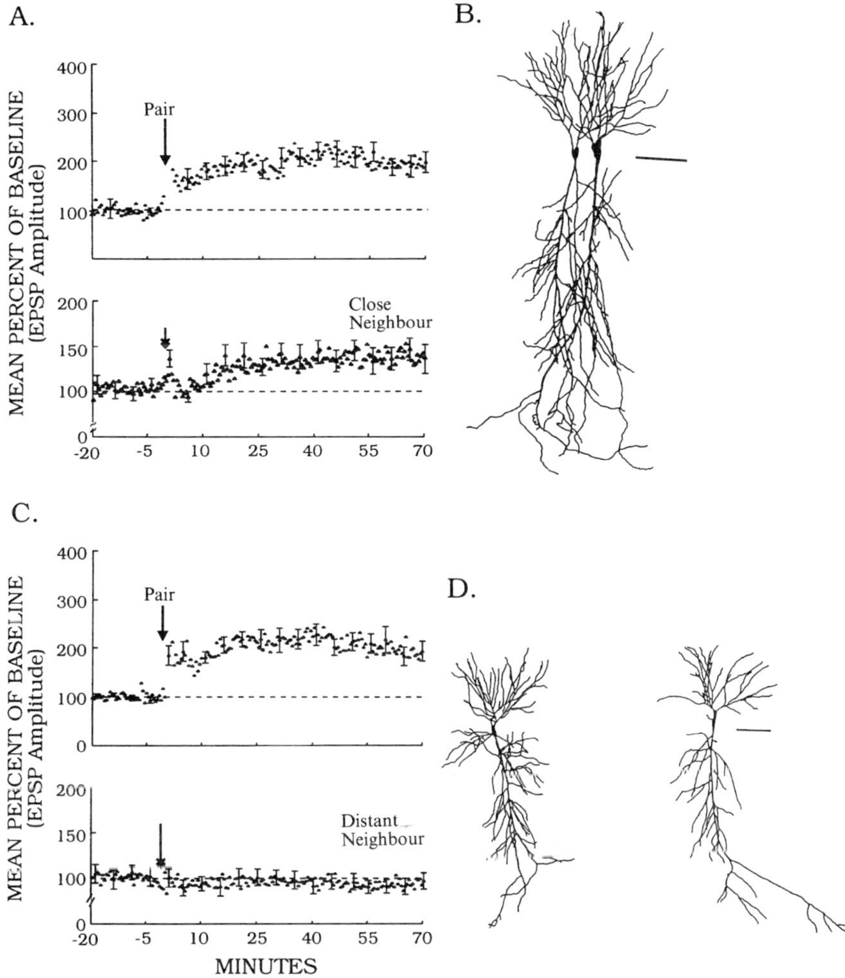

Figure 3 (A) Ensemble averages for dual intracellular recordings from nearby pyramidal neurons (n=20). Top plot shows average data for the 'paired' neurons in which LTP was induced by pairing postsynaptic depolarization with 1 Hz stimulation of afferent fibres (Pair). Bottom plot shows average data for the enhancement observed in the neighbouring neuron in response to LTP induction in the 'paired' neuron. (B) Camera lucida tracing of biocytin-filled neurons which were recorded from during an experiment which contributed to the ensemble averages shown in (A). In the example shown cell bodies are within 40 μm of each other; note that the dendritic trees overlap extensively. Scale bar=100 μm. (C) Ensemble averages for dual intracellular recordings from distant (~500 μm) CA1 pyramidal neurons (n=17). Top plot shows average data for the 'paired' neuron where LTP was induced by pairing (Pair). Bottom plot shows average data for the synaptic strength at synapses onto the distant neighbouring neuron. No enhancement of synaptic transmission was observed. (D) Camera lucida tracing of biocytin-filled neurons which were recorded from during an experiment which contributed to the ensemble averages shown in (A). The cell bodies are approximately 585 μm apart. Scale bar=100 μm.

strength was approximately 98% of baseline values (Figure 3C). Camera lucida reconstructions of the HRP processed neurons revealed that the pairs of cells used in these experiments were on average 595 μm apart (range 340–800 μm; Figure 3D). None of these pairs of neurons had any apparent regions of dendritic overlap. A comparison of the amount of synaptic enhancement exhibited by near and far neighbouring cells reveals that potentiation of synaptic transmission at neighbouring cell synapses appears to be restricted to synapses near the site of LTP induction. We combined the data from both near and far cells and examined the relationship between intersomatic distance and distributed potentiation. We found a significant negative correlation (r=–0.58) such that increasing intersomatic distance is associated with decreasing neighbouring synapse potentiation.

7.3.1.2 The role of nitric oxide

Does nitric oxide participate in the distributed potentiation? To begin to address this issue, we blocked NO production in the paired cell. The paired cell was filled with the NO synthase inhibitor L-methyl-arginine (L-Me-Arg), delivered through the recording microelectrode. As reported in previous studies (Schuman and Madison, 1991; O'Dell *et al.*, 1991) interfering with NO production in the paired cell blocked the formation of LTP (mean per cent of baseline transmission ~100%, n=18; Figure 4A). The injection of L-Me-Arg into the paired cell also prevented the enhancement of synaptic transmission at the neighbouring cell synapses (~102%; Figure 4A). The inability of the neighbouring synapses to exhibit potentiation was not due to distance, since these pairs of cells were on average just as close as the nearby cells (mean intersomatic distance ~143 μm; Figure 4B). In a few experiments the intracellular recordings were of a sufficient duration to examine whether LTP induction in the neighbouring cell could bring about potentiation at the synapses onto the NO synthase-inhibited cell. In three of five experiments we found that inducing LTP in the neighbouring cell could enhance synaptic transmission at the L-Me-Arg-injected cell synapses. These results suggest a role for NO in the production of distributed synaptic enhancement. While these data are consistent with the idea that NO itself mediates the potentiation at neighbouring synapses, it is still possible that a molecular process downstream from NO production is actually involved. In addition, the contribution of other diffusible signals such as arachidonic acid, carbon monoxide and PAF needs to be considered.

To further test the idea that neurons may provide signals to potentiate neighbouring synapses we conducted an experiment that takes advantage of a difference in two LTP induction protocols: pairing and tetanus. As discussed above, the pairing protocol involves depolarization of a single pyramidal neuron coupled with low frequency stimulation; thus, only the depolarized cell undergoes the standard LTP induction cascade. In contrast, LTP elicited by tetanic stimulation depolarizes a population of neurons in a local area and hence many cells undergo LTP induction. We filled individual CA1 pyramidal neurons with an NOS inhibitor and attempted to induce LTP by pairing (Figure 5). As shown in Figure 5, L-Me-Arg

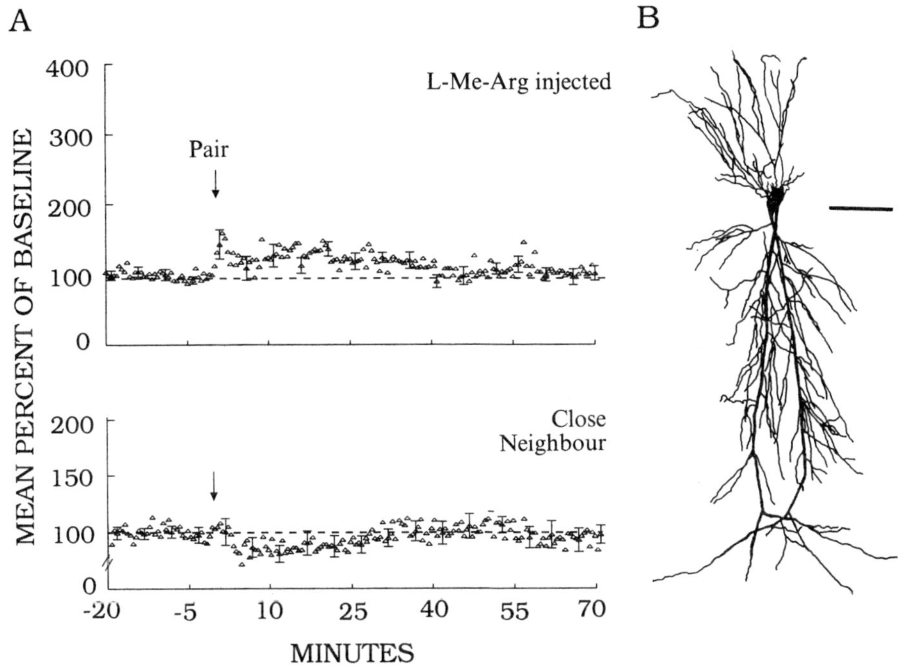

Figure 4 (A) Ensemble averages for dual intracellular recordings from two nearby pyramidal neurons; the producer cell (top) was filled with the NOS inhibitor L-Me-Arg (n=18). Postsynaptic injection of L-Me-Arg blocked not only pairing-induced LTP in the 'paired' cell, but also blocked the spread of enhancement in the neighbouring cell. Top plot shows average data for the LTP produced by pairing (Pair). Bottom plot shows average data for the enhancement observed in the neighbouring neuron in response to LTP induction in the producer neuron. Please note difference in y axis values in (A) and (B). (B) Camera lucida tracing of biocytin-filled neurons which were recorded from during an experiment which contributed to the ensemble averages shown in (A). In the example shown cell bodies are within 10 μm of each other. Scale bar=100 μm.

injections effectively prevented LTP induced by pairing. After failing to induce LTP by pairing, tetanic stimulation was then delivered to the same stimulating pathway. Surprisingly, high-frequency stimulation restored the ability of the NOS inhibited cell to exhibit LTP (Figure 5). Our interpretation of this experiment is that the uninhibited neighbouring cells become depolarized during tetanic stimulation and as a result provide a signal, perhaps NO itself, that can potentiate the synapses onto the inhibited neuron.

7.3.1.3 The contribution of the neighbouring cell

The simplest interpretation of the above experiments indicates that in order to exhibit synaptic potentiation pyramidal neurons need not actively participate;

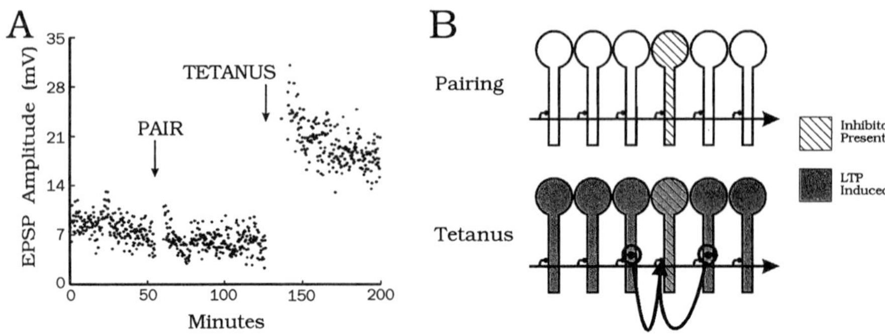

Figure 5 Representative experiment examining the potential contribution of neighbouring synapses to LTP induced at synapses onto an NOS-inhibited pyramidal neuron. (A) EPSP amplitude from a single CA1 pyramidal neuron filled with the NOS inhibitor L-Me-Arg. This cell exhibited no LTP in response to pairing postsynaptic depolarization with 1 Hz stimulation of afferents (PAIR). In contrast, when high frequency stimulation (100 Hz; TETANUS) was delivered to the same afferents, the cell exhibited robust LTP. (B) Schematic diagram depicting how neighbouring neurons may provide NO to induce LTP onto an NOS-inhibited neuron during LTP induction. Tetanic stimulation delivered to afferent fibres should induce LTP in many neurons, including the NOS-inhibited neuron; during tetanic stimulation the NO produced in the neighbouring neurons may diffuse to potentiate synapses onto the NOS-inhibited neuron.

rather, depolarized neighbouring neurons can provide the factor(s) required. This idea is supported by the demonstration that synapses onto an NOS inhibited cell can still exhibit potentiation if neighbouring cells experience tetanic stimulation. However, similar experiments conducted with Ca^{2+} chelators and membrane hyperpolarization indicate that this simple scenario cannot be correct. These studies found that injections of Ca^{2+} chelators (Lynch *et al.*, 1983; Malenka *et al.*, 1988) or hyperpolarizing current (Malinow and Miller, 1986) into individual CA1 pyramidal neurons could prevent LTP induced by tetanic stimulation. Clearly, depolarization of neighbouring cells cannot compensate for postsynaptic manipulations that alter Ca^{2+} and membrane potential, suggesting that there may be additional requirements for a neuron to exhibit synaptic potentiation.

Thus, the potential contribution of the neighbouring cell to distributed potentiation was assessed. To accomplish this, the effects of dialysis, Ca^{2+} chelation and membrane hyperpolarization were assessed. The neighbouring cell was recorded from in whole-cell voltage clamp mode using a pipette containing 10 mM BAPTA (free acid) to chelate intracellular Ca^{2+}. In addition, during the period of LTP induction of the paired cell, the neighbouring cell was hyperpolarized to −95 mV. The paired cell exhibited LTP that was comparable to that observed in previous experiments (mean per cent of baseline transmission ~190%, n=14; Figure 6). However, the combination of dialysis, hyperpolarization and Ca^{2+} chelation prevented the distributed enhancement usually observed at the neighbouring cell synapses (mean per cent of baseline transmission ~100%, Figure 6). Examination of HRP-processed

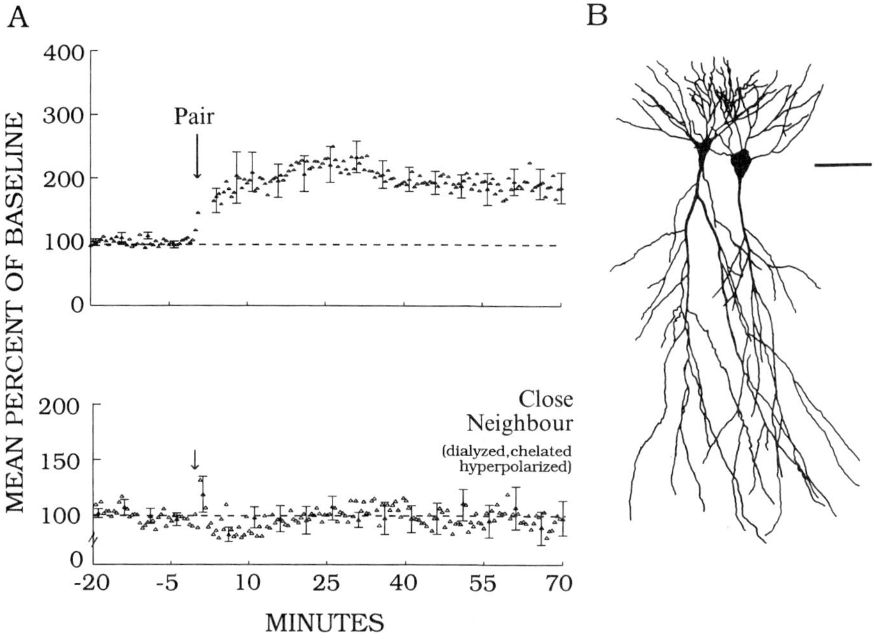

Figure 6 (A) Ensemble averages for dual recordings from two nearby pyramidal neurons (n=14); the paired cell was recorded from with a standard intracellular micropipette; the neighbouring cell (bottom) was recorded in whole-cell voltage-clamp mode and filled with the Ca^{2+} chelator BAPTA. During the induction of LTP in the 'paired' cell the neighbouring neuron was hyperpolarized to −95 mV. Top plot shows average data for the LTP produced by pairing (Pair). Bottom plot shows average data for the enhancement observed in the neighbouring neuron in response to LTP induction in the producer neuron. The combination of postsynaptic dialysis, Ca^{2+} chelation and hyperpolarization of the neighbouring neuron prevented the synaptic enhancement typically observed. (B) Camera lucida tracing of biocytin-filled neurons which were recorded from during an experiment which contributed to the ensemble averages shown in (A). In the example shown cell bodies are within 65 μm of each other. Scale bar=100 μm.

neurons revealed that the pairs of cells in this experiment were also anatomically close (mean intersomatic distance ~80 μm; Figure 6). These data indicate that under these experimental conditions the neighbouring cell actively contributes to the potentiation.

What is the nature of the neighbouring cell contribution? There are two general possibilities to account for the involvement of the neighbouring neuron in distributed potentiation. First, the signal generated in the paired cell may interact with a Ca^{2+}- and/or voltage-dependent postsynaptic process in the neighbouring cell to produce distributed potentiation. According to this idea, the distributed potentiation is both induced and expressed in postsynaptic neurons. Second, there may be a requirement for an additional retrograde signal generated in the neighbouring neuron. The generation of this signal may be either a constitutive process or one stimulated by LTP induction in the paired cell. For example, the constitutive release of an additional messenger(s) could be stimulated by tonic Ca^{2+} influx through either NMDA (Sah *et al.*, 1989) or voltage-dependent Ca^{2+} channels. Alternatively, the release of the additional messenger may be stimulated by LTP induction in the paired cell. This stimulation could take the form of a diffusible chemical messenger or alternatively be mediated by increases in extracellular K^{+} associated with low frequency stimulation or depolarization of the paired neuron during LTP induction. Elucidating the precise nature of the Ca^{2+}- and/or voltage-dependent signal which acts in concert with other postsynaptically generated messengers will be an extremely important area of future investigation.

7.3.2 Functional significance of distributed potentiation

Does this distribution of plasticity make sense given what we know about information processing in the hippocampus? *In vivo* studies of hippocampal function suggest the hippocampus should be regarded as a *supramodal* processor, encoding the significance of stimuli independent of particular sensory input (Eichenbaum *et al.*, 1992). This property is largely derived from the multimodal nature of the inputs to the hippocampus, namely afferents from the association areas of neocortex and olfactory cortex which are involved in higher order sensory and motor processing (VanHoesen *et al.*, 1972; Deacon *et al.*, 1983). Accordingly, the response properties of pyramidal neurons are not organized in a strict topographic fashion where functional characteristics are faithfully mapped in a point-to-point fashion onto the hippocampal structure. However, it does appear that there is a 'patchy' organization of function, such that clusters of neighbouring neurons have overlapping response properties and encode similar aspects of behavioural stimuli (Eichenbaum *et al.*, 1989). These neighbouring cells also send their axonal projections to roughly equivalent areas of entorhinal cortex. These types of observations indicate that in the hippocampus, behaviours are perhaps best described by the response properties of ensembles of neurons, rather than individual neurons (Eichenbaum, 1993). Indeed, a recent study suggests that this may be the case for the encoding of spatial representations in the rat hippocampus (Wilson and McNaughton, 1993). Using a multisite electrode recording array implanted in the hippocampus, the activity of 70–145 single hippocampal neurons was monitored while rats foraged for food in a chamber. Wilson and McNaughton found that increasing the number of cells sampled increased the ability of the population to accurately predict the rats' spatial location. Given the increasing evidence that simple behaviours are best described by ensembles of neuronal activity (Wu *et al.*, 1994), it seems particularly appropriate that certain forms of synaptic plasticity can be distributed across local populations of synapses. As described above, the distributed potentiation can spread to synapses that share common afferents, providing a mechanism that potentially allows all synapses which are active and near a site of messenger production to be enhanced.

7.4 Long-term depression in the hippocampus

Certain patterns of synaptic activity can also lead to persistent decreases in synaptic strength or long-term depression (LTD). In the hippocampus, LTD is most reliably produced by prolonged (5–15 min) 1 Hz stimulation of afferent fibres (Dudek and Bear, 1992; Mulkey and Malenka, 1992). LTD can be reversed by high frequency stimulation protocols typically used to induce LTP. Two groups have found that like LTP, LTD induction also requires the activation of NMDA receptors (Dudek and Bear, 1992; Mulkey and Malenka, 1992). In addition, membrane hyperpolarization can also prevent LTD (Mulkey and Malenka, 1992), presumably by preventing the activation of NMDA receptors. Mulkey and Malenka also found that chelating intracellular Ca^{2+} with BAPTA also blocks LTD (Mulkey & Malenka, 1992). These data indicate that the same intracellular transduction events are required for LTP and LTD. What then determines whether a synapse will be potentiated or depressed? It has been suggested that the level of postsynaptic depolarization, and hence the Ca^{2+} influx, will determine the phenotype of the plasticity (Artola and Singer, 1993). Thus, at intermediate ranges of depolarization and Ca^{2+} concentration synaptic depression will occur, whereas higher levels of depolarization and Ca^{2+} will produce potentiation. A test of this idea involves experimentally manipulating intracellular Ca^{2+} levels with the prediction that intermediate concentrations of Ca^{2+} should produce LTD.

Is LTD expressed by pre- or postsynaptic mechanisms? Mulkey and Malenka (1992) reported that the expression of LTD did not alter paired-pulse facilitation, a phenomenon considered to be due to changes in neurotransmitter release. Since it is often suggested that plasticity which utilizes a presynaptic mechanism should interact with paired-pulse facilitation, this result could be interpreted as evidence for a postsynaptic change mediating LTD. In contrast, utilizing quantal analysis techniques, Larkman *et al.* (1991) found that changes associated with synaptic depression in the hippocampus are largely presynaptic in nature. In addition, two recent studies have suggested that LTD is accompanied by an increase in the number of presynaptic failures to release neurotransmitter, further indicating a presynaptic change may contribute to LTD (Bolshakov and Siegelbaum, 1994; Stevens and Wang, 1994). If the data continue to support a postsynaptic locus for LTD induction and a presynaptic locus for LTD expression then, like LTP, there would be a need for a retrograde signal.

7.4.1 The role of nitric oxide in hippocampal LTD

Because the same initial transduction events involving NMDA receptors and intracellular Ca^{2+} appear to be involved in both LTP and LTD in the hippocampus, it is attractive to suggest that NO might also play a role in hippocampal LTD. In support of this idea, one group has shown that inhibitors of NOS or haemoglobin can block LTD induced by 1 Hz stimulation in hippocampal slices (Izumi and Zorumski, 1993). Consistent with the idea that the inhibitors are acting on the NOS, this block of LTD was relieved in the presence of L- but not D-arginine. In addition, Bliss and colleagues have also observed that photolysis of caged NO compounds can result in a short-lasting depression of synaptic transmission in the hippocampus (Williams *et al.*, 1993a). Another group has shown that application of two different NO donors, sodium nitroprusside (100 μM) or *S*-nitrosocysteine (100 μM), to hippocampal slices can induce a persistent depression of synaptic transmission (Izumi and Zorumski, 1993). Unlike stimulation-induced LTD, however, the NO donor-induced depression could not be reversed by LTP-inducing high frequency stimulation. This latter result suggests that NO-induced depression and synaptically induced depression may utilize different cellular mechanisms. Related to this issue, it has not yet been shown that NO-induced depression occludes synaptically induced depression, as would be expected if the two forms of LTD rely on similar mechanisms.

7.5 Long-term depression in the cerebellum

Long-term depression is also intensely studied in the cerebellum. A major source of information outflow in the cerebellum is the inhibitory synapses made by Purkinje cells onto cerebellar and vestibular nuclei. Since LTD involves a reduction in synaptic strength at synapses on postsynaptic Purkinje cells, the net effect of LTD in the cerebellum is a reduction in inhibition. LTD has been suggested to underlie a form of behavioural plasticity, known as the vestibular ocular reflex, or VOR (Ito, 1982). The VOR is an adaptive visual response that stabilizes retinal images when the head moves. The same cerebellar circuitry involved in LTD may also mediate some aspects of the VOR (Ito, 1990).

In the cerebellar cortex, the Purkinje cells receive synaptic input from two separate sets of afferents, the parallel fibres (granule cells axons) and the climbing fibres (inferior olive axons). Both the parallel fibre and the climbing fibre synapses use glutamate as a neurotransmitter. Conjunctive stimulation (1–4 Hz for 25 s to 10 min) of the parallel fibres and the climbing fibres produces a long-lasting depression of synaptic transmission at the synapses between the parallel fibres and Purkinje cells. The observed decrease in synaptic strength is due to a reduction in the sensitivity of postsynaptic AMPA receptors (Ito *et al.*, 1982; Crepel and Krupa, 1988; Hirano, 1991; Linden *et al.*, 1991).

Studies aimed at elucidating the molecular mechanisms which underlie LTD have outlined a series of events which include glutamate binding to postsynaptic receptors, rises in intracellular Ca^{2+}, and activation of protein kinases(s). Glutamate released from climbing fibres is thought to activate primarily the AMPA class of receptor. Because each climbing fibre possesses multiple synapses with each Purkinje cell, climbing fibre activity can potently depolarize Purkinje neurons resulting in the activation of voltage-dependent Ca^{2+} channels and the influx of Ca^{2+}. The glutamate released from parallel fibres has been proposed to activate both AMPA and metabotropic classes of glutamate receptors. Activation of both of these receptors is a necessary step in the induction of LTD as antagonists to either class of receptor will block LTD (Linden *et al.*, 1991). Activation of the metabotropic receptor activates phospholipase C resulting in generation of diacylglycerol and an IP_3-mediated rise in intracellular Ca^{2+}. The Ca^{2+} signal (also derived from climbing fibre-mediated activation of voltage-sensitive Ca^{2+} channels) as well as the generation of diacylglycerol serve to activate protein kinase C (PKC). PKC activity is required for LTD (Linden and Connor, 1991) and LTD can be mimicked by the application of phorbol esters (Crepel and Krupa, 1988). In addition to PKC, activation of cGMP-dependent protein kinase has also been implicated in LTD, as discussed in more detail below. The activity of these protein kinases has then been proposed to bring about alterations in the sensitivity of AMPA-type glutamate receptors by directly phosphorylating channel subunits or associated molecules.

Where might NO fit into the induction cascade outlined above? The brain NOS that was originally purified from cerebellum (Bredt *et al.*, 1990) is found in the granule cells as well as the inhibitory basket cells, but appears to be absent from the Purkinje cells (Bredt *et al.*, 1991). Thus, the Ca^{2+} signal in Purkinje cells that is required for LTD most likely does not function as an activator of NOS (unless an as yet unidentified isoform of NOS exists in Purkinje cells). However, NOS may be activated by an influx of Ca^{2+} in the granule cells or by granule cell-induced excitation of the basket cells. If NO plays a role in LTD, then a soluble guanylyl cyclase is a likely target since cGMP concentrations are much higher in the cerebellum than in other brain areas. Immunohistochemical studies indicate that guanylyl cyclase (Nakane *et al.*, 1983) and cGMP-dependent protein kinase (Lohmann *et al.*, 1981) are present in Purkinje cell bodies, dendrites and axons. In contrast, cGMP is found primarily in Bergmann fibres and cell bodies, and in astroglial cells in the granular layer and white matter, but appears to be absent from Purkinje cells (de Vente *et al.*, 1989). In addition, several different agonists including glutamate, kainate, as well as the NO donor sodium nitroprusside fail to elevate cGMP in Purkinje neurons (Garthwaite and Garthwaite, 1987). These findings may suggest that the guanylyl cyclase present in Purkinje cells may be stimulated by an as yet unidentified signal transduction cascade.

What is the experimental evidence that NO participates in cerebellar LTD? Crepel and colleagues (Crepel and Jaillard, 1990; Daniel *et al.*, 1993) have shown that extracellular application of the NOS inhibitor L-NMMA blocks LTD produced by pairing parallel fibre-mediated EPSPs with postsynaptic Purkinje cell Ca^{2+} spikes in cerebellar slices. However, LTD is not blocked when the NOS inhibitor is included in the whole-cell recording pipette, suggesting that NO production is not required in Purkinje cells (Daniel *et al.*, 1993). In contrast, inclusion of an NO donor in the Purkinje cell recording pipette resulted in a progressive decline in the amplitude of parallel fibre-mediated EPSPs (Daniel *et al.*, 1993). This NO-mediated decrease in the EPSP occluded the subsequent induction of LTD by pairing parallel fibre stimulation with Ca^{2+} spikes. LTD can be prevented by extracellular application of methylene blue, which has been reported to inhibit guanylyl cyclase (Crepel and Jaillard, 1990). In addition, bath or intracellular application of

8-bromo-cGMP also depressed the Purkinje cell EPSP. These results are consistent with a role for NO in LTD where NO is produced by parallel fibre stimulation and then diffuses into the Purkinje cells to activate guanylyl cyclase and depress the EPSP.

A role for NO has been explored in another LTD paradigm. Ito and Karachot (1989) have documented a quisqualate (QA)-induced desensitization of Purkinje cell glutamate receptors in grease gap recordings from Purkinje cell axons in cerebellar slices. QA is believed to induce the observed desensitization by acting upon two classes of glutamate receptors, both the ionotropic AMPA type as well as the metabotropic type. Application of AMPA alone does not induce the desensitization (presumably because it does not act at the metabotropic receptor) but the co-application of AMPA and the NO donor sodium nitroprusside or a membrane permeant cGMP analogue will produce desensitization (Ito and Karachot, 1989). Also, prior incubation with either the NOS inhibitor L-NMMA or haemoglobin blocked the QA-induced desensitization of responses. These results can be contrasted to the findings of Linden and Connor (1992) who showed that in cultured Purkinje neurons, NO is not important for the depression of glutamate currents produced by conjoint depolarization and glutamate iontophoresis. The differences in these two findings may be accounted for by different preparations (cerebellar slices vs cultured Purkinje neurons), or by different induction procedures (quisqualate applications vs glutamate iontophoresis coupled with depolarization).

Another study has shown that a correlate of LTD, alterations in extracellular K^+ concentration ($[K^+]_o$), is also influenced in a manner consistent with a role for NO. When the parallel fibres are stimulated in the molecular layer of a cerebellar slice an increase in $[K^+]_o$ can be recorded with an ion-sensitive electrode (Shibuki and Okada, 1991). LTD, produced by conjunctive stimulation, is accompanied by a depression of the parallel fibre-elicited K^+ response (Shibuki and Okada, 1991). The conjunctive stimulation-induced decrease in $[K^+]_o$ is blocked when cerebellar slices are bathed in the NOS inhibitor L-NMMA os haemoglobin. In addition, sodium nitroprusside or a cGMP analogue paired with parallel fibre stimulation significantly depressed the K^+ response. This same study showed that an NO-sensitive probe inserted in the molecular layer was able to detect increases in NO concentration following conjunctive stimulation.

Thus, several studies suggest that NO functions as a important signal in the cellular events which underlie cerebellar LTD. How might NO be incorporated into the anatomical and cellular circuitry important for LTD? The most parsimonious transduction scheme would most likely begin with NO generation in the basket or granule cells, induced by parallel fibre stimulation (Figure 2). NO would then diffuse to Purkinje cells to activate guanylyl cyclase, increase cGMP levels, and potentially activate a cGMP-dependent protein kinase. A cGMP-dependent protein kinase is one kinase that has been proposed to mediate the decreased postsynaptic responsiveness by phosphorylating a postsynaptic AMPA receptor or associated molecule (Ito, 1989). One caveat to the above sequence of events is the inability to observe NO stimulation of Purkinje cell guanylyl cyclase (Garthwaite and Garthwaite,

1987). In addition, the molecular underpinnings of the proposed down-regulation of the AMPA channels need to be further explored. It will be interesting to see whether the particular AMPA receptor subunits expressed in Purkinje cells possess consensus sites for phosphorylation by cGMP protein kinase and/or PKC, and whether NO can stimulate the phosphorylation of the receptor subunits by either of these kinases.

7.6 Conclusions

Many studies indicate that NO likely serves as a signalling molecule in two different forms of synaptic plasticity: LTP and LTD. However, many of the details of NO's origin and action are still not known. In LTP, the identity of the NOS responsible for producing NO during induction is unclear, although pharmacological studies suggest that it is located in postsynaptic neurons. Similarly, in the case of cerebellar LTD, the origin of NO is unclear, as NOS appears to be absent from Purkinje cells. The molecular targets of NO are still not known, although since cGMP has been implicated in cerebellar LTD, NO may activate a guanylyl cyclase. Because it is freely diffusible, NO may function to distribute plasticity across many synapses in a local domain (Schuman and Madison, 1994). Distributed plasticity has been observed in the hippocampus, and a potentially related phenomenon may exist in the cerebellum (Vincent and Marty, 1993). This distributed synaptic potentiation may be utilized in both developing and adult brain. During development, this locally distributed potentiation may participate in the formation of functionally segregated anatomical structures such as cortical columns (Bonhoeffer *et al.*, 1989; Montague *et al.*, 1991) and also could induce temporary functional domains in regions where no such anatomical specialization exists. In as much as NO has also been implicated in different forms of behavioural plasticity (Chapman *et al.*, 1992; Hölscher and Rose, 1992; Böhme *et al.*, 1993), it may serve the analogous function of coordinating ensembles of neurons that encode learned behaviours and memories.

References

Arai, A. & Lynch, G. (1992) *Eur. J. Neurosci.* **4**, 411–419.
Artola, A. & Singer, W. (1993) *Trends Neurosci.* **16**, 480–487.
Ascher, P. & Nowak, L. (1988) *J. Physiol. (Lond.)* **399**, 247–266.
Asztely, F., Wigstrom, H. & Gustafsson, B. (1992) *Eur. J. Neurosci.* **4**, 681–690.
Barrionuevo, G. & Brown, T.H. (1983) *Proc. Natl. Acad. Sci. USA* **80**, 7347–7351.
Bashir, Z.I., Alford, S., Davies, S.N., Randall, A.D. & Collingridge, G.A. (1991) *Nature* **349**, 156–158.
Bekkers, J.M. & Stevens, C.F. (1990) *Nature* **346**, 724–729.
Berretta, N., Berton, F., Bianchi, R., Brunelli, M., Capogna, M. & Francesconi, W. (1991) *Eur. J. Neurosci.* **3**, 850–854.

.s, T.V.P. & Collingridge, G.L. (1993) *Nature* **361**, 31–39.
.liss, T.V.P., Douglas, R.M., Errington, M.L. & Lynch, M.A. (1986) *J. Physiol. (Lond.)* **377**, 391–408.
Böhme, G.A., Bon, C., Stutzmann, J.-M., Doble, A. & Blanchard, J.-C. (1991) *Eur. J. Pharmacol.* **199**, 379–381.
Böhme, G.A., Bon, C., Lemaire, M., Reibaud, M., Piot, O., Stutzmann, J.-M., Doble, A. & Blanchard, J.-C. (1994) *Proc. Natl. Acad. Sci. USA* **90**, 9191–9194.
Bolshakov, Y. & Siegelbaum, S.A. (1994) *Science* **264**, 1148–1152.
Bonhoeffer, T., Staiger, V. & Aertsen, A. (1989) *Proc. Natl. Acad. Sci. USA* **86**, 8113–8117.
Bredt, D.S., Hwang, P.M. & Snyder, S.H. (1990) *Nature* **347**, 768–770.
Bredt, D.S., Glatt, C.E., Hwang, P.M., Fotuhi, M., Dawson, T.M. & Snyder, S.H. (1991) *Neuron* **7**, 615–624.
Chapman, P.F., Atkins, C.M., Allen, M.T., Haley, J.E. & Steinmetz, J.E. (1992) *NeuroReport* **3**, 567–570.
Chetkovich, D.M., Klann, E. & Sweatt, J.D. (1993) *NeuroReport* **4**, 919–922.
Clark, G.D., Happel, L.T., Zorumski, C.F. & Bazan, N.G. (1992) *Neuron* **9**, 1211–1216.
Cohen, N.J. & Eichenbaum, H. (1993) *Memory, Amnesia, and the Hippocampal System.* London, MIT Press.
Collingridge, G.L., Kehl, S.J. & McLennon, H. (1983) *J. Physiol. (Lond.)* **334**, 19–31.
Crepel, F. & Jaillard, D. (1990) *NeuroReport* **1**, 133–136.
Crepel, F. & Krupa, M. (1988) *Brain Res.* **458**, 397–401.
Daniel, H., Hemart, N., Jaillard, D. & Crepel, F. (1993) *Eur. J. Neurosci.* **5**, 1079–1082.
de Vente, J., Bol, J.G.J.M. & Steinbusch, H.W.M. (1989) *Brain Res.* **504**, 332–337.
Deacon, T.W., Eichenbaum, H., Rosenberg, P. & Eckman, K.W. (1983) *J. Comp. Neurol.* **220**, 168–190.
del Castillo, J. & Katz, B. (1954) *J. Physiol. (Lond.)* **124**, 560–573.
Del Cerro, S., Arai, A. & Lynch, G. (1990) *Behav. Neural Biol.* **54**, 213–217.
Dinerman, J.L., Dawson, T.M., Schell, M.J., Snowman, A. & Snyder, S.H. (1994) *Proc. Natl. Acad. Sci. USA* **91**, 4214–4218.
Dolphin, A.C., Errington, M.L. & Bliss, T.V.P. (1982) *Nature* **297**, 496–498.
Drapeau, C., Pellerin, L., Wolfe, L.S. & Avoli, M. (1990) *Neurosci. Lett.* **115**, 286–292.
Dudek, S.M. & Bear, M.F. (1992) *Proc. Natl. Acad. Sci. USA* **89**, 4363–4367.
Eichenbaum, H. (1993) *Science* **261**, 993–994.
Eichenbaum, H., Wiener, S.I., Shapiro, M. & Cohen, N.J. (1989) *J. Neurosci.* **9**, 2764–2775.
Eichenbaum, H., Otto, T. & Cohen, N.J. (1992) *Behav. Neural Biol.* **57**, 2–36.
Fatt, P. & Katz, B. (1952) *J. Physiol. (Lond.)* **117**, 109–128.
Fonnum, F. (1984) *J. Neurochem.* **41**, 1–11.
Foster, T.C. & McNaughton, B.L. (1991) *Hippocampus* **1**, 79–91.
Garthwaite, J. & Garthwaite, G. (1987) *J. Neurochem.* **48**, 29–39.
Grant, S.G.N., O'Dell, T.J., Karl, K.A., Stein, P.L., Soriano, P. & Kandel, E.R. (1992) *Science* **258**, 1903–1910.
Gribkoff, V.K. & Lum-Ragan, J.T. (1992) *J. Neurophysiol.* **68**, 639–642.
Haley, J.E., Wilcox, G.L. & Chapman, P.F. (1992) *Neuron* **8**, 211–216.
Haley, J.E., Malen, P.L. & Chapman, P.F. (1993). *Neurosci. Lett.* **160**, 85–88.
Hirano, T. (1991) *Synapse* **7**, 321–323.
Hollmann, M. & Heinemann, S. (1994) *Annu. Rev. Neurosci.* **17**, 31–108.
Hölscher, C. & Rose, S.P.R. (1992) *Neurosci. Lett.* **145**, 165–167.
Huang, P.L., Dawson, T.M., Bredt, D.S., Snyder, S.H. & Fishman, M.C. (1993) *Cell* **75**, 1273–1286.
Ito, M. (1982) *Annu. Rev. Neurosci.* **5**, 275–296.
Ito, M. (1989) *Annu. Rev. Neurosci.* **12**, 85–102.
Ito, M. (1990) *Semin. Neurosci.* **2**, 381–390.
Ito, M. & Karachot, L. (1989) *Neurosci. Res.* **458**, 397–401.

Ito, M., Sakurai, M. & Tongroach, P. (1982) *J. Physiol. (Lond.)* **324**, 113–124.
Izumi, Y. & Zorumski, C.F. (1993) *NeuroReport* **4**, 1131–1134.
Kato, K., Clark, G.D., Bazan, N.G. & Zorumski, C.F. (1994) *Nature* **367**, 175–179.
Kauer, J.A., Malenka, R.C. & Nicoll, R.A. (1988) *Neuron* **1**, 911–917.
Kelso, S.R., Ganong, A.H. & Brown, T.H. (1986) *Proc. Natl. Acad. Sci. USA* **83**, 5326–5330
Korn, H. & Faber, D.S. (1991) *Trends Neurosci.* **14**, 439–445.
Kullman, D.M. & Nicoll, R.A. (1992) *Nature* **357**, 240–244.
Kullman, D.M., Perkel, D.J., Manabe, T. & Nicoll, R.A. (1992) *Neuron* **9**, 1175–1183.
Larkman, A., Stratford, K. & Jack, J. (1991) *Nature* **350**, 344–347.
Larkman, A., Hannay, T., Stratford, K. & Jack, J. (1992) *Nature* **360**, 70–73.
Linden, D.J. (1994) *Neuron* **12**, 457–472.
Linden, D.J. & Connor, J.A. (1991) *Science* **254**, 1656–1659.
Linden, D.J. & Connor, J.A. (1992) *Eur. J. Neurosci.* **4**, 10–15.
Linden, D.J., Dickinson, M.H., Smeyne, M. & Connor, J.A. (1991) *Neuron* **7**, 81–89.
Lohmann, S.M., Walter, U., Miller, P.E., Greengard, P. & DeCammilli, P. (1981) *Proc. Natl. Acad. Sci. USA* **78**, 653–657.
Lovinger, D.M., Wong, K.L., Murakami, K. & Routtenberg, A. (1987) *Brain Res.* **436**, 177–183.
Lynch, G.A. & Baudry, M. (1984) *Science* **224**, 1057–1063.
Lynch, G., Larson, J., Kelso, S., Barrionuevo, G. & Schottler, F. (1983) *Nature* **305**, 719–721.
MacDermott, A.B., Mayer, M.L., Westbrook, G.L., Smith, S.J. & Barker, J.L. (1986) *Nature* **321**, 519–522.
Malenka, R.C., Kauer, J.A., Zucker, R.S. & Nicoll, R.A. (1988) *Science* **242**, 81–84.
Malenka, R.C., Kauer, J.A., Perkel, D.J., Mauk, M.D., Kelly, P.T., Nicoll, R.A. & Waxham, M.N. (1989) *Nature* **340**, 554–557.
Malgaroli, A. & Tsien, R.W. (1992) *Nature* **357**, 134–139.
Malinow, R. (1991) *Science* **252**, 722–724.
Malinow, R. & Miller, J.P. (1986) *Nature* **320**, 529–530.
Malinow, R. & Tsien, R.W. (1990) *Nature* **346**, 177–180.
Malinow, R., Madison, D.V. & Tsien, R.W. (1988) *Nature* **335**, 820–824.
Malinow, R., Schulman, H. & Tsien, R.W. (1989) *Science* **245**, 862–866.
Manabe, T., Renner, P. & Nicoll, R.A. (1992) *Nature* **355**, 50–55.
Mayer, M.L., Westbrook, G.L. & Guthrie, P.B. (1984) *Nature* **309**, 261–263.
Montague, P.R., Gally, J.A. & Edelman, G.M. (1991) *Cerebral Cortex* **1**, 199–220.
Morris, E.G.M., Garrud, P., Rawlins, J.N.P. & O'Keefe, J. (1982) *Nature* **297**, 681–683.
Morris, R.G.M. (1989) *J. Neurosci.* **9**, 3040–3057.
Morris, R.G.M., Anderson, E., Lynch, G.S. & Baudry, M. (1986) *Nature* **319**, 774–776.
Mulkey, R.M. & Malenka, R.C. (1992) *Neuron* **9**, 967–975.
Muller, D. & Lynch, G. (1988) *Proc. Natl. Acad. Sci. USA* **85**, 6997–7000.
Muller, D., Arai, A. & Lynch, G. (1992) *Hippocampus* **2**, 29–38.
Musleh, W.Y., Shahi, K. & Baudry, M. (1993) *Synapse* **13**, 370–375.
Nakane, M., Ichikawa, M. & Deguchi, T. (1983) *Brain Res.* **273**, 9–15.
O'Dell, T.J., Hawkins, R.J., Kandel, E.R. & Arancio, O. (1991) *Proc. Natl. Acad. Sci. USA* **88**, 11285–11289.
O'Dell, T.J., Huang, P.L., Dawson, T.M., Dinerman, J.L., Snyder, S.H., Kandel, E.R. & Fishman, M.C. (1994) *Science* **265**, 542–545.
Perkel, D.J. & Nicoll, R.A. (1993) *J. Physiol. (Lond.)* **471**, 481–500.
Pollock, J.S., Klinghofer, V., Förstermann, U. & Murad, F. (1992) *FEBS Lett.* **309**, 402–404.
Sah, P., Hestrin, S. & Nicoll, R.A. (1989) *Science* **246**, 815–818.
Sastry, B.R., Goh, J.W. & Auyeung, A. (1986) *Science* **232**, 988–990.
Schuman, E.M. & Madison, D.V. (1991) *Science* **254**, 1503–1506.
Schuman, E.M. & Madison, D.V. (1994) *Science* **263**, 532–536.

., D. (1991) *Nature* **349**, 326–328.
R., Wehner, J.M. & Tonegawa, S. (1992a) *Science* **257**, 206–211.
.ns, C.F., Tonegawa, S. & Wang, Y. (1992b) *Science* **257**, 201–206.
. Zola-Morgan, S. (1991) *Science* **253**, 1380–1386.
. & Wang, Y. (1993) *Nature* **364**, 147–149.
.F. & Wang, Y. (1994) *Nature* **371**, 704–707.
on, A. (1991) *Eur. J. Neurosci.* **3**, 587–601.
.chanoff, J.G., Weinberg, R.J., Kharazia, V.N., Nakane, M. & Schmidt, H.H.H.W. (1993) *J. Comp. Neurol.* **331**, 111–121.
VanHoesen, G.W., Pandya, D.N. & Butters, N. (1972) *Science* **175**, 1471–1473.
Vincent, P. & Marty, A. (1993) *Neuron* **11**, 885–893.
Vincent, S.R. & Kimura, H. (1992) *Neuroscience* **46**, 755–784.
Waxham, M.N., Malenka, R.C., Kelly, P.T. & Mauk, M.D. (1993) *Brain Res.* **609**, 1–8.
Wendland, B., Schweizer, F.E., Ryan, T.A., Nakane, M., Murad, F., Scheller, R.H. & Tsien, R.W. (1994) *Proc. Natl. Acad. Sci. USA* **91**, 2151–2155.
Wieraszko, A., Li. G., Karnecki, E., Hogan, M.V. & Ehrlich, Y.H. (1993) *Neuron* **10**, 553–557.
Williams, J.H. & Bliss, T.V.P. (1989) *Neurosci. Lett.* **107**, 301–306.
Williams, J.H., Errington, M.L., Lynch, M.A. & Bliss, T.V.P. (1989) *Nature* **341**, 739–742.
Williams, J.H., Bettache, N., Trentham, D.R. & Bliss, T.V.P. (1993a) *J. Physiol.* **467**, 166P.
Williams, J.H., Li, Y.-G., Nayak, A., Errington, M.L., Murphy, K.P.S.J. & Bliss, T.V.P. (1993b) *Neuron* **11**, 877–884.
Wilson, M.A. & McNaughton, B.L. (1993) *Science* **261**, 1055–1058.
Witter, M.P., Groenewegen, H.J., Lopes da Silva, F.H. & Lohman, A.H.M. (1989) *Progress in Neurobiology* **33**, 161–253.
Wu, J., Cohen, L.B. & Falk, C.X. (1994) *Science* **263**, 820–823.
Wyllie, D.J.A., Manabe, T. & Nicoll, R.A. (1994) *Neuron* **12**, 127–138.
Xie, X., Berger, T. & Barrionuevo, G. (1992) *J. Neurophysiol.* **67**, 1009–1013.
Zhuo, M., Small, S.A., Kandel, E.R. & Hawkins, R.D. (1993) *Science* **260**, 1946–1950.
Zucker, R.S. (1989) *Annu. Rev. Neurosci.* **12**, 13–32.

CHAPTER 8

NITRIC OXIDE SIGNALLING IN TH HYPOTHALAMUS

Shimon Amir

Center for Studies in Behavioral Neurobiology, Concordia University, 1455 de Maisonneuve Boulevard West, Montreal, Quebec H3G 1M8, Canada

Table of Contents

8.1 Introduction 151
8.2 NOS in the hypothalamus 152
8.3 Corticotropin-releasing hormone neurosecretory system 152
8.4 Luteinizing hormone-releasing hormone neurosecretory system 155
8.5 Vasopressin neurosecretory system 155
8.6 Somatostatin release 156
8.7 Autonomic regulation 156
8.8 Functional considerations 158
8.9 Conclusions 160

Acknowledgements 160
References 160

8.1 Introduction

The mammalian hypothalamus plays an important role in the control of autonomic and endocrine processes such as cardiovascular activity and secretion of stress hormones; it is also implicated in diverse physiological and behavioural functions such as reproduction and maternal behaviour, temperature regulation, energy balance and circadian rhythmicity. Comprised of a network of anatomically discrete nuclei, the hypothalamus functions to alter metabolic, endocrine and autonomic activities in anticipation of, or in response to, specific organismic or environmental challenges, by transducing sensory signals to the activation of inhibition of the appropriate effector mechanisms. It is now well recognized that an important mechanism underlying the functional integrity of the hypothalamic surveillance system is neurochemical diversity. Transduction of signals to and from various hypothalamic regulatory systems involves disparate classes of excitatory and inhibitory transmitters, hormones, peptides and second messengers. An important new addition to this

Nitric Oxide in the Nervous System
ISBN 0-12-721985-4

, nitric oxide (NO), a highly reactive free radical gas produced y a calcium- and calmodulin-dependent and NADPH-requiring .ne, nitric oxide synthase (NOS) (Garthwaite, 1991; Bredt and -). Neuronal NOS is a constitutive protein that can rapidly and tran- .thesize and release small amounts of NO in response to trans-synaptic .hat increase intracellular calcium concentration (Schmidt *et al.*, 1989; Bredt ..nyder, 1990; Mayer *et al.*, 1990; Algarsamy *et al.*, 1994). Once released, NO .n interact with various target proteins in the generating cell or in adjacent cells to influence ion channels, transmitter receptors, enzymes and second messenger systems (Knowles *et al.*, 1989; Bredt and Snyder, 1989; Schmidt *et al.*, 1993; Schuman and Madison, 1994; Vincent, 1994). Here I will review recent studies of NO as a messenger molecule within the hypothalamus, focusing on the involvement of this messenger in the regulation of hypothalamic systems involved in neuroendocrine and autonomic control.

8.2 NOS in the hypothalamus

Studies of the distribution of neuronal NOS and NOS mRNA within the mammalian central nervous system have been conducted using both immunohistochemical and NADPH diaphorase histochemical methods (Bredt *et al.*, 1990, 1991; Dawson *et al.*, 1991; Hope *et al.*, 1991). These studies have revealed the presence of a large number of NOS containing neurons in the magnocellular and parvocellular subdivisions of the hypothalamic paraventricular nucleus (PVN), supraoptic nucleus (SON), nucleus circularis and lateral hypothalamus. Additional groups of neurons within the hypothalamus and associated structures expressing NOS have been demonstrated in the preoptic nucleus, periventricular nucleus and portions of the ventromedial nucleus (VMH), beneath and around the fornix, lateral anterior nucleus, bed nucleus of the stria terminalis, magnocellular neurons in the horizontal limb of the diagonal band, supraoptic decussation, tuber cinereum and the medial and lateral tuberal nuclei (see Chapter 5). Nerve endings containing NOS have been demonstrated in the internal portion of the median eminence (Arévalo *et al.*, 1992; Vincent and Kimura, 1992).

8.3 Corticotropin-releasing hormone neurosecretory system

The synthesis and release of glucocorticoids from the adrenal cortex is regulated by the anterior pituitary hormone adrenocorticotropin (ACTH), whose secretion is controlled by the hypothalamic corticotropin-releasing hormone (CRH) neurosecretory system. Cell bodies containing CRH as well as other ACTH-releasing factors such as vasopressin are located in the parvocellular subdivision of the PVN;

these cells project axons to the portal capillary plexus in the external zone of the median eminence. Because of the similarity in localization of NOS-containing neurons and CRH-containing neurons within the PVN (Torres *et al.*, 1993; Siaud *et al.*, 1994), it has been suggested that NO might be involved in the control of CRH release. In a series of *in vitro* experiments on hypothalamic slices, it has been shown that treatment with either the NO precursor L-arginine or the NO donors molsidomine or sodium nitroprusside has no effect on basal release of CRH. Similar treatments were found to block the stimulated release of CRH induced by KCl or interleukin-1β. The inhibitory effect of L-arginine could be reversed by treatment with the competitive NOS blocker N^G-monomethyl-L-arginine (L-NMMA); the inhibitory effect of the NO donors was blocked by treatment with haemoglobin, a scavenger of NO. On the basis of these findings it has been suggested that NO exerts an inhibitory effect on stimulated CRH release by hypothalamic neurons (Costa *et al.*, 1993). Contrary to these findings, Karanth *et al.* (1993) reported evidence suggesting that NO exerts a permissive or facilitatory effect on CRH release. In their experiments, prolonged treatment with L-arginine elevated the basal release of CRH as well as enhanced the interleukin-2-induced stimulation of CRH release from hypothalamic slices. Treatment with the NOS blocker L-NMMA completely suppressed the interleukin-2-induced release of CRH. L-NMMA had no effect on noradrenaline (NA)-induced CRH release. Additional data supporting a permissive role of NO on CRH release has been reported by Brunetti *et al.* (1993). Their experiments showed that treatment with the competitive NOS blocker N^G-nitro-L-arginine (L-NOArg) inhibits interleukin-1β-induced release of CRH and ACTH from cultured hypothalamic neurons *in vitro*.

The data from *in vitro* studies with slices and cultured cells, suggesting that NO is involved in the stimulated release of CRH from hypothalamic neurons, prompted studies of the role of NO on stress- and cytokine-induced release of ACTH and adrenal corticosterone *in vivo*. Earlier studies by Calza *et al.* (1993) had demonstrated an increase in NOS mRNA in both the magnocellular and parvocellular subdivisions of the PVN in rats following immobilization stress. Using the c-*fos* gene product Fos as a marker of neuronal activation, we have found that immobilization stress is associated with the activation of many PVN cells that express the NOS histochemical marker, NADPH diaphorase (Rackover *et al.*, 1994). Further, it was found that systemic treatment with the NOS blocker N^G-nitro-L-arginine methyl ester (L-NAME) significantly attenuates immobilization stress-induced release of adrenal corticosterone in rats; pretreatment with D-NAME, an inactive isomer of L-NAME, had no effect. Systemic pretreatment with 7-nitro-indazole, a NOS competitive blocker devoid of cardiovascular actions (Moore *et al.*, 1993), or intracerebroventricular administration of L-NAME, also attenuates the immobilization stress-induced corticosterone release. Pretreatment with L-NAME at a dose that blocked the immobilization-induced release of corticosterone had no effect on the stress-induced expression of Fos in the NOS-containing PVN neurons, suggesting that the inhibitory effect of L-NAME on corticosterone release was mediated at the level of, or downstream from, the activated NOS-containing PVN cells. Importantly, NO has been implicated in basal steroidogenesis (Adams *et al.*, 1991) as well as in the control of basal adrenal vascular tone in the rat isolated adrenal gland preparation, but was found not to be involved in either the vascular or steroidogenic response to ACTH stimulation (Cameron and Hinson, 1993). This further supports the conclusion that the inhibitory effect of L-NAME on stress-induced corticosterone release is mediated centrally and may not involve peripheral actions at the level of the adrenal gland.

Contrary to the finding suggesting that NO has a permissive role in the control of stress-induced corticosterone release, Rivier and Shen (1994) have obtained data suggesting an inhibitory effect of NO on interleukin-1β-, vasopressin- and oxytocin-induced ACTH release *in vivo*. Specifically, they found that peripheral administration of the NOS blocker L-NAME enhances the stimulatory effect of peripherally administered interleukin-1β, vasopressin and oxytocin on ACTH release, whereas treatment with L-arginine blunts the effect of interleukin-1β on ACTH release. Furthermore, systemic treatment with L-NAME enhanced lipopolysaccharide-induced ACTH and corticosterone release. In contrast, systemic L-NAME had no effect on central interleukin-1β-induced release of ACTH, suggesting that the facilitatory effects of L-NAME on ACTH and corticosterone release following systemic treatment with the different secretagogues was mediated at a level of the pituitary, downstream from the hypothalamus.

While the studies conducted by Costa *et al.* (1993) and Rivier and Shen (1994) suggest an inhibitory influence of NO on cytokine-induced release of CRH, ACTH and corticosterone, all other studies cited above are consistent with the view that NO exerts a facilitatory effect on the release of these stress hormones. Specifically, they suggest that NOS-containing neurons in the PVN are activated in response to stress, and that upon activation these neurons are induced to synthesize and release NO, which has a permissive effect on CRH secretion. This ultimately leads to enhanced release of ACTH from the anterior pituitary and, consequently, to enhanced synthesis and secretion of corticosterone. It remains to be determined precisely how NO influences CRH secretion from PVN neurosecretory neurons, and what are the neurochemical signals that trigger NO production in NOS-containing PVN cells during stress. One hypothesis, advanced by Karanth *et al.* (1993), suggests that production of NO in NOS-containing PVN neurons is coupled to the activation of muscarinic receptors by the neurotransmitter acetylcholine. Such activation would trigger calcium influx into the NOS-containing cell, leading to the rapid activation of the enzyme and release of NO. NO diffuses to the CRH neuron where it stimulates prostaglandin-E_2 production by activating the cyclo-oxygenase enzyme. Prostaglandin-E_2, through activation of adenylate cyclase, stimulates the generation of cAMP which induces exocytosis of CRH secretory granules (Karanth *et al.*, 1993). The mechanisms by which NO suppresses CRH release from hypothalamic neurons or inhibits secretagogue-stimulated ACTH release from the pituitary (Costa *et al.*, 1993; Rivier and Shen, 1994) remains to be elucidated.

8.4 Luteinizing hormone-releasing hormone neurosecretory system

Pulsatile release of luteinizing hormone (LH) from the anterior pituitary is controlled by the hypothalamic releasing peptide luteinizing hormone-releasing hormone (LHRH). LHRH is synthesized by neurosecretory cells in the medial hypothalamus and preoptic area and released from terminals in the median eminence into the hypophyseal portal system. The possibility that NO might be involved in the regulation of the hypothalamic LHRH neurosecretory system is consistent with the following observations. NOS-containing neurons are located in close proximity to LHRH neurons (Grossman *et al.*, 1994). Treatment with the NOS blocker L-NMMA inhibits NA- and glutamate-induced LHRH release from medial basal hypothalami *in vitro*; L-NMMA has no effect on basal release of LHRH. Basal release is enhanced by treatment with the NO donor sodium nitroprusside; this effect is blocked by the NO scavenger haemoglobin. Finally, injection of L-NMMA into the third cerebral ventricle blocks pulsatile LH release in rats (Rettori *et al.*, 1993, 1994). In another study, L-NMMA was found to block *N*-methyl-D-aspartate (NMDA)-induced LHRH release from median eminence-arcuate nucleus fragments *in vitro*, and systemic injection of L-NMMA was found to block progesterone-induced stimulation of LH release in oestrogen-primed ovariectomized rats (Bonavera *et al.*, 1993). It has been suggested that the effect of NO on LHRH release is linked to its stimulatory action on prostaglandin-E_2 synthesis within the LHRH-secreting neurons (Rettori *et al.*, 1992, 1993, 1994). Others have suggested that the effect of NO on LHRH secretion is mediated by cGMP (Moretto *et al.*, 1993). In these experiments, treatment with sodium nitroprusside stimulated LHRH release as well as induced an elevation in both extra- and intracellular cGMP levels in hypothalamic explants *in vitro*. The effect of sodium nitroprusside on LHRH release was blocked by Rp-8-Br-cGMP, a cGMP analogue which blocks cGMP-dependent protein kinase, implicating cGMP as a mediator of the stimulatory effect of NO on LHRH release. It should be noted that other studies have demonstrated an inhibitory action of NO on basal and stimulated LHRH release from GT1-1 neuronal cell line (Sortino *et al.*, 1994) and on LHRH-induced release of pituitary LH (Ceccatelli *et al.*, 1993).

8.5 Vasopressin neurosecretory system

Co-localization mapping studies have shown that a proportion of NOS-containing magnocellular neurons in the PVN and SON also express the peptides vasopressin (Calka and Block, 1993a) and oxytocin (Miyagawa *et al.*, 1994). It has also been shown that nerve terminals in the posterior pituitary express NOS, and that stimuli which increase vasopressin synthesis and release (i.e. hypertonic saline) increase the amount of NOS mRNA and protein in the PVN, SON and posterior pituitary

(Sagar and Ferriero, 1987; Pow, 1992; Calca *et al.*, 1994; Villar *et al.*, 1994; Kadowaki *et al.*, 1994). Studies of the involvement of NO in the control of vasopressin release have shown that treatment with the NO precursor L-arginine reduces KC1- and interleukin-1β evoked vasopressin release from incubated rat hypothalami *in vitro* (Yasin *et al.*, 1993). Similar inhibitory effects on KC1-evoked vasopressin release were noted after treatment with the NO donor sodium nitroprusside. Goyer *et al.* (1994) have shown that intravenous infusion of the NOS blocker L-NAME increases vasopressin concentration in conscious rabbits. A similar treatment had no effect on the increase in vasopressin induced by hypertonic saline or sodium nitroprusside infusion, indicating that NO is involved in the control of basal vasopressin secretion, but not in vasopressin responses to hyperosmolality and hypotension. In contrast, Ota *et al.* (1993) reported that central injection of the NO donor *S*-nitroso-*N*-acetylpenicillamine or the NO biosynthetic precursor L-arginine stimulates vasopressin release *in vivo.*

8.6 Somatostatin release

Co-existence of NOS and somatostatin (SRIF) has been demonstrated in some PVN neurons (Alonso *et al.*, 1992). The possibility that NO is involved in the control of SRIF synthesis and release has been assessed using periventricular nuclei of male rats *in vitro.* Treatment with the NO donor sodium nitroprusside stimulated SRIF mRNA levels and SRIF release; a similar effect was observed after treatment with the cGMP analogue dibutyryl cGMP (Aguila, 1994). Treatment with growth hormone-releasing factor (GRF) increased cGMP formation and SRIF mRNA level and stimulated SRIF release. These effects were blocked by L-NMMA but not D-NMMA, implicating the NO/cGMP pathway in the control of SRIF synthesis and release.

8.7 Autonomic regulation

The localization of NOS within the hypothalamic nuclei involved in autonomic control suggests a potential role for NO as a modulator of neuronal activity within these centres. To study the direct effect of NO on the sympathetic outflow from the PVN, Horn *et al.* (1994) injected the NO donor sodium nitroprusside into the PVN of urethane-anaesthetized rats and evaluated the effect on blood pressure and release of amino acids using microdialysis. Intra-PVN injection of sodium nitroprusside caused a decrease in blood pressure and a rise in extracellular glutamate within the PVN. In another study, Bains and Ferguson (1994) have shown that treatment with the NOS blocker L-NAME enhances the pressor response elicited by electrical stimulation of the subfornical organ, an effect which is mediated by angiotensin II release in the PVN. These studies suggest that NO acts in the PVN to down-regulate neural activity leading to increased sympathetic tone and a rise in blood pressure. Interestingly, angiotensin (1–7) has been shown to be co-localized with NOS in the PVN as well as other hypothalamic nuclei such as the SON and nucleus circularis (Calka and Block, 1993b).

Glutamate is an important excitatory transmitter in the hypothalamic centres involved in neuroendocrine and autonomic regulation (Brann and Mahesh, 1992; van den Pol and Trombley, 1993; Meeker *et al.*, 1993, 1994; Patchev *et al.*, 1994). It is also a primary stimulus for the activation of NOS in central neurons (Southam *et al.*, 1991; Luo *et al.*, 1993, 1994). The effect of glutamate on NO production is mediated by NMDA receptors; activation of NMDA receptors leads to an increase in intracellular calcium level and consequent activation of NOS. NMDA receptors have been implicated in the transmission of photic signals from the retina to the suprachiasmatic nucleus (SCN), site of a circadian pacemaker in mammals (Abe *et al.*, 1991; Ohi *et al.*, 1991; Colwell *et al.*, 1991; Castel *et al.*, 1993; Gannon and Rea, 1993; Reghunandanan *et al.*, 1993). They have also been implicated in the mechanism that couples photic stimulation to sympathetic nervous system activation, a process mediated by the SCN (Amir, 1989, 1992; Amir *et al.*, 1989; Niijima *et al.*, 1992, 1993). Earlier immunohistochemical and NADPH diaphorase histochemical studies have reported on the absence of any NOS containing neurons in the SCN (Vincent and Kimura, 1992), arguing against the possibility that light-induced NMDA-mediated changes in SCN neuronal activity involve NO production by NOS. We have observed, however, that a small number of neurons located medial to the SCN, between the floor of the third cerebral ventricle and the optic chiasma, consistently label with NADPH diaphorase stain. To test the possibility that the NMDA-mediated link between photic stimulation and sympathetic nervous system activation involves release of NO in the SCN region, we examined the effect of infusing L-NAME into the SCN on light-induced cardiac acceleration in urethane-anaesthetized rats. L-NAME infusion into the SCN region attenuated the light-induced increase in heart rate; the inactive analogue D-NAME had no effect. Similarly, infusion of methylene blue, a blocker of NOS and the cGMP synthesizing enzyme soluble guanylate cyclase which is activated by NO, attenuated the light-induced increase in heart rate. Infusion of L-NAME or methylene blue into the SCN region had no effect on cardiac acceleration induced by a non-photic stimulus, tail pinch, pointing to a selective role of this NO/cGMP signalling mechanism in neural system that transduces photic stimulation to cardiac acceleration (Amir, 1992). Interestingly, both glutamate and cGMP have been shown to induce a light-like phase shift of the circadian rhythm of cell firing in the SCN *in vitro* (Prosser *et al.*, 1989; Ding and Gillette, 1993); treatment with L-NAME was found to inhibit the phase-shifting effect of glutamate (Ding and Gillette, 1993; Watanabe *et al.*, 1994), implicating NO production in glutamate-induced phase shift in this system. However, the effect of L-NAME on light-induced phase shift of circadian rhythms *in vivo* has not been determined.

In the SON, activation of NMDA receptors leads to a rapid increase in sympathetic activity and heart rate (Amir, 1994). Studies of the role of the NO/cGMP signalling system revealed that infusion into the SON of compounds that either block NO synthesis by NOS (L-NAME) or inhibit NO-induced stimulation of cGMP production by soluble guanylate cyclase (methylene blue) attenuate the increase in heart rate induced by intra-SON injection of NMDA. Furthermore, intra-SON injection of the calcium ionophore A23187, which stimulates NOS by increasing intracellular calcium directly, or injection of a membrane permeating analogue of cGMP, 8-Br-cGMP, mimicked the effect of NMDA injection on cardiac acceleration. Thus, activation of NMDA receptors in the SON leads to a sympathetically mediated increase in heart rate; the signal from NMDA receptor stimulation to cardiac acceleration involves NO production by NOS and consequent increase in cGMP, which is mediated by NO-induced activation of soluble guanylate cyclase.

Neurons containing NOS have been demonstrated in the preoptic area of the anterior hypothalamus (POAH), a structure involved in diverse physiological functions, including temperature regulation and cold-induced thermogenesis. Prostaglandin-E_2 injection into the POAH increases body temperature in urethane-anaesthetized rats by stimulating heat production in brown adipose tissue. This effect is blocked by co-injection of the NOS blocker L-NMMA, suggesting that the elaboration or transmission of the thermogenic signal involves NO production (Amir *et al.*, 1991).

8.8 Functional considerations

The studies reviewed above are consistent with the view that NO is involved in the regulation of specific hypothalamic neurosecretory and autonomic neuronal systems. Specifically, these studies show that many hypothalamic neurons which are involved in neuroendocrine and autonomic regulation express NOS. Further, pharmacological interventions that influence NO production, that interfere with the action of NO on target proteins, or that mimic the action of endogenous NO on these proteins, can influence the function of these hypothalamic neurons. Notwithstanding, some key questions on the function of the NO signalling system within the hypothalamus remain to be answered. For example, what are the physiological and chemical stimuli that activate NOS in hypothalamic neurosecretory and autonomic neurons? Neurosecretion by hypothalamic neurosecretory cells can be triggered by diverse types of stimuli; some stimuli may be specific to a particular neurosecretory system, others may have a non-specific effect leading to the activation of multiple neurosecretory and autonomic pathways. Although evidence now exists that some stimuli that activate hypothalamic neurosecretory neurons can influence the expression of NOS mRNA as well as bring about an increase in the number in neurons expressing NOS, it is not known whether or by which mechanism these and other stimuli trigger NO synthesis by NOS. Clearly, excitatory amino acids are the most likely transmitter candidates in these systems; this is consistent with the wide distribution of glutamate terminals and receptors in the hypothalamus, the role they play in the regulation of neurosecretory neurons and autonomic neurons, and the key role played by NMDA receptors in NO production by NOS throughout the nervous system (Garthwaite *et al.*, 1988; Garthwaite, 1991). Mediation by cholinergic neurons via muscarinic cholinergic receptors is also possible (Hu and El-Fakahany, 1993), and has been discussed in relation to the mechanism by which interleukins influence CRH neurosecretion from hypothalamic slices *in vitro* (Karanth *et al.*, 1993). The role of other hypothalamic transmitter systems, such as the serotonergic system, is less well defined.

Another question concerns the mechanism by which NO influences the activity of hypothalamic neurons. The best known target of NO is the soluble guanylate cyclase (Knowles *et al.*, 1989; Southam and Garthwaite, 1993; Luo *et al.*, 1994; see Chapter 3), and it is most likely that some of the NO-mediated changes in hypothalamic neuronal function involve an increase in intracellular cGMP in the cells targeted by NO. NO may also influence neuronal function by stimulating the ADP-ribosylation of proteins (Brune and Lapetina, 1989). These NO-mediated cellular actions are likely to influence transmitter release and/or neurotransmitter receptor function, as has been reported in other brain regions (Lonart *et al.*, 1992; Prast and Philippu, 1992; Zu Zhu and Luo, 1992; Manzoni *et al.*, 1992).

A further question concerns the functional nature of NO signalling within the hypothalamus. Because of its unique biophysical and chemical characteristics, NO most likely functions as a diffusible intercellular messenger rather than as a classical chemical synaptic transmitter. As such, NO could serve to facilitate communication between cells or groups of cells that lack direct synaptic access to each other. This mode of spatial communication may provide a means by which activities within different neurosecretory and autonomic hypothalamic systems are integrated and modulated.

Finally, NO might be implicated in the processes underlying neural plasticity within the hypothalamus. Release of NO has been suggested to be a necessary step in the expression of some forms of activity-dependent synaptic plasticity, such as hippocampal long-term potentiation (Zorumski and Izumi, 1993; see Chapter 7). It has been proposed that NO serves as a retrograde messenger, allowing transmission of signals from the postsynaptic membrane to the presynaptic membrane, signals that ultimately bring about increases in neurotransmitter release. Activity-dependent changes in structure and function of neurons and glia have also been described in the hypothalamus (Theodosis and Poulain, 1993). Neurosecretory cells and glia in the PVN and SON undergo structural remodelling, including changes in conformation and synaptic rearrangement, following exposure to stimuli such as suckling and osmotic stress. NOS is expressed in many of these hypothalamic neurosecretory neurons, and osmotic stimuli have been shown to increase NOS mRNA and protein in PVN and SON neurosecretory cells (Sagar and Ferriero, 1987; Pow, 1992; Kadowaki *et al.*, 1994). Thus, it is possible that some physiologically linked changes in structure or function of hypothalamic neurosecretory cells and associated glia may be related to NO release and action.

8.9 Conclusions

The idea that NO functions as an intercellular messenger molecule in hypothalamic systems involved in neuroendocrine and autonomic regulation is rapidly gaining acceptance. The studies reviewed here are consistent with a general view that NO participates in the control of many different neurosecretory and autonomic processes. There is less agreement, however, as to precisely how NO influences these processes. This is best illustrated in studies on the role of NO in the control of the CRH neurosecretory system. Available evidence indicates that NO can exert either an inhibitory influence or a facilitatory influence on the release of CRH *in vitro*, and on the activity of the hypothalamic–pituitary–adrenal system *in vivo*. These apparent contradictions may be due merely to differences in experimental models and procedures used in different studies. It may, however, also represent the true nature of this elusive messenger molecule, namely an ability to exert diametrically opposed effects within discrete systems. It is possible, for example, that the effect exerted by NO within a given system will change from one which is inhibitory to one which is facilitatory on the system, or vice versa, depending on the level of basal activity within the system, the type of stimulus presented to the system, or both. Indeed, such a multifaceted mode of action has been invoked to account for the ability of NO to exert both neuroprotective and neurodestructive effects in the brain (Lipton *et al.*, 1993). This and many other questions on the nature of NO signalling within the hypothalamus remain to be answered. Given the vigour of research and rate of progress in the field, it is most likely that some of these queries will be resolved by the time this paper is published.

Acknowledgements

This work was supported by grants from the 'Fonds pour la Formation de Chercheurs et l'Aide a la Recherche du Quebec' and the 'Natural Sciences and Engineering Research Council of Canada'.

References

Abe, H., Rusak, B. & Robertson, H.A. (1991) *Neurosci. Lett.* **127**, 9–12.
Adams, M.L., Nock, B., Truong, R. & Cicero, T.J. (1991) *Life Sci.* **50**, PL35–PL40.
Aguila, M.C. (1994) *Proc. Natl. Acad. Sci. USA* **91**, 782–786.
Algarsamy, S., Lonart, G. & Johnson, K.M. (1994) *J. Neurochem.* **62**, 400–403.
Alonso, J.R., Sanchez, F., Arévalo, R., Carretero, J., Vazquez, R. & Aijon, J. (1992) *Neurosci. Lett.* **148**, 101–104.
Amir, S. (1989) *Brain Res.* **503**, 163–166.
Amir, S. (1992) *Brain Res.* **586**, 336–339.

Amir, S. (1994) *Brain Res.* **645**, 330–334.
Amir, S., Shizgal, P. & Rompré, P.-P. (1989) *Brain Res.* **498**, 140–144.
Amir, S., De Blasio, E. & English, A. (1991) *Brain Res.* **556**, 157–160.
Arévalo, R., Sanchez, F., Alonso, J.R., Carretero, J., Vazquez, R. & Aijon, J. (1992) *Brain Res. Bull.* **28**, 599–603.
Bains, J.S. & Ferguson, A.V. (1994) *Regulatory Peptides* **50**, 53–59.
Bonavera, J.J., Sahu, A., Kalra, P.S. & Kalra, S.P. (1993) *Endocrinology* **133**, 2481–2487.
Brann, D.W. & Mahesh, V.B. (1992) *Trends Endocrinol. Metab.* **3**, 122–126.
Bredt, D. & Snyder, S.H. (1989) *Proc. Natl. Acad. Sci. USA* **86**, 9030–9033.
Bredt, D.S. & Snyder, S.H. (1990) *Proc. Natl. Acad. Sci. USA* **87**, 682–685.
Bredt, D. & Snyder, S.H. (1992) *Neuron* **8**, 3–11.
Bredt, D.S., Hwanb, P.M. & Snyder, S.H. (1990) *Nature* **347**, 768–770.
Bredt, D.S., Glatt, C.E., Hwang, P.M., Fotuhi, M., Dawson, T.M. & Snyder, S.H. (1991) *Neuron* **7**, 615–624.
Brune, B. & Lapetina, E.G. (1989) *J. Biol. Chem.* **264**, 8455–8458.
Brunetti, L., Preziosi, P., Ragazzoni, E. & Vacca, M. (1993) *Life Sci.* **53**, 219–222.
Calka, J. & Block, C.H. (1993a) *Brain Res. Bull.* **32**, 207–210.
Calka, J. & Block, C.H. (1993b) *Brain Res. Bull.* **30**, 677–685.
Calca, J., Wolf, G. & Brosz, M. (1994) *Brain Res. Bull.* **34**, 301–308.
Calza, L., Giardino, L. & Ceccatelli, S. (1993) *NeuroReport* **4**, 627–630.
Cameron, L.A. & Hinson, J.P. (1993) *J. Endocrinol.* **139**, 415–423.
Castel, M., Belenky, M., Cohen, S., Ottersen, O.P. & Storm-Mathisen, J. (1993) *Eur. J. Neurosci.* **5**, 368–381.
Ceccatelli, S., Hulting, A.-L., Zhang, X., Gustafsson, L., Villar, M. & Hokfelt, T. (1993) *Proc. Natl. Acad. Sci. USA* **90**, 11292–11296.
Colwell, C.S., Foster, R.G. & Menaker, M. (1991) *Brain Res.* **554**, 105–110.
Costa, A., Trainer, P., Besser, M. & Grossman, A. (1993) *Brain Res.* **605**, 187–192.
Dawson, T.M., Bredt, D.S., Fotuhi, M., Hwang, P.M. & Snyder, S.H. (1991) *Proc. Natl. Acad. Sci. USA* **88**, 7797–7801.
Ding, J.M. & Gillette, M.U. (1993) *Soc. Neurosci. Abs.* **19**, 1815.
Gannon, R.L. & Rea, M.A. (1993) *Brain Res.* **622**, 337–342.
Garthwaite, J. (1991) *Trends Neurosci.* **14**, 60–67.
Garthwaite, J., Charles, S.L. & Chess-Williams, R. (1988) *Nature* **336**, 385–388.
Goyer, M., Bui, H., Chou, L., Evans, J., Keil, L.C. & Reid, I.A. (1994) *Am. J. Physiol.* **266**, H822–H828.
Grossman, A.B., Rossmanith, W.G., Kabigting, E.B., Cadd, G., Clifton, D. & Steiner, R.A. (1994) *J. Endocrinol.* **140**, R5–R8.
Hope, B.T., Michael, G.J., Knigge, K.M. & Vincent, S.R. (1991) *Proc. Natl. Acad. Sci. USA* **88**, 2811–2814.
Horn, T., Smith, P.M., McLaughlin, B.E., Bauce, L., Marks, G.S., Pittman, Q.J. & Ferguson, A.V. (1994) *Am. J. Physiol.* **266**, R306–R313.
Hu, J. & El-Fakahany, E.E. (1993) *J. Neurochem.* **61**, 578–585.
Kadowaki, K., Kishimoto, J., Leng, G. & Emson, P.C. (1994) *Endocrinology* **134**, 1011–1017.
Karanth, S., Lyson, K. & McMann, S.M. (1993) *Proc. Natl. Acad. Sci. USA* **90**, 3383–3387.
Knowles, R.G., Palacios, M., Palmer, R.M.J. & Moncada, S. (1989) *Proc. Natl. Acad. Sci. USA* **86**, 5159–5162.
Lipton, S.A., Choi, Y-B., Pan, Z.-H., Lei, S.Z., Chen, H.-S. V., Sucher, N.J., Loscalzo, J., Singel, D.J. & Stamler, J.S. (1993) *Nature* **364**, 626–632.
Lonart, G., Wang, J. & Johnson, K.M. (1992) *Eur. J. Pharmacol.* **220**, 271–272.
Luo, D., Knezevich, S. & Vincent, S.R. (1993) *Neuroscience* **57**, 897–900.
Luo, D., Leung, E. & Vincent, S. (1994) *J. Neurosci.* **14**, 263–271.
Manzoni, O., Prezeau, L., Marin, P., Deshager, S., Bockaert, J. & Fagni, L. (1992) *Neuron* **8**, 653–662.

Mayer, B., John, M. & Bohme, E. (1990) *FEBS Lett.* **277**, 215–219.
Meeker, R.B., Swanson, D.J., Greenwood, R.S. & Hayward, J.N. (1993) *Brain Res.* **600**, 112–122.
Meeker, R.B., Greenwood, R.S. & Hayward, J.N. (1994) *Endocrinology* **134**, 621–629.
Miyagawa, A., Okamura, H. & Ibata, Y. (1994) *Neurosci. Lett.* **171**, 13–16.
Moore, P.K., Babbedge, R.C., Wallace, P., Gaffen, Z.A. & Hart, S.L. (1993) *Br. J. Pharmacol.* **108**, 296–297.
Moretto, M., Lopez, F.J. & Negro-Vilar, A. (1993) *Endocrinology* **133**, 2399–2402.
Niijima, A., Nagai, K., Nagai, N. & Nakagawa, H. (1992) *J. Autonom. Nerv. Syst.* **40**, 155–160.
Niijima, A., Nagai, K., Nagai, N. & Nakagawa, H. (1993) *Physiol. Behav.* **54**, 555–561.
Ohi, K., Takashima, M., Nishikawa, T. & Takahashi, K. (1991) *Neuroendocrinology* **53**, 344–348.
Ota, M., Crofton, J.T., Festavan, G.T. & Share, L. (1993) *Neuroendocrinology* **57**, 955–959.
Patchev, V.K., Karalis, K. & Chrousos, G.P. (1994) *Brain Res.* **633**, 312–316.
Pow, D.V. (1992) *J. Neuroendocrinol.* **4**, 377–380.
Prast, H. & Philippu, A. (1992) *Eur. J. Pharmacol.* **216**, 139–140.
Prosser, R.A., McArthur, A.J. & Gillette, M.U. (1989) *Proc. Natl. Acad. Sci. USA* **86**, 6812–6815.
Rackover, M., Funk, D. & Amir, S. (1994) *Soc. Neurosci. Abs.* **20**.
Reghunandanan, V., Reghunandanan, R. & Singh, P.I. (1993) *Prog. Neurobiol.* **41**, 647–655.
Rettori, V., Gimeno, M., Lyson, K. & McCann, S.M. (1992) *Proc. Natl. Acad. Sci. USA* **89**, 11543–11546.
Rettori, V., Belova, N., Dees, W.L., Nyberg, C.L., Gimeno, M. & McCann, S.M. (1993) *Proc. Natl. Acad. Sci. USA* **90**, 10130–10134.
Rettori, V., Kamat, A. & McCann, S.M. (1994) *Brain Res. Bull.* **33**, 501–503.
Rivier, C. & Shen, G.H. (1994) *J. Neurosci.* **14**, 1985–1993.
Sagar, S.M. & Ferriero, D.M. (1987) *Brain Res.* **400**, 348–352.
Schmidt, H.H.H.W., Lohmann, S.M. & Walter, U. (1993) *Biochim. Biophys. Acta* **1178**, 153–175.
Schmidt, K., Mayer, B. & Kukovetz, W.R. (1989) *Eur. J. Pharmacol.* **170**, 157–166.
Schuman, E.M. & Madison, D.V. (1994) *Annu. Rev. Neurosci.* **17**, 153–183.
Siaud, P., Mekaouche, M., Balmefrezol, M., Givalois, L., Barbanel, G. & Assenmacher, I. (1994) *Neurosci. Lett.* **170**, 51–54.
Sortino, M.A., Aleppo, G., Scapagnini, U. & Canonico, P.L. (1994) *Endocrinology* **134**, 1782–1787.
Southam, E. & Garthwaite, J. (1993) *Neuropharmacology* **32**, 1267–1277.
Southam, E., East, S.J. & Garthwaite, J. (1991) *J. Neurochem.* **56**, 2072–2081.
Theodosis, D.T. & Poulain, D.A. (1993) *Neuroscience* **57**, 501–535.
Torres, G., Lee, S. and Rivier, C. (1993) *Mol. Cell. Neurosci.* **4**, 155–163.
van den Pol, A.N. & Trombley, P.Q. (1993) *J. Neurosci.* **13**, 2829–2836.
Villar, M.J., Ceccatelli, S., Ronnqvist, M. & Hokfelt, T. (1994) *Brain Res.* **644**, 273–281.
Vincent, S.R. (1994) *Prog. Neurobiol.* **42**, 129–160.
Vincent, S.R. & Kimura, H. (1992) *Neuroscience* **46**, 755–784.
Watanabe, A., Hamada, T., Shibata, S. & Watanabe, S. (1994) *Brain Res.* **646**, 161–164.
Yasin, S., Costa, A., Trainer, P., Windle, R., Forsling, M.L. & Grossman, A. (1993) *Endocrinology* **133**, 1466–1469.
Zorumski, C.F. & Izumi, Y. (1993) *Biochem. Pharmacol.* **46**, 777–785.
Zu Zhu, X. & Luo, L.G. (1992) *J. Neurochem.* **59**, 932–935.

CHAPTER 9

GLIAL CELLS AS NITRIC OXIDE SOURCES AND TARGETS

Sean Murphy, Dana M. Grzybicki* and Martha L. Simmons†

Departments of Pharmacology and Pathology, University of Iowa College of Medicine, Iowa City, IA 52242, USA and† Department of Pathology, The University of California at San Francisco, San Francisco, CA 94143, USA*

Table of Contents

9.1 Introduction 164
9.2 Evidence for a constitutive, calcium-activated nitric oxide synthase in astrocytes 165
9.3 Induction of NOS in glia *in vitro* 167
9.3.1 Microglia 167
9.3.2 Astrocytes 168
9.3.3 Gliomas 169
9.3.4 Muller cells 170
9.3.5 Mechanisms of NOS induction in glial cells 170
9.4 Regulation of the inducible NOS in glia 172
9.4.1 Regulation of NOS transcription and translation 174
9.4.2 Post-translational regulation of NOS 175
9.5 Induction of NOS *in vivo* 176
9.5.1 Viral infection 177
9.5.2 Toxoplasmosis 179
9.5.3 Autoimmune disease 179
9.5.4 Trauma 180
9.5.5 Ischaemia 180
9.5.6 Spinal hyperalgesia 182
9.5.7 Opioid tolerance 182
9.6 Glia as targets for NO 183
9.7 Summary 185

Acknowledgements 186
References 186

Nitric Oxide in the Nervous System
ISBN 0-12-721985-4

9.1 Introduction

Appreciation of glial biology and of the diverse roles these cells play *in vivo* is growing exponentially. Indeed, we should no longer divide these cells simply into 'microglia' and 'macroglia' (astrocytes and oligodendrocytes), for that belies the functional and spatial heterogeneity which is now so evident. However, for simplicity, we retain the convention here.

The anatomical location of astrocytes, intimately associated with cerebral vessels and with neurons, predicts important functional roles. These cells contribute not only to CNS development and response to injury, but also to normal cell–cell signalling. The demonstration that astrocytes *in vitro* can release neuro- and vasoactive molecules (peptides, eicosanoids) strengthens this prediction (see Murphy, 1993). Astrocytes are coupled *in situ* via gap junctions and communicate information through a glial network (Finkbeiner, 1993). Astrocytes display a wide variety of receptors for neuroligands (Hosli and Hosli, 1993), both *in vitro* and *in situ*. For a number of these receptors the details of the signalling pathways that they recruit have been worked out and we know something of the regulatory functions of these receptors. In addition, astrocytes respond to factors from infiltrating lymphocytes and macrophages, as well as activated resident microglia, and can release a variety of cytokines. This suggests a role for astrocytes in CNS–immune interactions and an involvement in neuropathologies (Benveniste, 1992).

Microglial lineage remains a controversial issue. The origin of ramified microglia has been described as a series of successive transformations of blood monocytes to amoeboid microglia, to process-bearing cells (Ling and Wong, 1993). In contrast, Lassman *et al.* (1993) maintain that parenchymal ramified microglia are not continuously replaced by bone marrow-derived elements. Richardson *et al.* (1993), based on *in vitro* studies, insist that there are microglial progenitor cells indistinguishable from astrocytes. However this question of lineage may be resolved, we can say that microglia express an array of monocyte-like surface antigens, that these cells become activated by cytokines and release a host of factors including cytokines and low molecular weight products which may be harmful or beneficial to surrounding cells (Giulian, 1993). Amongst these products is nitric oxide (NO; Banati *et al.*, 1993). More recently, it has become clear that microglia, like the macroglia, express a variety of receptors for conventional transmitters (for example, see Kettenmann *et al.*, 1993).

Signalling in the nervous system, therefore, is not restricted to neurons. All parenchymal cells can receive and transmit information, signalling between each other, with the vasculature, and with infiltrating cells. The release of NO may be one mechanism by which glia achieve this level of communication. In this chapter we describe the regulation of NO production in glia and discuss the potential effects of NO on glial cell function.

9.2 Evidence for a constitutive, calcium-activated nitric oxide synthase in astrocytes

The first studies of nitric oxide synthase (NOS) in the CNS, using a polyclonal antiserum for immunocytochemistry and oligonucleotides for *in situ* hybridization to constitutive cNOS mRNA, ascribed localization to discrete mammalian brain regions (Bredt *et al.*, 1991a). These findings suggested the distribution amongst parenchymal cells was selective and entirely neuronal, and that it coincided with the distribution of NADPH diaphorase (Bredt *et al.*, 1991b), an enzymatic activity now routinely taken (after appropriate aldehyde fixation) as a histochemical indicator of NOS (Vincent and Kimura, 1992). However, although all forms of NOS display NADPH diaphorase activity, the two are not necessarily coincident (Tracey *et al.*, 1993; Matsumoto *et al.*, 1993). A neuronal localization is further supported by the distribution of argininosuccinate synthetase activity (citrulline to arginine conversion) which is restricted to neurons (Arnt-Ramos *et al.*, 1992), and immunocytochemistry indicates that citrulline is only detected in neurons which are NADPH diaphorase positive (Pasqualloto *et al.*, 1991).

The immunocytochemical localization of arginine, however, reveals a predominantly astrocytic distribution (Aoki *et al.*, 1991). Though Murad's group confirmed the neuronal distribution of NOS with an antiserum raised against the enzyme from rat cerebellum (Schmidt *et al.*, 1992), labelling was not confined to neurons but was widely distributed over several non-neuronal cell types and tissues, including Bergmann glia and astrocytes. Furthermore, NADPH diaphorase was found to colocalize with cNOS in both neuronal and glial cells. An electron microscopy study (Wolf *et al.*, 1992) has revealed that, although highly concentrated in neurons, NADPH diaphorase is ubiquitously distributed in cells of the brain. A non-neuronal expression of NADPH diaphorase has also been reported in perivascular glia (Poeggel *et al.*, 1992).

Arising from our interest in astrocyte-derived vasorelaxant factors we asked the question whether secondary cultures highly enriched in astroglial cells could produce NO. Conditioned medium from astrocytes acutely stimulated with particular agonists caused relaxation in endothelium-denuded rings of coronary artery (Murphy *et al.*, 1990). Detection by chemiluminescence of NO in the conditioned medium, coupled with the observed sensitivity of the vasodilatation to NOS inhibitors (N-substituted arginine analogues) applied to the astrocytes, and to haemoglobin (Hb) added to the conditioned medium, suggested that astrocytes released either NO or an NO-containing factor. Shortly thereafter it was reported that an astrocytoma cell line released a similar factor which inhibited the aggregation of platelets (Mollace *et al.*, 1990). Recently, Minc-Golomb and Schwartz (1992, 1993) have used RNA-specific reverse transcription followed by polymerase chain reaction (RT-PCR) and *in situ* hybridization, with primers and probes directed against the constitutive NOS found in neurons, to demonstrate expression of mRNA transcripts in astrocytes cultured from the cerebral cortex, striatum and cerebellum. Additionally, the expression of an active constitutive

NOS enzyme could be confirmed by measurements of L-citrulline formation.

Using either chemiluminescence detection of NO or activation of soluble guanylyl cyclase (sGC), a number of agonists have been found which activate NOS in astrocytes. The most effective are noradrenaline (NA) and quisqualate. It has been calculated that 75% of the NA effect is mediated by α_1- and the rest by β-adrenoceptors (Agullo and Garcia, 1991, 1992a). The NA effect depends upon extracellular calcium and remains under conditions that abolish mobilization of intracellular calcium. This is supported by the effectiveness of calcium ionophores but not thapsigargin (which elevates intracellular calcium at the expense of stores), suggesting that intracellular calcium mobilization is insufficient (either in magnitude or location) to activate this NOS (Murphy *et al.*, 1991). As occurs for other α_1-receptor-mediated responses that involve calcium entry, the cGMP response is largely inhibited by pretreatment of astrocytes with either pertussis toxin or phorbol ester. The effect of quisqualate is mimicked by ibotenate but not D-L-α-amino-3-hydroxy-5-methyl-4-isoxalone propionate (AMPA), suggesting the involvement of a metabotropic glutamate receptor. The best responses to NA were found in astrocytes cultured from the cerebellum (3–4-fold that in the cerebrum), with intermediate responses in cultures from the hippocampus and hypothalamus (Garcia and Agullo, 1992). Glutamate responses were detected in hippocampal but not cerebellar astrocyte cultures. However, Kiedrowski *et al.* (1992), using both the citrulline assay and cGMP accumulation, were unable to find calcium ionophore activation of NOS in cultures of cerebellar astrocytes. A larger number of receptor agonists which raise intracellular calcium in astrocytes (McCarthy and Salm, 1991) do not activate this NOS (Murphy *et al.*, 1991; Agullo and Garcia, 1992b), suggesting heterogeneity of the astrocytes in culture in terms of NOS localization and receptor expression.

Such functional and localization studies suggest that NOS distribution in the brain parenchyma is not exclusively neuronal. There is no doubt that NOS expression is high in particular neurons, but the histochemical and immunocytochemical data indicate some level of expression in astrocytes. To date, there is no evidence that oligodendrocytes or microglia have a calcium-activated, constitutive NOS. The low level of expression of NOS in cultured astrocytes might imply that the enzyme is restricted to a subpopulation of cells. If so, then such heterogeneity might hold true *in vivo*. Given the immaturity of astrocytes when taken for culture, NOS expression might have some meaning in cell differentiation. As yet we do not know if the astrocyte constitutive NOS is identical to that found in neurons. The *in situ* hybridization studies might indicate a product of alternate splicing. Oguro *et al.* (1993) isolated and sequenced the entire region of cDNAs encoding constitutive NOS in the mouse and have demonstrated a variant formed by alternate splicing. The same structural diversity was also demonstrated in human cerebellum by PCR.

9.3 Induction of NOS in glia *in vitro*

9.3.1 Microglia

The inducible form of NOS was first described in macrophages. In contrast to the constitutive NOS, it is not present in the cell under basal conditions but needs to be induced with toxins or cytokines, a process which requires both transcription and translation. However, once it is present, it is continually active and does not require receptor-mediated, calcium/calmodulin-dependent activation (see Nathan, 1992, for review). Since microglia carry out many of the macrophage functions of the brain it is not surprising that they too express an inducible NOS.

Numerous reports have described microglial NOS activity *in vitro* (Boje and Arora, 1992; Chao *et al.*, 1992; Zielasek *et al.*, 1992; Banati *et al.*, 1993; Corradin *et al.*, 1993; Merrill *et al.*, 1993). As an indicator of NO production, most of these studies have measured the accumulation of nitrite in culture medium over a period of 18–24 h. If this accumulation is sensitive to L-arginine analogues known to be inhibitors of NOS, it most likely results from the oxidation of the NO produced. Due to the low sensitivity of this method it is more suited to measuring activity of the inducible enzyme. Since this NOS is continually active once induced, stable nitrite can accumulate over a period of hours to measurable levels.

Rat microglia in primary culture were found to produce nitrite in response to treatment with lipopolysaccharide (LPS) and/or interferon (IFN)-γ (Chao *et al.*, 1992; Zielasek *et al.*, 1992). Tumour necrosis factor (TNF)-α enhanced nitrite accumulation induced by IFN-γ but not that induced by LPS (Merrill *et al.*, 1993). Boje and Arora (1992) demonstrated the highest level of induction in rat microglia after treatment with a combination of LPS, TNF-α, interleukin (IL)-1β, IFN-γ and IL-6. Three clonal lines of mouse microglia were found to produce NO after treatment with combinations of either IFN-γ/TNF-α or IFN-γ/LPS and this induction was enhanced when the microglia phagocytosed zymogen particles (Corradin *et al.*, 1993).

Generally, NO production by microglia required transcription and translation and needed L-arginine and NADPH but not calcium and calmodulin. Indeed, the microglial form of the enzyme shared most of the characteristics of the macrophage NOS, including the ability to cause cytotoxicity when microglia are co-cultured with oligodendrocytes (Merrill *et al.*, 1993) and neurons (Chao *et al.*, 1992; Boje and Arora, 1992).

In contrast, Lee *et al.* (1993), using similar methods, were not able to demonstrate NO production in cultures of human foetal microglia in response to IL-1β, LPS, IFN-γ or TNF-α alone or in combination. This is consistent with the observation that NOS in human macrophages cannot be induced by the same stimuli that induce rodent macrophages. In fact evidence for a human NOS has been very difficult to obtain and induction seems to require a multitude of cytokines in combination (Geller *et al.*, 1993). Hence, although there is considerable *in vitro* evidence for a microglial NOS in rodent systems, *in vitro* evidence for a human microglial NOS is lacking.

9.3.2 Astrocytes

The first evidence for an inducible NOS in astrocytes *in vitro* came from studies in our laboratory of cGMP production by neonatal rat Type 1 astrocytes (Simmons and Murphy, 1992). These cells, when exposed to LPS, produced high levels of intracellular cGMP in a dose-dependent manner. This cGMP production could be blocked by inhibitors of NOS such as L-NAME or L-NMMA and was enhanced by the presence of L-arginine. The induction was blocked by cycloheximide and actinomycin D and required at least 4 h of exposure to LPS to become evident. These observations suggested that LPS caused *de novo* production of an NOS and that the resulting NO was activating the abundant sGC in the cells. Hence L-arginine-dependent cGMP production indicated transcriptional induction of NOS.

To characterize further the NOS enzyme we studied the ability of cytosolic preparations from LPS-treated astrocyte cultures to produce increases in cGMP in a fibroblast cell line (Simmons and Murphy, 1992). This method allowed demonstration of the calcium/calmodulin independence and NADPH dependence of the enzyme. Additionally, the ability of the astrocyte cytosol to activate sGC in fibroblasts was enhanced by the presence of superoxide dismutase (SOD), suggesting the involvement of an unstable mediator such as NO. Hence, LPS-treated astrocytes appeared to express an NOS activity similar to that found in macrophages and microglia.

These observations were quickly confirmed by Galea *et al.* (1992) who estimated NO production via both an assay for the L-citrulline co-product and by accumulation of nitrite. Again the enzyme activity was seen only after hours of exposure to LPS, was calcium/calmodulin independent, required NADPH and tetrahydrobiopterin (BH4), and was sensitive to the presence of L-arginine analogues.

Both groups recognized and addressed the confounding factor of contaminating microglia in the astrocyte cultures. Indeed both Boje and Arora (1992) and Zielasek *et al.* (1992) had looked at nitrite production by astrocyte cultures as well as microglial cultures and had concluded that the small activity seen in the astrocyte cultures was due simply to contaminating microglia. We addressed the problem by demonstrating that microglial contamination could be reduced from 7% to 1% with more than 50% NOS activity remaining in the more homogeneous cultures (Simmons and Murphy, 1992). Galea *et al.* (1992) studied pure microglial cultures and compared quantity of NO production on a per cell basis with the astrocyte cultures. They calculated that the 3–5% microglial contamination in their astrocyte cultures could not account for the amount of NOS activity measured in these cultures. Additionally, they were able to demonstrate NADPH diaphorase staining in a subset of astrocytes in their cultures. Astrocytes may produce less NO on a per cell basis than microglia, but since astrocytes far outnumber microglia in brain, their contribution could be substantial.

Cytokines can also influence induction of NOS in astrocyte cultures. Using the same L-arginine-dependent cGMP assay we found that astrocytes exposed to IFN-γ or IL-1β alone did not produce NO but when the two cytokines were applied together they represented a better stimulus than LPS (Simmons and Murphy, 1993). TNF-α had some activity on its own and this was enhanced by the presence of LPS (Galea *et al.*, 1992). Co-presentation of either IL-6 or IL-2 did not enhance or inhibit induction of NOS by other cytokines or by LPS. Dexamethasone, in contrast, strongly inhibited induction of NOS mRNA (Lin and Murphy, unpublished observations) and NOS activity by LPS and TNF-α, while not affecting induction by IFN-γ/IL-1β (Simmons and Murphy, 1993; Demerle-Pallardy *et al.*, 1993).

Cultured human astrocytes, in contrast to human microglia, have recently been found to express an inducible NOS (Lee *et al.*, 1993). These cells produce nitrite and become NADPH diaphorase positive in response to IL-1β in combination with either IFN-γ or TNF-α. Hence, at least three different cytokines can contribute to induction of NOS in cultured rat and human astrocytes.

Most recently, Galea *et al.* (1993) have cloned the astrocyte inducible NOS cDNA. They used a cDNA library from astrocytes exposed to LPS for 5 h and a probe which covered the well-conserved NADPH binding sites of the macrophage sequence. Degenerate primers to other highly conserved sequences and PCR were used to clone the upstream regions. They found that the rat astrocyte iNOS cDNA is 92% identical in sequence to the mouse macrophage sequence in the coding region and 81% identical in the untranslated regions. The deduced amino acid sequence is 93% identical to that of the mouse macrophage enzyme and most of the substitutions are conservative. Hence the macrophage and astrocyte enzymes in rodents are all but identical in sequence, although explanations for differences in the control of their expression may be found once the genetic sequences are known and promoter analysis carried out. A partial promoter analysis of the macrophage NOS has appeared (Xie *et al.*, 1993).

9.3.3 Gliomas

Many of the studies in astrocyte cultures cited above were also carried out using C6 glioma cells, a rat cell line known to express astrocytic properties. These cells were initially used to verify NOS induction in a homogeneous astrocyte-like population and their study was continued due to ease of handling and high expression of NOS. We (Simmons and Murphy, 1992, 1993) found that LPS and single cytokines (IL-1β, IFN-γ, TNF-α) were able to induce NOS in C6 cells to a much higher level than that found in primary or secondary astrocyte cultures. Vigne *et al.* (1993) also demonstrated NO-dependent cGMP production in C6 cells induced by IL-1β and TNF-α. Again dexamethasone was able to inhibit the induction by LPS and TNF-α but not that by IL-1β or IFN-γ (Simmons and Murphy, 1993; Demerle-Pallardy *et al.*, 1993). In contrast, Feinstein *et al.* (1994a) found that C6 cells could not be induced to express NOS by LPS alone but required a combination with one of the three cytokines. These variable findings may be due to differences in the passages of cells used or in the methods used to estimate NO production.

Cultured human astrocytoma cells also contain a macrophage-type NOS. LPS, IL-1β and TNF-α were all able to increase nitrite production and L-citrulline formation in T67 astrocytoma cells (Mollace *et al.*, 1993). These observations required at least 4 h to see and were again independent of calcium and calmodulin. Therefore, human astrocytoma cells appear to express an inducible NOS similar to that found in rodent and human astrocytes.

9.3.4 Muller cells

Cultured Muller cells, a specialized form of glia found only in the retina, have very recently been found to express an inducible NOS *in vitro*. Using nitrite accumulation, L-citrulline formation and detection of mRNA, Goureau *et al.* (1994) found that NOS in Muller cells was not induced by LPS, IFN-γ or TNF-α individually but could be induced by a combination of any two. Much higher levels of induction were seen when all three compounds were used together. Additionally, they found that the induction could be blocked by TGF-β, an effect which has been observed in a number of other cell types.

In summary, there is ample evidence *in vitro* that a variety of CNS glial cell types, in addition to glioma cells, can express an NOS in response to LPS and cytokines. As yet, the inducibility of oligodendrocytes to express NOS has not been demonstrated.

9.3.5 Mechanisms of NOS induction in glial cells

During the course of these studies it became clear to us that the mechanisms involved in the induction of NOS by LPS in comparison to IFN-γ/IL-1β might be quite different. In both astrocytes and C6 cells induction by the cytokines had a more rapid time course than that by LPS. Additionally, as previously mentioned, induction by LPS was inhibited by dexamethasone whereas induction by IFN-γ/IL-1β was not. Mollace *et al.* (1993) noticed differences in the time course of LPS and cytokine induction in their human astrocytoma cells as well but found that the LPS had a more rapid time course than IL-1β or TNF-α. LPS appears in both cell types to be acting through a different pathway than the cytokines. One explanation is that the LPS is causing release of a cytokine such as IL-1β or TNF-α from the astrocytes which then goes on to induce NOS more directly. Such LPS-induced cytokine production from astrocytes has been demonstrated previously (Chung and Beneviste, 1990; Fontana *et al.*, 1982).

We have looked at the effect of the presence of IL-1 neutralizing antibodies on NOS induction by LPS in cultured cells (Figure 1). There is a small but significant titre-dependent decrease in the level of induction by LPS in the presence of these antibodies. However, induction by IFN-γ/phorbol 12-myristate 13-acetate (PMA) was unaffected. Hence dexamethasone may serve to prevent LPS-stimulated release

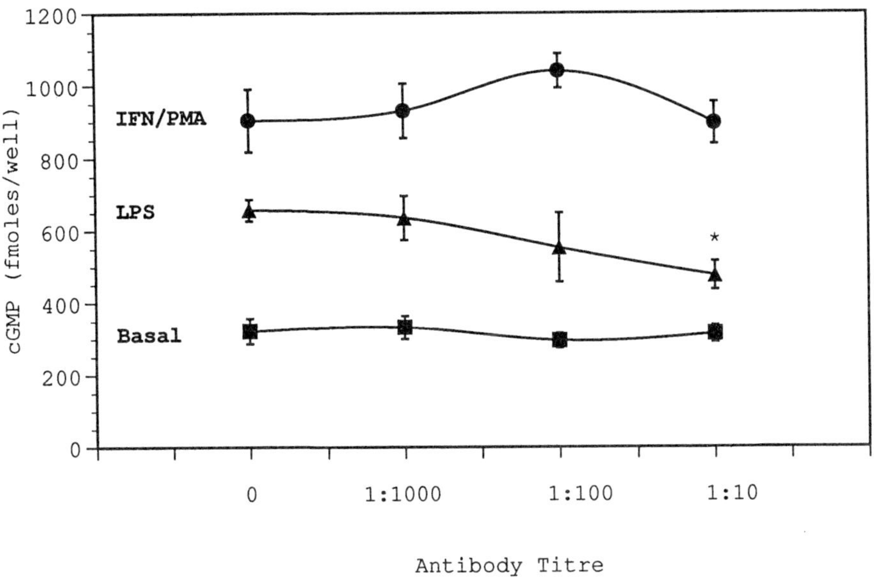

Figure 1 Role of IL-1 in astrocyte NOS induction by LPS. Neutralizing antibodies to rat IL-1α and IL-1β at various titres were added just prior to LPS (10 μg/ml) or PMA (0.1 μm)/IFN-γ (50 U/ml). After 8 h, NOS activity was assayed by cGMP. Values represent mean±s.e.m., n=4. *p<0.05 compared to cells with no antibodies added.

of IL-1 (or TNF-α) from the cultures and thus prevent induction. Similarly, Mollace *et al.* (1993) found that induction by LPS in the presence of antibodies directed against the type 2 IL-1β receptor reduced the level of NOS induction by over 25% in the human astrocytoma cells. They suggested that LPS in their cells is working at least partially through IL-1β.

The second messenger systems by which cytokines and LPS induce NOS in glial cells may also explain observed differences in response to these compounds. What these systems are for cytokine signalling is not known, but there is now evidence in glial cells for involvement of protein kinases in NOS induction by LPS and cytokines.

Simmons and Murphy (1994) found that the PKC activator PMA could only induce astrocyte NOS to a very small extent by itself. However, when combined with IFN-γ it produced induction as strong as the IFN-γ/IL-1β combination. Again we used L-arginine-dependent cGMP production as a bioassay for the NO production. Merrill *et al.* (1993) have shown that PKC activation by PMA can induce NOS in microglia. PKC has also been implicated in NOS induction in macrophages and hepatocytes through the use of PKC inhibitors (Hortelano *et al.*, 1992, 1993; Severn *et al.*, 1993). In contrast, we found in astrocytes that only the non-specific PKC inhibitor H-7 prevented induction by LPS and cytokines. Staurosporine and the more specific PKC inhibitor calphostin C had no effect, and phorbol ester-induced depletion of PKC in the cells also did not prevent induction. These observations suggest that while PKC could contribute to the induction of NOS in astrocytes, its activity is not essential for the induction by cytokines or LPS. Indeed, PMA is known to cause the release of TNF-α from astrocytes (Chung *et al.*, 1992) and in that way might activate NOS induction.

Tyrosine kinase has also been implicated recently in the induction of NOS by LPS, IL-1β and IFN-γ in islets of Langerhans, macrophages, retinal epithelial cells and smooth muscle cells (Dong *et al.*, 1993; Corbett *et al.*, 1993; Goureau *et al.*, 1993; Marczin *et al.*, 1993). We have found that the tyrosine kinase inhibitor genistein inhibits induction of NOS by LPS, IFN-γ/IL-1β, and IFN-γ/PMA in astrocyte cultures (Simmons and Murphy, 1994). However, a more specific tyrosine kinase inhibitor tyrphostin 47 inhibited only the induction by LPS. Feinstein *et al.* (1994b) have also found that genistein inhibits induction by LPS and TNF-α in both primary astrocyte cultures and C6 cells. They also reported that tyrphostin worked equally as well to supress induction by LPS or LPS/IFN-γ. These effects were shown not only with accumulation of nitrite but also in levels of NOS mRNA. In all these studies the tyrosine kinase inhibitors needed to be present prior to the inducing agents to have an effect. There is evidence for tyrosine kinase activation by LPS, IFN-γ and IL-1β in other cell types but exactly what type of tyrosine kinase is involved in the induction of NOS in glia remains to be determined.

In summary, although there is ample evidence for induction of NOS in glial cells *in vitro*, the mechanisms by which this induction occurs in any cell type are complex and have yet to be completely described. One trend throughout different cell types, but particularly in primary astroctyes, is that the highest level of induction is generally achieved by a combination of inducing agents. These inducing agents may employ a variety of different second messenger systems (Otero and Merrill, 1994) and the induction may be prevented by interference with these systems or by cotreatment with other physiological mediators (Figure 2). Such complex regulation of NOS expression allows very close control over the production of a highly reactive product.

9.4 Regulation of the inducible NOS in glia

Evidently, the activity of induced NOS in glia is retained for hours to days, at least *in vitro* where arginine is not limiting. Apart from the central question as to the conditions under which glia can be induced to express NOS *in vivo* (see Section 9.5), it is important to know what determines whether a glial cell can be induced, and if the expression of NOS can be regulated at the level of transcription, translation and/or via post-translational modification. In essence, are there mechanisms to prevent or limit glia from becoming sources of prolonged and potentially deleterious NO in the CNS?

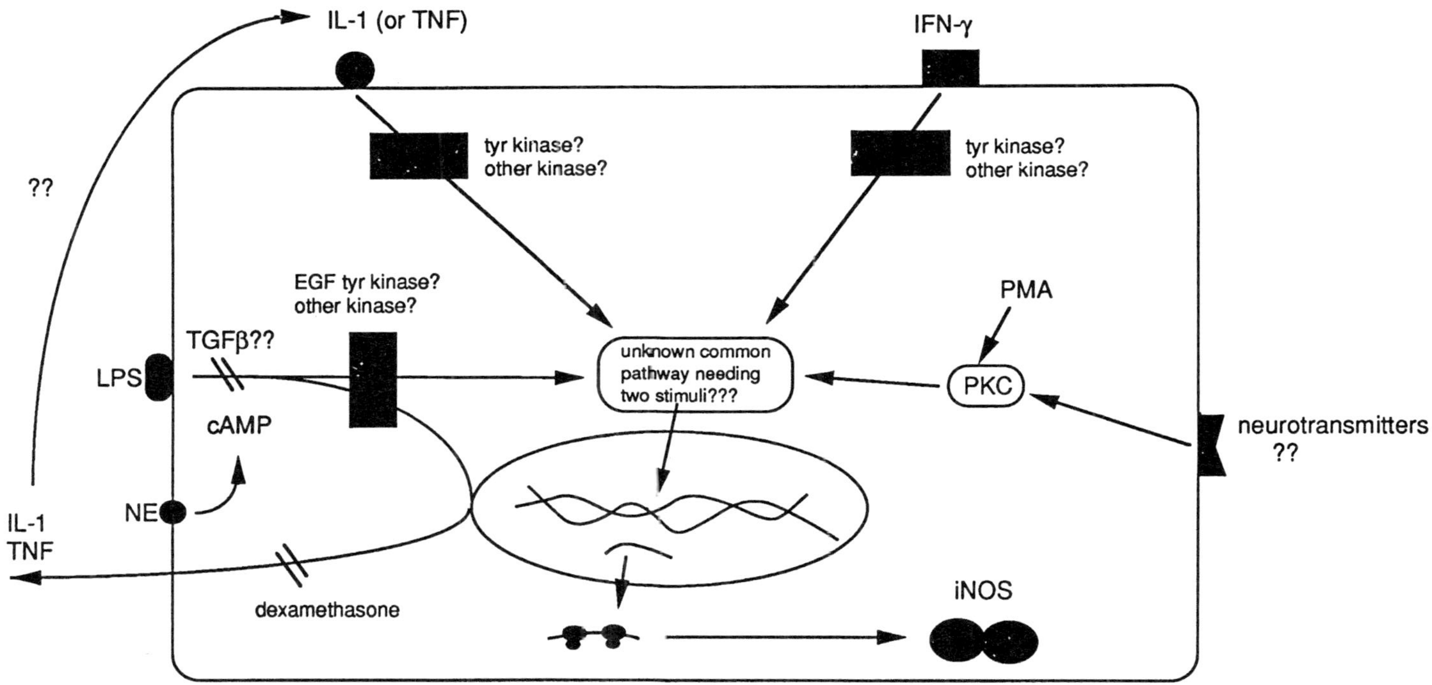

Figure 2 Mechanisms of induction of NOS in astrocytes *in vitro*. Maximal induction appears to require two stimuli (e.g. IFN/IL-1 or IFN/PMA). LPS may evoke induction of NOS via cytokines and the effects of LPS can be inhibited by dexamethasone, cAMP and possibly other cytokines such as TGF-β. See text for details.

9.4.1 Regulation of NOS transcription and translation

If astrocytes are exposed to NA, selective β-adrenergic receptor agonists, or dbcAMP, then they are much less responsive to endotoxin in terms of transcriptional induction of NOS (Feinstein *et al.*, 1993). As little as 100 pM NA is sufficient for as short a period as 5 min, and the effect is not directly on NOS catalytic activity. A similar effect has been reported in macrophages (Bulut *et al.*, 1993). Thus, there must be a step in the pathway of transcriptional induction of NOS which involves cAMP and probably protein kinase A (PKA). This could be either suppression of induction of activating cytokines (such as TNF-α), or the cAMP-dependent induction of an inactivating cytokine. One candidate could be IL-6 which inhibits LPS-induced TNF production (Aderka *et al.*, 1989), and which Norris and Benveniste (1993) have shown to be released from astrocytes upon exposure to NA. We have investigated whether protein synthesis (transcription factor or cytokine) is required for the induction of NOS mRNA. As seen in Figure 3, cycloheximide markedly inhibited (>60%) induction by various routes.

Deactivation by cytokines is an important mechanism of counterbalancing immunoregulation in macrophages. In astrocytes, related glioma cells, and microglia, the cytokines IL-4 and IL-10, together with TGF-β, and also dexamethasone, diminish the responsiveness of the cells to LPS in terms of NOS induction (Simmons and Murphy, 1993; Merrill *et al.*, 1993). Similar effects are seen in endothelial cell NOS induction (Schneemann *et al.*, 1993). The effects of deactivating cytokines in astrocytes are at the NOS transcriptional level and probably reflect a block on LPS-evoked transcription of NOS-inducing cytokines such as TNF-α.

Cells can be made refractory to cytokine-evoked transcriptional induction of NOS. Pretreatment of macrophages with low doses of LPS and TNF-α suppresses NOS activity and destabilizes mRNA following induction (Bogdan *et al.*, 1993; Zhang and Morrison, 1993). Endotoxin tolerance of cells has been suggested to be due partly to a decrease in TNF (Severn *et al.*, 1993), but this did not seem to be the case here (Bogdan *et al.*, 1993). Thus, preactivation of pathways normally contributing to synergistic induction of NOS may deplete cells of factors needed for its expression. This implies that regulation of NOS induction *in vivo* may depend on the relative timing of presentation of cytokines to cells.

It is clear that NO, via interaction with the haem moiety in NOS, regulates catalytic activity (Rogers and Ignarro, 1992; Buga *et al.*, 1993; Assreuy *et al.*, 1993; Rengasamy and Johns, 1993). However, we have found that in cells undergoing transcriptional induction of NOS, removal of the NO being formed (using haemoglobin or haematin) or prevention of its formation (e.g. with NAME) will amplify the transcription of NOS evoked by cytokines (Park, Lin and Murphy, 1994). This is clearly an effect of NO, as the protection by Hb can be reversed by a slow-releasing NO donor such as spermine NONOate. The site of NO action is transcriptional rather than on mRNA stability as the effect is sensitive to actinomycin D treatment. We have found a similar effect in endothelial cells where induction of NOS is inhibited if the cells are first exposed to NO, either from chemical or cellular

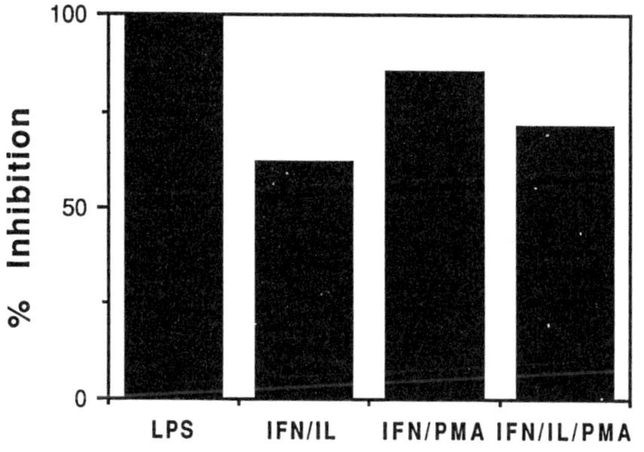

Figure 3 Inhibition of transcriptional induction of astrocyte NOS by cycloheximide. Treatment with cycloheximide (10 μg/ml) was for 60 min before addition of LPS (10 μg/ml), IFN-γ (50 U/ml), IL-1β (500 pg/ml), PMA (0.1 μM). After 5 h, total RNA was separated by gel electrophoresis, transferred to nylon membrane and NOS mRNA detected using a ^{32}P-labelled macrophage NOS cDNA probe (Xie *et al.*, 1992) and autoradiography. For quantitation, NOS mRNA bands were scanned by densitometry and compared with non-cycloheximide-treated cultures run in parallel.

(astrocyte) donors (Borgerding and Murphy, in preparation). The regulation of NOS transcriptional induction by NO is a good way of limiting a response *in vivo* to just a few cells.

As yet we do not know of conditions which may shorten or lengthen NOS mRNA life in glia. Using endothelial cells, Yoshizumi *et al.* (1993) have shown that the constitutive NOS mRNA is normally very stable (half-life of 48 h) but that in the presence of TNF-α this decreases to 3 h, an effect which is dependent on protein synthesis. This could explain an often seen event which is that induction of NOS in cells correlates with a decrease (via NO formed?) in the expression of constitutive NOS (Lamas *et al.*, 1992; MacNaul and Hutchinson, 1993). The functional benefit of such a reciprocal relationship between inducible and constitutive NOS remains obscure.

9.4.2 Post-translational regulation of NOS

An amino acid sequence for the inducible form of NOS in rat astroglial cells has been proposed based on cloning studies (Galea *et al.*, 1993) and it is 92% identical with that in rat macrophages and smooth muscle cells. While we do not know the sequences for constitutive NOS in glia, nor for the inducible NOS in microglia, we assume these are similar to that for neuronal and macrophage NOS, respectively. From the predicted amino acid sequence and from biochemical studies, it is clear that NOS can be phosphorylated by protein kinases. Both the constitutive and inducible forms of NOS display a weak phosphorylation sequence for PKA. In the former it is in the N-terminal region while in the latter it is in the consensus binding site for NADPH (Lowenstein *et al.*, 1992). Brune and Lapetina (1991) showed phosphorylation of the neuronal NOS by PKA, but not by PKC or CaM kinase. However, there was no change in enzyme activity, and they suggest that phosphorylation might affect compartmentation, association with other proteins, or turnover. However, Nakane *et al.* (1991) described phosphorylation of this NOS by CaM kinase II and PKC, and found the former resulted in marked inhibition while the latter in a moderate increase in NOS activity. Bredt *et al.* (1992), using a stably transfected cell line, showed that PKA, PKC and CaM kinase II all phosphorylated NOS but on different serine sites. Only PKC activation caused a marked inhibition in NOS activity. Later work from the same laboratory suggested that all four kinases (PKA, PKG, PKA and CaM kinase) inhibit the cNOS (Dawson *et al.*, 1993). From the work of Feinstein *et al.* (1993) it does not seem that PKA affects cNOS catalytic activity in astrocytes. As yet we do not know about potential regulatory effects of other protein kinases on NOS in glia. If there are any, then it will be important to discover whether activation of these kinases through receptors modulates NOS once induced.

As mentioned above there is feedback inhibition on NOS catalytic activity by NO (but not by oxidation products such as nitrite, nitrate, peroxynitrite) which serves not only to modulate NO production in the cell of origin but also in near-neighbour cells. The suggestion is that NO binds to the haem moiety in NOS and interferes with electron transport and substrate oxygenation (Rogers and Ignarro, 1992; Buga *et al.*, 1993). The implications of this form of regulation are numerous. Where NO is acting as a transmitter, it would limit the duration of signal production. In the presence of superoxide anion NO combines to form peroxynitrite (see Chapter 10), thus blocking the negative feedback by NO on NOS. This could be an important amplification mechanism in the response to invading pathogens. Alternatively, it may represent an uncontrolled 'wind-up' with disastrous consequences for CNS integrity.

Availability of substrate and cofactors are again factors which would limit NOS activity. While it is unlikely that arginine is limiting for the constitutive NOS because this is only transiently turned on, it is very possible that arginine availability would determine the duration of NO production by the induced, tonically active NOS (for discussion, see Stuehr and Griffith, 1992). The synthesis of BH4, an essential factor for NOS activity, appears to be highly regulated in macrophages by activating and deactivating cytokines and, interestingly, much of it is released from the cell (Schoedon *et al.*, 1993).

9.5 Induction of NOS *in vivo*

The conditions under which a specific cell type is induced to express NOS seem to vary; likewise, how much NOS activity is expressed after induction also depends on the specific cell type expressing the enzyme. However, common to all cells presently identified which express inducible NOS is the requirement for exposure to specific cytokines and/or LPS. Since cytokines and LPS are not present in the normal brain in the quantities that cause induction of the enzyme *in vitro*, studies looking for NOS expression in the CNS, and specifically in glial cells, are based on the assumption that NOS expression will occur *in vivo* only under pathological conditions such as infection or injury, when inducing molecules such as IL-1 are expressed in higher concentrations than in normal brain. Evidence for the transcriptional induction of NOS in the CNS *in vivo* has only recently been sought and is provided by models of acute CNS infection, autoimmune reactions, ischaemia, traumatic injury, spinal hyperalgesia, and opioid tolerance (see Table 1). In a number of these studies, especially those using inhibitors of NOS, it is not clear whether the NOS induced is of the calcium-independent or calcium-dependent type.

9.5.1 Viral infection

During acute and chronic viral infection in the CNS, a variety of cytokines are produced in increased amounts, including IFN-γ and IL-1 (Frei *et al.*, 1988; Benveniste, 1992; Schneider-Schaulies *et al.*, 1993). Such a CNS environment containing increased concentrations of cytokines would be conducive to NOS induction. Additionally, a possible cytotoxic role for NO under such conditions has been suggested by the fact that macrophages expressing NOS and producing NO have been shown to be cytotoxic for tumour cells (Stuehr and Nathan, 1989) and for cells infected with intracellular pathogens (Adams *et al.*, 1990). Recently, murine macrophages induced with IFN-γ to express NOS have been shown to inhibit replication of ectromelia, vaccinia, and herpes simplex-1 viruses *in vitro* (Karupiah *et al.*, 1993). *In vivo*, inhibition of NOS converted resolving ectromelia virus infection into clinically fulminant mousepox. Since astrocytes and microglia can be induced to express NOS these cells could play a toxic role *in vivo* during acute viral infection. This toxicity could result in promoting the clearance of virus from infected cells and/or contribute to tissue damage through cytotoxic effects on uninfected neighbour cells.

Using RT-PCR and Southern blot analyses, inducible NOS mRNA has been demonstrated in the brains of rats infected with borna disease and rabies virus, and in the brains of mice with herpes simplex virus, either before or coincident with the onset of clinical symptoms (Koprowski *et al.*, 1993). No message for inducible NOS could be detected in control animals, supporting the idea that it is not normally expressed in the CNS. The primers used for PCR in these studies were based on the murine macrophage cDNA sequence which is nearly identical to the astrocyte sequence. Although a microglial NOS has not yet been cloned, the *in vitro* activity is so similar that cross-hybridization of the primers with microglial message seems likely. Therefore the NOS mRNA found in these *in vivo* studies could arise from invading macrophages or from resident glia.

This same group has recently extended their studies on borna disease virus-infected rats to time course RT-PCR and immunohistochemical studies, and to *in*

Table 1 Induction of NOS *in vivo*

Pathologic model	Animal model	Method	Reference
Acute viral infection			
Rabies*	Rat	RT/PCR	Koprowski *et al.* (1993)
Herpes simplex*	Mouse	RT/PCR	Koprowski *et al.* (1993)
Borna disease	Rat	RT/PCR, immunocyt. *In situ* hybrid.	Zheng *et al.* (1993)
Toxoplasmosis*	Mouse	RT/PCR	Gazzinelli *et al.* (1993)
Autoimmune disease* (EAE)	Rat	RT/PCR	Koprowski *et al.* (1993)
Trauma			
Stab wound	Mouse, rat	NADPHd	Wallace and Fredens (1992) Regidor *et al.* (1993)
Ventral root avulsion†	Rat	NADPHd, immunocyt. *In situ* hybrid.	Wu *et al.* (1993)
Spinal hyperalgesia*†	Rat	NOS inhibition	Meller *et al.* (1992)
Opioid tolerance*†	Mouse	NOS inhibition	Kolesnikov *et al.* (1993) Thorat *et al.* (1993)
Ischaemia†	Rat, mouse, cat	NOS inhibition	Buisson *et al.* (1993)* Nowicki *et al.* (1991)* Nishikawa *et al.* (1993)* Fujisawa *et al.* (1993)* Yamamoto *et al.* (1992)* Kuluz *et al.* (1993)* Wallace and Fredens (1992) Endoh *et al.* (1993a,b)

* Cell type expressing NOS not identified.
† In these studies the isoform of NOS is uncertain.

situ hybridization (Zheng *et al.*, 1993). The PCR data show that NOS mRNA peaks when neurological signs and inflammatory reactions are also at a peak. Film autoradiograms of *in situ* hybridization using a specific probe for inducible NOS 21 days post-infection show that NOS mRNA was detected in only some of the acutely virus-infected regions of the brain, and no message was detected in the brains of rats with chronic infection. Immunohistochemical studies revealed a variable intensity of NOS protein in only a percentage of cells that also stained for a macrophage marker.

Both of these studies demonstrate that inducible NOS message is expressed *in vivo* in the brain during acute viral infection. Liquid emulsion autoradiograms of *in situ* hybridizaton for NOS mRNA are needed to identify definitively the cell types responsible. While the borna studies seem to point to infiltrating macrophages, expression by glia cannot be ruled out. Film autoradiography was performed at a single time point late in infection and single cell resolution is not possible with this method. The immunohistochemistry data also suggest NOS protein production only in macrophages, but the authors do not indicate the time point in the disease process at which these studies were done, and no additional cell type-specific labels were used. Thus, while evidence exists for NOS induction during some acute viral infections in the brain, definitive information regarding cellular source(s) is still lacking.

9.5.2 Toxoplasmosis

Evidence for transcriptional induction of NOS *in vivo* in another CNS infection, acute cerebral toxoplasmosis, has recently been provided by Gazzinelli *et al.* (1993). Using RT-PCR, these authors demonstrated the up-regulation of NOS mRNA from 2 to 4 weeks after infection with *Toxoplasma gondii.* This declined at 8 weeks and correlated with enhanced pathology. The authors suggest a possible toxic role for NO, since the decline of mRNA was associated with progression of disease. Although it is not stated from which cDNA their PCR primers were designed, NOS expression in this study is taken as a marker of macrophage activation. Cross-hybridization with a glial NOS mRNA cannot be ruled out and *in situ* hybridization studies will identify the cellular location of NOS *in vivo.*

9.5.3 Autoimmune disease

The studies by Koprowski *et al.* (1993) also included RT-PCR data for NOS mRNA expression during experimental allergic encephalitis (EAE), an autoimmune demyelinating process considered a model for human multiple sclerosis (MS). As in acute viral infection, brains of animals with EAE have been shown to express increased levels of cytokines (Benveniste, 1992). Additionally, cytokine-activated microglial cytotoxicity for oligodendrocytes *in vitro* has recently been shown to be mediated by NO (Merrill *et al.*, 1993). Therefore, a glial source and possible cytotoxic role for NO produced by inducible NOS during CNS demyelination *in vivo* is suggested. Koprowski *et al.* (1993) found increased NOS expression in rats developing EAE which preceded detectable clinical signs and symptoms, and this remained present after animals had recovered. To date, only PCR studies have been reported. Again, *in situ* hybridization studies are needed to define the cell type(s) expressing inducible NOS under conditions of autoimmune reaction in the CNS.

9.5.4 Trauma

Akin to inflammatory processes in the CNS, traumatic injury to the brain will result in an increased expression of cytokines such as IL-1 by resident cells (and infiltrating macrophages), creating an environment conducive for induction of NOS *in vivo*. Two models of traumatic injury that have been reported to result in NOS expression *in vivo* are cerebral stab wound and ventral spinal root avulsion.

Wallace and Fredens (1992) utilized NADPH diaphorase staining in aldehyde-fixed tissue as a marker for the potential presence of NOS. In this study, stereotaxic injections of kainic acid or saline were made into the mouse retrosplenial cortex. Mice were then sacrificed after 5 or 6 days. In three of five sham-operated controls, activated astrocytes surrounding the needle track displayed an NADPH diaphorase reaction, and this might suggest NOS expression. No NADPH diaphorase-positive astrocytes were found deeper in the hippocampus at the site of saline injection, suggesting that the tissue damage resulting from the stab wound itself caused induction of astrocyte NADPH diaphorase. These findings have been reproduced by Regidor *et al.* (1993) who injected cysteamine or artificial cerebrospinal fluid into the cerebral cortex, hippocampus and neostriatum of rats. At time points ranging from 1 to 21 days post-injection, reactive astrocytes around the needle track in all three structures of all animals displayed NADPH diaphorase activity.

Similar studies of NOS induction have been made after avulsion injury to motoneurons in the rat spinal cord (Wu *et al.*, 1993). There is an induction of NOS in astrocytes surrounding injured motoneurons 6–8 weeks post-surgery. This occurs at a time of maximal motoneuron loss after injury and the authors suggest that glial NO may be cytotoxic for motoneurons.

The results from these studies demonstrate that direct injury to the cortex and to deeper structures can induce glia to express NADPH diaphorase. While NOS is an NADPH diaphorase, not all NADPH diaphorase activity in a cell results from NOS (Tracey *et al.*, 1993). Therefore, although NADPH diaphorase inducibility in astrocytes suggests possible induction of NOS *in vivo*, *in situ* hybridization studies using specific probes are needed. A glial scar forms at the site of injury after a stab wound to the CNS. One possible role for glial NO production via NOS at a site of injury may be to regulate blood flow to the area during wound repair.

9.5.5 Ischaemia

Excitotoxic injury in the CNS is thought to underlie the pathologic changes occurring with ischaemia and seizure activity. In light of the *in vitro* work linking NO production by constitutive NOS to excitatory amino acid neurotoxicity, multiple studies in the last few years have attempted to demonstrate increased NO production in the brain during ischaemia, and have investigated the effects of NOS inhibitors in animal models of ischaemia *in vivo*. Three different groups, utilizing different methods, have demonstrated that NO levels do increase in rat brain after focal ischaemia (Kader *et al.*, 1993; Malinski *et al.*, 1993; Tominaga *et al.*, 1993). The timing of the peak measurements in these studies (minutes) suggests that the increase in NO is from a constitutive form of NOS in resident cells of the CNS.

Many other studies, demonstrating the effects of NOS inhibition on lesion formation, have utilized methods in which lesion size is evaluated hours or days after the ischaemic event. In these cases, it is not clear which form(s) of the enzyme is responsible. Ischaemia produces an increase in cytokines such as IL-1β (Liu *et al.*, 1993). Therefore, the inducible form of the enzyme in glia might contribute to the observed changes. The majority of these studies have revealed that inhibition of NOS is protective during ischaemia, resulting in a decrease in lesion size (e.g. Nowicki *et al.*, 1991; Buisson *et al.*, 1993).

In addition, a recent study in which the effect of L-NAME inhibition on cortical excitotoxic injuries produced by direct delivery of glutamate reveals that L-NAME is protective, reducing lesion volume by 30% (Fujisawa *et al.*, 1993). Rats in this study were sacrificed 4 h after the beginning of glutamate injection, making it possible that inhibition of glial NOS could be involved in the effects.

However, two inhibition studies have produced disparate results from those above. Both Yamamoto *et al.* (1993) and Kuluz *et al.* (1993) have shown that NOS inhibition in models of focal ischaemia results in an increase in lesion size. The timing of both of these studies leaves open the possibility that inhibition of an induced NOS may be involved in the resulting pathology.

The disparate results concerning the neurotoxic versus neuroprotective effects of NO in ischaemia could be due to the many physiological variables inherent in different animal models of ischaemia (Ginsberg and Busto, 1989), variations in different methods used, the different physiological effects of NOS inhibitors besides NOS inhibition (Fujisawa *et al.*, 1993), and the fact that NO itself has multiple physiologic and potentially pathologic effects. Despite the contradictory outcomes of the inhibition studies, they do suggest that production of NO *in vivo* is involved in some way in the pathology occurring with ischaemia. Sustained production via a glial NOS could be one source of this NO.

The induction of NADPH diaphorase and NOS protein demonstrates more specifically glial NOS expression *in vivo* after ischaemic insult. Wallace and Fredens (1992) found that retrosplenial cortical injection of the potent neurotoxin kainic acid in mice resulted in the expression of hippocampal astrocyte NADPH diaphorase bilaterally 5 or 6 days post-injury. This NADPH diaphorase staining is only suggestive of glial NOS activity *in vivo* induced locally and distantly in areas of excitotoxic injury. More definitive evidence is provided by Endoh *et al.* (1993a,b) who have produced transient global ischaemia in rats and looked at both NADPH diaphorase staining and immunohistochemistry using an antibody to macrophage

NOS. These studies show NADPH diaphorase and NOS immunoreactivity in reactive hippocampal astrocytes bilaterally, starting at 1 day post-ischaemia. Double staining for glial fibrillary acidic protein has confirmed the identity of these positive cells as astrocytes.

These recent studies directly demonstrate glial expression of NOS *in vivo* after ischaemia. However, global and focal ischaemia vary in their pathological effects on the CNS (Choi, 1993). Therefore, results obtained with one form of ischaemia cannot necessarily be extrapolated to the other. *In situ* hybridization studies with specific probes and immunohistochemistry are needed to confirm glial expression. Possible roles of glial NO production during ischaemia are many, some potentially protective and some potentially toxic. For example, glial NO could significantly reduce platelet aggregation and endothelial adhesion in local vessels, thereby helping to prevent further vascular occlusion and ischaemia. However, local vascular dilatation by NO could promote continued oxygen delivery to the area, resulting in oxygen radical formation and cytotoxicity.

9.5.6 Spinal hyperalgesia

Some forms of nociceptive processing in the spinal cord, specifically mechanisms underlying the pathology of thermal hyperalgesia, have been shown to involve NMDA receptor activation followed by NO production via constitutive NOS and production of cGMP in target cells (Meller and Gebhart, 1993). This facilitation of thermal reflexes involving NOS is acute in nature. Recently, a rat model of neuropathic pain has been described which may be mediated by the sustained production of NO (Meller *et al.*, 1992). Ligation of the sciatic nerve in rats resulted in a marked thermal hyperalgesia 3 days post-surgery. Intrathecal injection of the NOS inhibitor L-NAME or the NOS and soluble guanylyl cyclase inhibitor methylene blue, blocked the thermal hyperalgesia for 2 and 4 h, respectively.

These results suggest involvement of an inducible form of NOS *in vivo* which could be produced by glial cells in the spinal cord. Studies directly demonstrating the presence of inducible NOS mRNA and protein *in vivo* are needed in this model to confirm whether glial iNOS is present during the production of thermal hyperalgesia or other forms of hyperalgesia.

9.5.7 Opioid tolerance

Related to studies of nociceptive processing are those implicating NO production in the mechanisms of opioid antinociception and opioid tolerance. Two recent reports show possible NO effects on opioid antinociception; however, the time courses of both of these studies, which report disparate results, suggest that if NO is involved in opioid antinociception, it is produced by a constitutive NOS (Xu and Tseng, 1993; Przewlocki *et al.*, 1993). This is in contrast to the studies on opioid tolerance, where the time courses suggest that a sustained production of NO by inducible NOS may have effects on development of tolerance (Thorat *et al.*, 1993;

Kolesnikov *et al.*, 1993). Similar to the recent studies regarding opioid antinociception, the tolerance reports are disparate regarding the effects of NOS inhibition *in vivo* on *k*-opioid tolerance in mice, but they do suggest that NO production *in vivo* is involved in the development of some opioid tolerances. This NO may be produced in glia, and further inhibition, *in situ* hybridization, and immunohistochemical studies will support or reject this possibility.

In summary, studies from the last several years are beginning to produce evidence that NOS is induced *in vivo* in glial cells under certain pathologic conditions. The most convincing evidence to date is provided by immunohistochemistry, demonstrating the appearance of NOS-positive astrocytes 1 day after transient, global ischaemia in rats. Evidence in models of traumatic injury are highly suggestive but not definitive, due to the possible non-specificity of NADPH diaphorase staining. The models of acute encephalitis and demyelinating disease convincingly demonstrate NOS expression *in vivo*, but the cell type responsible is still incompletely defined. Finally, the production of NO in certain hyperalgesia models, and in models of opioid tolerance, are only suggestive. What we lack thus far are *in situ* hybridization and immunohistochemical studies to demonstrate glial expression of NOS message and protein *in vivo* in pathologic tissues. Only through such demonstration will clues regarding the cytotoxic and/or cytoprotective role(s) of glial NO *in vivo* be obtained.

9.6 Glia as targets for NO

The primary chemical reactions of NO that have biological significance are those with redox metals to form NO–metal complexes, and reaction with superoxide (O_2^-) to form peroxynitrite ($ONOO^-$). The latter decomposes rapidly to OH^-, a powerful oxidant that reacts indiscriminately.

The relevance of the first reaction is probably the best known. The most common redox metal involved is iron, either in haem or iron–sulphur centres. Hence, NO binds to the haem iron in soluble guanylyl cyclase to activate the enzyme and elevate cGMP, and this then can result in a myriad of effects (Schmidt *et al.*, 1993). Astrocytes show elevated cGMP in response to NO (De Vente and Steinbusch, 1992) and yet little is known of the consequences of this cyclic nucleotide specifically in these cells. In other cells types, cGMP regulates specific ion channels, phosphodiesterases (and hence the level of cellular cAMP) and specific protein kinases (Lincoln and Cornwell, 1993; see Chapter 4). Recently, Billiar *et al.* (1993) demonstrated that a significant amount of cGMP was released from hepatocytes which had been induced to express NOS. In our studies (Simmons and Murphy, unpublished) induced astrocytes extrude cGMP to a significant level (60% of the total cGMP produced). Whether this has any functional significance in the CNS is not clear.

Recently, cGMP has been shown to mobilize intracellular calcium by stimulating cyclic ADP ribose synthesis which in turn activates calcium-induced calcium release via the ryanodine receptor (Galione *et al.*, 1993). The existence of ryanodine-sensitive pools in glia is not clear (Charles *et al.*, 1993) but the resulting rise in intracellular calcium in glial cells could initiate events ranging from enzyme activation to cytoskeletal rearrangements (Finkbeiner, 1993). Peunova and Enikolopov (1993) report that, in PCI2 cells, NO acts as an amplifier of calcium signals with subsequent PKA-dependent activation of the transcription factor CREB. This suggests that NO could potentiate phosphorylation of proteins by PKA. Calcium and NO have to act within a very narrow time window for this enhancement to occur. Thus, small calcium transients in glia generated by receptor activation (Finkbeiner, 1993) would then become functionally significant in the presence of NO. Such mechanisms could explain why NO appears to modulate cytokine release in some cell types. While NO suppresses LPS-stimulated TNF synthesis in monocytes, NO enhances IL-1-induced TNF synthesis (Eigler *et al.*, 1993). While glia are proven sources of particular cytokines (Benveniste, 1992) it is not known whether their expression is regulated by NO.

It is suggested that NO is a tonic inhibitor of cell proliferation *in vivo*, and NO does inhibit DNA synthesis in fibroblasts (Firnhaber and Murphy, 1993) and mitogenesis in cerebellar glia (Garg *et al.*, 1992). In a study of C6 glioma cells, their PKG-dependent proliferation in response to TNF-α and IFN-γ did not occur if NOS activity was blocked (Munoz-Fernandez and Fresno, 1993). Considering the relatively high levels of NO produced in gliomas after induction by cytokines it is possible that some gliomas may limit their own growth, or that of surrounding reactive astrocytes, via production of NO. Alternatively, enough NO could be produced from astrocytomas to be toxic to surrounding neurons and oligodendrocytes.

Cyclo-oxygenase and lipoxygenase enzymes, important in the production of eicosanoids, both have an iron haem centre at their active sites. The effects of NO on these enzymes are unclear. Salvemini *et al.* (1993) find that NO enhances cyclo-oxygenase activity via a mechanism that is cGMP independent. Indeed, treatment of astrocytes with stable cGMP analogues inhibits prostanoid release (Lin and Murphy, unpublished observations). However, Kanner *et al.* (1992) suggest that NO has an inhibitory effect on lipoxygenase and cyclo-oxygenase. As microglia and astrocytes are sources of a variety of eicosanoids (primary prostaglandins, thromboxane, leukotrienes, hydroxyeicosatetraenoic acids), then activation or inhibition of their synthesis could have various effects on vasculature, synaptic transmission and on energy supply (Schaad *et al.*, 1991).

Likewise, NO can bind to and inhibit such enzymes as aconitase (Liew and Cox, 1991; Welsh and Sandler, 1992) and other P-450 enzymes (Wink *et al.*, 1993). Defects in mitochondrial energy metabolism might underlie the pathology of neurodegenerative diseases such as Parkinson's and Alzheimer's (Coyle and Puttfarcken, 1993). Components of the respiratory chain susceptible to damage by free radicals (Zhang *et al.*, 1990) are thought to explain the tumour cytotoxicity caused by activated macrophages (Hibbs *et al.*, 1988; Stuehr and Nathan, 1989). Bolanos *et al.* (1994) assessed the activities of the mitochondrial respiratory chain components in astrocytes in which NOS had been induced. Cytochrome *c* oxidase and succinate–cytochrome *c* reductase activities were inhibited but not NADH-ubiquinone-1 reductase or citrate synthase. There was no effect on the survival of the cells as there was a switch to anaerobic glycolysis. Mitrovic *et al.* (1994) have looked at the effects of NO donors on mitochondria in glial cells. Mitochondrial damage was evident in all three glial cell types. Astrocytes and oligodendrocytes suffered a decrease in activity of the iron–sulphur-containing succinate dehydrogenase while microglia were not much affected. The activity of isocitrate dehydrogenase, which has no iron–sulphur centre and should not be susceptible to NO, was unaffected. A large proportion of oligodendrocytes (44%) were destroyed by NO but only 20% of the astrocytes and <5% of microglia, and single-strand DNA breaks were found only in oligodendrocytes. It is important to determine how microglia and astrocytes protect themselves from their own NO. Mitrovic *et al.* (1994) suggest this could result either from the oxidation of NO or the coupling of NO with iron–sulphur-containing proteins. The activity of NOS produces O_2^- and/or H_2O_2 in addition to NO (Heinzel *et al.*, 1992; Pou *et al.*, 1992). The formation of peroxynitrite after combination with superoxide ion is most important in terms of the potential toxicity of NO. In a series of studies, Beckman and colleagues (Beckman *et al.*, 1990; Radi *et al.*, 1991a,b; Ischiropoulous *et al.*, 1992) suggest that peroxynitrite is a key player in the toxicity resulting from NO production (see Chapter 10).

There is evidence that NO production from induced microglia is capable of causing toxicity to neurons (Boje and Arora, 1992; Chao *et al.*, 1992) and oligodendrocytes (Merrill *et al.*, 1993), but similar toxicity was not caused by astrocytes in these studies. The sheer number of astrocytes versus microglia (the latter make up only 10% of the glial population), and the intimate contacts astrocytes make with both neurons and oligodendrocytes, may permit even low amounts of NO produced by astrocytes to pose a potential threat to adjacent cells. Such toxicity might not be seen in monolayers of co-cultured cells which do not display the close *in situ* interactions.

9.7 Summary

Astrocytes *in vitro* express constitutive NOS activity which is activated via a discrete group of receptors. Whether, *in vivo*, these cells can be activated by signal molecules from adjacent active neurons to contribute NO remains to be seen. The location of their processes around synapses and against the cerebrovascular wall places astrocytes in a perfect position to modulate synaptic transmission and vessel dynamics through NO release.

Both astrocytes and microglial cells can be induced by cytokines *in vitro* to express an isoform of NOS. The mRNA is short-lived (hours) but the amount of NO released is potentially high. Regulation of the transcriptional induction of this NOS, and post-transcriptional control of enzyme activity are beginning to be understood.

There is clearly induction of NOS in the CNS *in vivo* in a variety of pathological states. However, the stimuli that provoke transcriptional induction, the cell types involved (resident or infiltrating cells), and the precise isoform of NOS being induced, all await a more complete description. Finally, determining what role(s) glial-derived NO has in the CNS, either in the progress of such pathology or to recovery, promises an interesting future.

Acknowledgements

We recognize the continuing contributions of Song kyu Park, Rose Borgerding, Sherry Kardos and Hsin-Lee Lin to the work from this laboratory. We thank Jean Merrill for providing preprints, the NIH for programme and project grant support (to S.M.) and for training grant support (D.M.G.), and the Life and Health Insurance Medical Research Fund (M.L.S.).

References

Adams, L.B., Hibbs, J.B., Taintor, R.R. & Krahenbuhl, J.L. (1990) *J. Immunol.* **144**, 2725–2729.

Aderka, D., Le, J. & Vilcek, J. (1989) *J. Immunol.* **143**, 3517–3523.

Agullo, L. & Garcia, A. (1991) *Eur. J. Pharmacol.* **206**, 343–346.

Agullo, L. & Garcia, A. (1992a) *Biochem. J.* **288**, 619–624.

Agullo, L. & Garcia, A. (1992b) *Biochem. Biophys. Res. Commun.* **182**, 1362–1368.

Aoki, E., Semba, R., Mikoshiba, K. & Kashiwamata, S. (1991) *Brain Res.* **547**, 190–192.

Arnt-Ramos, L.R., O'Brien, W.E. & Vincent, S.R. (1992) *Neuroscience* **51**, 773–789.

Assreuy, J., Cunha, F.Q., Liew, F.Y. & Moncada, S. (1993) *Br. J. Pharmacol.* **108**, 833–837.

Banati, R.B., Gehrmann, J., Schubert, P. & Kreutzberg, G.W. (1993) *Glia* **7**, 111–118.

Beckman, J.S., Beckman, T.W., Chen, J., Marshall, P.A. & Freeman, B.A. (1990) *Proc. Natl. Acad. Sci. USA* **87**, 1620–1624.

Benveniste, E.N. (1992) *Am.J. Physiol.* **263**, C1–C16.

Billiar, T.R., Curran, R.D., Harbrecht, B.G., Stadler, J., Wiliiams, D.L., Ochoa, J.B., Di Silvio, M., Simmons, R.L. & Murray, S.A. (1993) *Am. J. Physiol.* **262**, C1077–C1082.

Bogdan, C., Vodovotz, Y., Paik, J., Xie, Q-W. & Nathan, C. (1993) *J. Immunol.* **151**, 301–309.

Boje, K.M. & Arora, P.K. (1992) *Brain Res.* **587**, 250–256.

Bolanos, J.P., Peuchen, S., Heales, S.J.R., Land, J.M. & Clark, J.B. (1994) *J. Neurochem.* **63**, 910–916.

Bredt, D.S., Hwang, P.M., Glatt, C.E., Lowenstein, C., Reed, R.R. & Snyder, S.H. (1991a) *Nature* **351**, 714–718.

Bredt, D.S., Glatt, C.E., Hwang, P.M., Fotuhi, M., Dawson, T.M. & Snyder, S.H. (1991b) *Neuron* **7**, 615–624.

Bredt, D.S., Ferris, C.D. & Snyder, S.H. (1992) *J. Biol. Chem.* **267**, 10976–10981.

Brune, B. & Lapetina, E.G. (1991) *Biochem. Biophys. Res. Commun.* **181**, 921–926.

Buga, G.M., Griscavage, J.M., Rogers, N.E. & Ignarro, L.J. (1993) *Circ. Res.* **73**, 808–812.

Buisson, A., Margaill, I., Callebert, J., Plotkine, M. & Boulu, R.G. (1993) *J. Neurochem.* **61**, 690–696.

Bulut, V., Severn, A. & Liew, F.Y. (1993) *Biochem. Biophys. Res. Commun.* **195**, 1134–1138.

Chao, C.C., Shuxian, H., Molitor, T.W., Shaskan, E.G. & Peterson, P. (1992) *J. Immunol.* **149**, 2736–2741.

Charles, A.C., Dirksen, E.R., Merrill, J.E. & Sanderson, M.J. (1993) *Glia* **7**, 134–145.

Choi, D.W. (1993) *Proc. Natl. Acad. Sci. USA* **90**, 9741–9743.

Chung, I.Y. & Benveniste, E.N. (1990) *J. Immunol.* **144**, 2999–3007.

Chung, I.Y., Kwon, J. & Benveniste, E.N. (1992) *J. Immunol.* **149**, 3894–3902.

Corbett, J.A., Sweetland, M.A., Lancaster, J.R. & McDaniel, M.L. (1993) *FASEB J.* **7**, 369–374.

Corradin, S.B., Manuel, J., Donini, D., Quattrocchi, E. & Ricciardi-Castagnoli, P. (1993) *Glia* **7**, 255–262.

Coyle, J.T. & Puttfarcken, P. (1993) *Science* **262**, 689–695.

Dawson, V.L., Dawson, T.M., Bartley, D.A., Uhl, G.R. & Snyder, S.H. (1993) *J. Neurosci.* **13**, 2651–2661.

Demerle-Pallardy, C., Lonchampt, M-O., Chabrier, P-E. & Braquet, P. (1993) *Life Sci.* **52**, 1883–1890.

De Vente, J. & Steinbusch, H.W.M. (1992) *Acta Histochem.* **92**, 13–38.

Dong, Z., Qi, A., Xie, K. & Fidler, I.J. (1993) *J. Immunol.* **151**, 2717–2724.

Eigler, A., Sinha, B. & Endres, S. (1993) *Biochem. Biophys. Res. Commun.* **196**, 494–501.

Endoh, M., Maiese, K., Pulsinelli, W.A. & Wagner, J.A. (1993a) *Abstr. Soc. Neurosci.* **19**, 1660.

Endoh, M., Maiese, K., Pulsinelli, W.A. & Wagner, J.A. (1993b) *Neurosci. Lett.* **154**, 125–128.

Feinstein, D.L., Galea, E. & Reis, D.J. (1993) *J. Neurochem.* **60**, 1945–1948.

Feinstein, D.L., Galea, E., Roberts, S., Berquist, H., Wang, H. & Reis, D.J. (1994a) *J. Neurochem.* **62**, 315–321.

Feinstein, D.L., Cermak, J., Chugh, P., Lyandvert, L., Galea, E. & Reis, D.J. (1994b) *J. Neurochem.* **62**, 811–814.

Finkbeiner, S. (1993) *Glia* **9**, 83–104.

Fontana, A., Kristensen, F., Dubs, R., Gemsa, D. & Weber, E. (1982) *J. Immunol.* **129**, 2413–2419.

Frei, K., Leist, T.P., Meager, A., Gallo, P., Leppert, D., Zinkernagel, R.M. & Fontana, A. (1988) *J. Exp. Med.* **168**, 449–453.

Fujisawa, H., Dawson, D., Browne, S.E., MacKay, K.B., Bullock, R. & McCulloch, J. (1993) *Brain Res.* **629**, 73–78.

Galea, E., Feinstein, D.L. & Reis, D.J. (1992) *Proc. Natl. Acad. Sci. USA* **89**, 10945–10949.

Galea, E., Reis, D.J. & Feinstein, D.L. (1993) *Abstr. Soc. Neurosci.* **19**, 1121.

Galione, A., White, A., Willmot, N., Turner, M., Potter, B. & Watson, S.P. (1993) *Nature* **365**, 456–459.

Garcia, A. & Agullo, L. (1992) *Pharmacol. Res.* **26** (Suppl. 1), 207.

Garg, U.C., Devi, L., Tundorf, H., Goldfrank, L.R. & Bansinath, M. (1992) *Brain Res.* **592**, 208–212.

Gazzinelli, R.T., Eltoum, I., Wynn, T.A. & Sher, A. (1993) *J. Immunol.* **151**, 3672–3681.

Geller, D.A., Lowenstein, C.J., Shapiro, R.A., Nussler, A.K., Di Silvio, M., Wang, S.C., Nakayama, D.K., Simmons, R.L., Snyder, S.H. & Billiar, T.R. (1993) *Proc. Natl. Acad. Sci. USA* **90**, 3491–3495.

Ginsberg, M.D. & Busto, R. (1989) *Stroke* **20**, 1627–1642.

Giulian, D. (1993) *Glia* **7**, 102–110.

Goureau, O., Lepoivre, M., Becquet, F. & Courtois, Y. (1993) *Proc. Natl. Acad. Sci. USA* **90**, 4276–4280.

Goureau, O., Hicks, D., Courtois, Y. & Dekozak, Y. (1994) *J. Neurochem.* **63**, 310–317.

Heinzel, B., John, M., Klatt, P., Bohme, E. & Mayer, B. (1992) *Biochem. J.* **281**, 627–630.

Hibbs, J.B., Taintor, R.R., Vavrin, Z. & Rachlin, E.M. (1988) *Biochem. Biophys. Res. Commun.* **157**, 87–94.

Hortelano, S., Genaro, A.M. & Bosca, L. (1992) *J. Biol. Chem.* **267**, 24937–24940.
Hortelano, S., Genaro, A.M. & Bosca, L. (1993) *FEBS Lett.* **320**, 135–139.
Hosli, E. & Hosli, L. (1993) *Prog. Neurobiol.* **40**, 477–506.
Ischiropoulous, H., Zhu, L. & Beckman, J.S. (1992) *Arch. Biochem. Biophys.* **298**, 446–451.
Kader, A., Frazzini, V.I., Solomon, R.A. & Trifiletti, R.R. (1993) *Stroke* **24**, 1709–1716.
Kanner, J., Harel, S. & Granit, R. (1992) *Lipids* **27**, 46–49.
Karupiah, G., Xie, Q., Buller, M.L., Nathan, C., Duarte, C. & MacMicking, J. (1993) *Science* **261**, 1445–1448.
Kettenmann, H., Banati, R. & Walz, W. (1993) *Glia* **7**, 93–101.
Kiedrowski, L., Costa, E. & Wroblewski, J.T. (1992) *Neurosci. Lett.* **135**, 59–61.
Kolesnikov, Y.A., Pick, C.G., Ciszewska, G. & Pasternack, G.W. (1993) *Proc. Natl. Acad. Sci. USA* **90**, 5162–5166.
Koprowski, H., Zheng, Y.M., Heber-Katz, E., Fraser, N., Rorke, L., Fu, Z.F., Hanlon, C. & Dietzchold, B. (1993) *Proc. Natl. Acad. Sci. USA* **90**, 3024–3027.
Kuluz, J.W., Prado, R.J., Dietrich, D., Schleien, C.L. & Watson, B.D. (1993) *Stroke* **24**, 2023–2029.
Lamas, S., Marsden, P.A., Li, G.K., Tempst, P. & Michel, T. (1992) *Proc. Natl. Acad. Sci. USA* **89**, 6348–6352.
Lassmann, H., Schmied, M., Vass, K. & Hickey, W.F. (1993) *Glia* **7**, 19–24.
Lee, S.C., Dickson, D.W., Liu, W. & Brosnan, C.F. (1993) *J. Neuroimmunol.* **46**, 19–24.
Liew, F.Y. & Cox, F.E.G. (1991) *Immunol. Today* **12**, A17–A21.
Lincoln, T.M. & Cornwell, T.L. (1993) *FASEB J.* **7**, 328–338.
Ling, E-A. & Wong, W-C. (1993) *Glia* **7**, 9–18.
Liu, T., McDonell, P.C., Young, P.R., White, R.F., Siren, A.L., Hallenbeck, J.M., Barone, F.C. & Feuerstein, G.Z. (1993) *Stroke* **24**, 1746–1751.
Lowenstein, C.J., Glatt, C.S., Bredt, D.S. & Snyder, S.H. (1992) *Proc. Natl. Acad. Sci. USA* **89**, 6711–6715.
MacNaul, K.L. & Hutchinson, N.I. (1993) *Biochem. Biophys. Res. Commun.* **196**, 1330–1334.
Malinski, T., Bailey, F., Zhang, Z.G. & Chopp, M. (1993) *J. Cereb. Blood Flow Metab.* **13**, 355–358.
Marczin, N., Papapetropoulos, A. & Catravas, J.D. (1993) *Am. J. Physiol.* **265**, H1014–H1018.
Matsumoto, T., Nakane, M., Pollock, J.S., Kuk, J.E. & Forstermann, U. (1993) *Neurosci. Lett.* **155**, 61–64.
McCarthy, K.D. & Salm, A.K. (1991) *Neuroscience* **41**, 325–333.
Meller, S.T. & Gebhart, G.F. (1993) *Pain* **52**, 127–136.
Meller, S.T., Pechman, P.S., Gebhart, G.F. & Maves, T.J. (1992) *Neuroscience* **50**, 7–10.
Merrill, J.E., Ignarro, L.J., Sherman, M.P., Melinek, J. & Lane, T.E. (1993) *J. Immunol.* **151**, 2132–2141.
Minc-Golomb, D. & Schwartz, J.P. (1992) *Abstr. Soc. Neurosci.* **18**, 1004.
Minc-Golomb, D. & Schwartz, J.P. (1993) *Abstr. Soc. Neurosci.* **19**, 1175.
Mitrovic, B., Ignarro, L.J., Montestruque, S., Smoll, A. & Merrill, J.E. (1994) *Neurosci.* **61**, 575–585.
Mollace, V., Salvemini, D., Anggard, E. & Vane, J. (1990) *Biochem. Biophys. Res. Commun.* **172**, 564–569.
Mollace, V., Colasanti, M., Rodino, Massoud, R., Lauro, G.M. & Nistico, G. (1993) *Biochem. Biophys. Res. Commun.* **191**, 327–334.
Munoz-Fernandez, A. & Fresno, M. (1993) *Biochem. Biophys. Res. Commun.* **194**, 319–325.
Murphy, S. (1993) *Astrocytes: Pharmacology and Function.* San Diego, Academic Press.
Murphy, S., Minor, R.L., Welk, G. & Harrison, D.G. (1990) *J. Neurochem.* **55**, 349–351.
Murphy, S., Minor, R.L., Welk, G. & Harrison, D.G. (1991) *J. Cardiovasc. Pharmacol.* **17**, S265–S268.

Nakane, M., Moitchell, J., Forstermann, U. & Murad, F. (1991) *Biochem. Biophys. Res. Commun.* **180**, 1396–1402.
Nathan, C. (1992) *FASEB J.* **6**, 3051–3064.
Nishikawa, T., Kirsch, J.R., Koehler, R.C., Bredt, D.S., Snyder, S.H. & Traystman, R.J. (1993) *Stroke* **24**, 1717–1724.
Norris, J.G. & Benveniste, E.N. (1993) *J. Neuroimmunol.* **45**, 137–146.
Nowicki, J.P., Duval, D., Poignet, H. & Scatton, B. (1991) *Eur. J. Pharmacol.* **204**, 339–340.
Oguro, T., Yokoyama, T., Fujisawa, H., Kurashima, Y. & Esumi, H. (1993) *Biochem. Biophys. Res. Commun.* **193**, 1014–1022.
Otero, G.C. & Merrill, J.E. (1994) *Glia* **11**, 117–128.
Park, S.K., Lin, H.L. & Murphy, S. (1994) *Biochem. Biophys. Res. Commun.* **201**, 762–768.
Pasqualotto, B.A., Hope, B.T. & Vincent, S.R. (1991) *Neurosci. Lett.* **128**, 155–160.
Peunova, N. & Enikolopov, G. (1993) *Nature* **364**, 450–453.
Poeggel, G., Muller, M., Seidel, I., Rechardt, L. & Bernstein, H-G. (1992) *J. Cardiovasc. Pharmacol.* **20** (supp. 12), S76–S79.
Pou, S., Pou, W.S., Bredt, D.S., Snyder, S.H. & Rosen, G.M. (1992) *J. Biol. Chem.* **267**, 24173–24176.
Przewlocki, R., Machelska, H. & Przewlocka, B. (1993) *Life Sci.* **53**, PL1–5.
Radi, R., Beckman, J.S., Bush, K.M. & Freeman, B.A. (1991a) *J. Biol. Chem.* **266**, 4244–4250.
Radi, R., Beckman, J.S., Bush, K.M. & Freeman, B.A. (1991b) *Arch. Biochem. Biophys.* **288**, 481–487.
Regidor, J., Montesdeoca, J., Ramirez-Gonzalez, J.A., Hernandez-Urquia, C.M. & Divac, I. (1993) *Brain Res.* **631**, 171–174.
Rengasamy, A. & Johns, R.A. (1993) *Mol. Pharmacol.* **44**, 124–128.
Richardson, A., Hao, C. & Fedoroff, S. (1993) *Glia* **7**, 25–33.
Rogers, N.E. & Ignarro, L.J. (1992) *Biochem. Biophys. Res. Commun.* **189**, 242–249.
Salvemini, D., Misko, T.P., Masferrer, J.L., Seibert, K., Currie, M.G. & Needleman, P. (1993) *Proc. Natl. Acad. Sci. USA* **90**, 7240–7244.
Schaad, N.C., Magistretti, P.J. & Schorderet, M. (1991) *Neurochem. Int.* **18**, 303–322.
Schmidt, H.H.H.W., Gagne, G.D., Nakane, M., Pollock, J.S., Miller, M.F. & Murad, F. (1992) *J. Histochem. Cytochem.* **40**, 1439–1456.
Schmidt, H.H.H.W., Lohmann, S.M. & Walter, U. (1993) *Biochim. Biophys. Acta* **1178**, 153–175.
Schneemann, M., Schoedon, G., Frei, K. & Schaffner, A. (1993) *Immunol. Lett.* **35**, 159–162.
Schneider-Schaulies, J., Schneider-Schaulies, S. & Ter Meulen, V. (1993) *Virology* **195**, 219–228.
Schoedon, G., Schneemann, M., Blau, N., Edgell, C.-J.S. & Schaffner, A. (1993) *Biochem. Biophys. Res. Commun.* **196**, 1343–1348.
Severn, A., Xu, D., Doyle, J., Leal, L.M.C., O'Donnell, C.A., Brett, S.J., Moss, D.W. & Liew, F.Y. (1993) *Eur. J. Immunol.* **23**, 1711–1714.
Simmons, M.L. & Murphy, S. (1992) *J. Neurochem.* **59**, 897–905.
Simmons, M.L. & Murphy, S. (1993) *Eur. J. Neurosci.* **5**, 825–831.
Simmons, M.L. & Murphy, S. (1994) *Glia* **11**, 227–234.
Stuehr, D.J. & Nathan, C.F. (1989) *J. Exp. Med.* **169**, 1543–1555.
Thorat, S.N., Reddy, P.L. & Bhargava, H.N. (1993) *Brain Res.* **621**, 171–174.
Tominaga, T., Sato, S., Ohnishi, T. & Ohnishi, S.T. (1993) *Brain Res.* **614**, 342–346.
Tracey, W.R., Nakane, M., Pollock, J.S. & Forstermann, U. (1993) *Biochem. Biophys. Res. Commun.* **195**, 1035–1040.
Vigne, P., Damais, C. & Frelin, C. (1993) *Brain Res.* **606**, 332–334.
Vincent, S.R. & Kimura, H. (1992) *Neuroscience* **46**, 755–784.
Wallace, M.N. & Fredens, K. (1992) *NeuroReport* **3**, 953–956.
Welsh, N. & Sandler, S. (1992) *Biochem. Biophys. Res. Commun.* **182**, 333–340.

Wink, D.A., Osawa, Y., Darbyshire, J.F., Jones, C.R., Eshenaur, S.C. & Nims, R.W. (1993) *Arch. Biochem. Biophys.* **300**, 115–123.
Wolf, G., Wurdig, S. & Schunzel, G. (1992) *Neurosci. Lett.* **147**, 63–66.
Wu, W., Liuzzi, F.J., Schinco, F.P., Dawson, T.M. & Snyder, S.H. (1993) *Abstr. Soc. Neurosci.* **19**, 440.
Xie, Q., Cho, H.J., Calaycay, J., Mumford, R.A., Swiderek, K.M., Lee, T.D., Ding, A., Troso, T. & Nathan, C. (1992) *Science* **256**, 225–228.
Xie, Q., Whisnant, R. & Nathan, C. (1993) *J. Exp. Med.* **177**, 1779–1784.
Xu, J.Y. & Tseng, L.F. (1993) *Eur. J. Pharmacol.* **236**, 137–142.
Yamamoto, R., Bredt, D.S., Dawson, T.M., Synder, S.H. & Stone, R.A. (1993) *Brain Res.* **631**, 83–88.
Yoshizumi, M., Perrella, M.A., Burnett, J.C. & Lee, M. (1993) *Circ. Res.* **73**, 205–209.
Zhang, X. & Morrison, D.C. (1993) *J. Exp. Med.* **177**, 511–516.
Zhang, Y., Marcillat, O., Giulivi, C., Ernster, L. & Davies, K.J.A. (1990) *J. Biol. Chem.* **265**, 16330–16336.
Zheng, Y.M., Schafer, M.K., Weihe, E., Sheng, H., Corisdeo, S., Fu, Z.F., Koprowski, H. & Dietzschold, B. (1993) *J. Virol.* **67**, 5786–5791.
Zielasek, J., Tausch, M., Toyka, K.V. & Hartung, H.-P. (1992) *Cell. Immunol.* **141**, 111–120.

CHAPTER 10

NITRIC OXIDE TOXICITY IN NEURONAL INJURY AND DEGENERATION

Pamela D. Varner* and Joseph S. Beckman†

*Departments of *Anesthesiology and †Biochemistry, The University of Alabama at Birmingham, Birmingham, AL 35233, USA*

Table of Contents

10.1	Introduction	191
10.2	The short half-life of nitric oxide	193
10.3	Nitric oxide is not highly reactive or toxic	194
10.4	Nitric oxide and transition metals	195
10.5	Diffusion of nitric oxide	196
10.6	Haemoglobin is a major trap for nitric oxide *in vivo*	197
10.7	Superoxide in cerebral ischaemic injury	197
10.8	Peroxynitrite as an oxidant	198
10.9	The interaction of superoxide and nitric oxide in cerebral ischaemia	199
10.10	Inhibition of nitric oxide synthesis in cerebral ischaemia	200
10.11	Superoxide dismutase and cerebral ischaemia	200
10.12	Superoxide dismutase-catalysed nitration	201
10.13	Motor neuron disease, superoxide dismutase and peroxynitrite	202
	Acknowledgements	204
	References	204

10.1 Introduction

Although nitric oxide (NO) can be produced by an amazing variety of tissues throughout the body, the brain is the richest source of nitric oxide synthase (NOS) in non-diseased states. NO is an important cellular messenger that helps regulate cerebral blood flow and can modulate synaptic plasticity (Garthwaite *et al.*, 1988; Iadecola *et al.*, 1994). Only 2% of neurons on average contain NOS, but these

Nitric Oxide in the Nervous System
ISBN 0-12-721985-4

neurons are highly branched in such a manner that all CNS neurons are within a few microns of an NO source (Bredt *et al.*, 1990). The olfactory bulb and cerebellum are the richest source of NO but the overall variation in NOS activity varies by at most 8-fold over all of the brain regions (Förstermann *et al.*, 1990; East and Garthwaite, 1991). Shibuki and Okada (1991) measured between 70 and 100 nM NO being produced in cerebellar slices after brief electrical stimulation. After the induction of focal cerebral ischaemia in rats, Malinski *et al.* (1993) measured 2–4 μM NO in the cortex. Thus, all regions of the brain have the potential to produce surprisingly high concentrations of NO in a biological system, though these concentrations are far lower than what has been used *in vitro* to show toxicity.

During cerebral ischaemia, a subtype of glutamate receptor stimulated by the agonist *N*-methyl-D-aspartate (NMDA) may contribute to the pathogenesis of ischaemia. The NMDA receptor is activated by glutamate and glycine, but will only open its channel when the membrane has been depolarized. The NMDA channel allows an influx of calcium into cells, which is known to play a significant role in long-term potentiation (LTP) and other forms of synaptic plasticity. Among the many processes controlled by elevated intracellular calcium is the production of NO. Because the production of NO by macrophages has been linked with their microbicidal and tumouricidal activity, it is reasonable to expect that over-production of NO by activation of the NMDA receptor could lead to neuronal injury. Indeed, inhibition of NO synthesis can reduce infarct volume in various stroke models, sometimes to a greater extent than do inhibitors of the NMDA receptor (Nowicki *et al.*, 1991). Studies in primary cerebral cultures also show that NMDA neurotoxicity *in vitro* can be attenuated by NOS inhibitors such as *N*-nitro-L-arginine (Dawson *et al.*, 1991, 1993a,b). Furthermore, removal of the NO precursor L-arginine or scavenging NO with haemoglobin are also protective.

Only 7 years have passed since NO was identified as a biological molecule. Because biologists are unfamiliar with an inorganic hydrophobic gas, many assumptions have become ingrained in the literature that bear re-examination. NO is usually described as a short-lived and highly reactive species. We will describe how these generalizations have arisen and why NO is not nearly as reactive as commonly thought in a broad sense at the concentrations produced in brain. NO becomes highly reactive and toxic by conversion to more strongly oxidizing species, but NO itself is selective and limited in the biological molecules that it reacts with. Because NO does not react readily with most cellular components, it can diffuse into several surrounding neurons. NO is relatively long-lived compared to many neural processes such as synaptic transmission or the propagation of action potentials, which allows NO to help integrate neuronal activity on a local scale. These simple points have important implications for understanding how the production of NO contributes to cerebral ischaemic injury and to neurodegeneration.

10.2 The short half-life of nitric oxide

The reputation for the high reactivity of NO results in part from its short half-life in perfusion cascades. NO was originally detected as the endothelium-derived relaxing factor (EDRF), which is released from endothelium and causes the relaxation of vascular smooth muscle. The life time of EDRF is only 6–8 s in Krebs–Henseleit buffer solutions bubbled with 95% oxygen and of the order of 30–50 s when the solutions are bubbled with 20% oxygen (Moncada *et al.*, 1987; Palmer *et al.*, 1987; Furchgott and Vanhoutte, 1989). Hypoxic conditions stabilize EDRF (Rubanyi and Vanhoutte, 1986).

When EDRF was recognized to be NO, the inactivation by oxygen was proposed to be due to the following third-order reaction:

$$2NO + O_2 \xrightarrow{k_3} 2NO_2$$

which converts NO into the powerful and toxic oxidant, nitrogen dioxide. Fortunately, the formation of NO_2 is too slow to be physiologically significant at the concentrations of NO produced for signal transduction or else NO would be far more toxic. The reason is that the rate of reaction depends upon the square of NO concentration since two NO molecules are required in the reaction.

$$\frac{d[NO_2]}{dt} = k_3[O_2][NO]^2$$

where k_3 is the rate constant of $6–8\times10^6$ M^{-1} s^{-1} (Zafiriou and McFarland, 1980; Ford *et al.*, 1993).

NO_2 formation, visible as an orange gas, appears instantaneous if a cylinder of pure NO leaks into room air and must be handled carefully to prevent lethal exposure to NO_2. NO at 1 atmosphere of pressure will have a concentration of approximately 40 mmol/l. In water, a saturating concentration of NO is about 1.9 mM at room temperature (Seidell, 1919); but the concentration of NO produced for signal transduction is less than 0.1 μM, or about 10 000–fold lower than a saturating solution. Because the reaction depends upon the square of NO concentration, the rate of NO decomposition slows by a factor of a hundred million at physiological concentrations of NO. The half-life of a 0.1 μM solution of nitric oxide in phosphate buffer is about 80 min. Consequently, other reactions must be responsible for inactivation of NO *in vitro* and *in vivo*.

The chemical basis for the short half-life of NO in perfusion cascades remains unexplained. Clearly, the formation of NO_2 does not explain the short half-life of NO. Spontaneous oxidation of glucose and other compounds in the buffers may generate superoxide, which reacts rapidly with NO. Addition of superoxide dismutase generally doubles the half-life of NO in such perfusion cascades, strongly implicating superoxide as a major route of decomposition *in vitro*. The buffers are also likely to be contaminated with micromolar concentrations of transition metals, which can also bind NO and might catalyse secondary reactions that will remove small amounts of nitric oxide.

10.3 Nitric oxide is not highly reactive or toxic

Perhaps, the strongest evidence that NO is not highly toxic is the current clinical use of inhaled NO for the treatment of respiratory failure from adult respiratory distress syndrome (ARDS) and in premature or newborn infants with meconium aspiration syndrome or persistent foetal circulation. There remain substantial issues to be addressed about long-term consequences of inhaled NO, but it is clear that breathing 20–40 ppm (about 0.8–1.6 μmol/l) for several days does not cause obvious toxicity in humans.

NO reacts at near diffusion-limited rates with other free radicals and with transition metals, but is remarkably unreactive with most biological molecules. While NO is a free radical, free radicals are not necessarily highly reactive or destructive. For example, molecular oxygen (O_2) has two unpaired electrons in separate orbitals, making it a biradical. Yet, we survive in an atmosphere of 21% oxygen. From a chemical perspective, the reactivity of NO is quite similar to that of molecular oxygen (Janzen, 1994).

Both NO and O_2 become far more reactive by secondary reactions that produce more toxic species. However, their toxicity is limited by a kinetic barrier because the transformation to more reactive species depends upon one-electron transfer reactions. If NO were to oxidize most biological molecules, it would leave a highly unstable organic radical as an intermediate.

$$R{-}H + {\cdot}NO \rightarrow R{\cdot} + H^{+} + NO^{-}$$

Because NO is only a mild oxidant, with a reduction potential at pH 7 of 0.39 V, most biological molecules tend not to react with NO. Molecular oxygen is even worse as a one-electron oxidant than NO with a reduction potential of –0.1 V at physiological conditions. Strongly oxidizing free radicals such as hydroxyl radical (HO·) or nitrogen dioxide ($NO_2{\cdot}$) are sufficiently strong oxidants, with reduction potentials of 2.0 and 1.0 V, respectively, to overcome the kinetic barrier and are far more reactive.

Even thiols are resistant to direct oxidation by NO. The oxidation of cysteine by NO requires 6 h (Pryor *et al.*, 1982). While nitrosothiols are certainly produced *in vivo* (Stamler *et al.*, 1992a,b) they are not in a simple reversible equilibrium with NO (Stamler *et al.*, 1992c). The mechanisms resulting in the formation of nitrosothiols *in vivo* remain to be determined. NO will react rapidly with thiyl radicals by a near diffusion-limited reaction to form nitrosothiols, but this requires that a thiol first be oxidized by one electron to a thiyl radical.

$$R{-}SH + X \rightarrow R{-}S{\cdot} + X{\cdot}^{-}$$

$$R{-}S{\cdot} + {\cdot}NO \rightarrow R{-}S{-}NO$$

For a nitrosothiol to release NO, there must be a balancing reaction to provide an electron for the thiyl radical.

The toxicity of NO has been overestimated by *in vitro* systems. Because of the inconvenience of delivering NO at a slow continuous rate, many studies have utilized nitrovasodilators as a source of NO. Paradoxically, many of these agents are often more reactive than NO itself. Sodium nitroprusside is an excellent oxidant of sulfhydryls and also releases cyanide. Nitrosothiols can modify key protein sulfhydryls by a transfer of the nitrosyl group between thiols. Lipton *et al.* (1993) have shown that the NMDA receptor can be down-regulated by nitrosylation or oxidation of a redox active sulfhydryl group. Another commonly used vasodilator is SIN-1, which is known to generate both superoxide and NO simultaneously, which then form the much stronger oxidant peroxynitrite (Hogg *et al.*, 1992).

In other cases, NO has been simply bubbled into containers with a target molecule or added as bolus of 100 μM to 1 mM. Under these conditions, the formation of NO_2 is rapid and the target molecule will be rapidly oxidized. However, the situation is far different if a steady state concentration of NO in the micromolar range is maintained, because the rate of formation of NO_2 is so slow.

In many cases, what has been shown is that the production of NO leads to various forms of injury rather than NO itself being the causal agent. For example, macrophages and neutrophils can produce superoxide simultaneously and will be producing peroxynitrite (Ischiropoulos *et al.*, 1992a).

10.4 Nitric oxide and transition metals

Transition metals also contain orbitals with unpaired electrons, making them potentially good acceptors for both oxygen and NO. NO binds with high affinity to transition metals like the iron in haem proteins. Indeed, the major signal transduction pathway for NO is binding to the ferrous haem cofactor of guanylate cyclase, which then activates the enzyme to produce cGMP (Ignarro, 1991; Southam and Garthwaite, 1993; see Chapter 3).

The production of cGMP by NO could amplify calcium signals in neurons. Peunova and Enkolopov (1993) have shown that NO amplifies the effects of calcium on the protein kinase A-dependent activation of the transcription factor for cAMP-responsive element. Such indirect effects of NO mediated through cGMP are certainly possible and need to be further explored.

NO is unusual in that it can bind strongly to both ferrous and ferric iron (Fe^{2+} and Fe^{3+}; Traylor and Sharma, 1992). Binding to ferrous haem is readily reversible. However, ferric iron will oxidize NO to nitrosonium ion (NO^+), which can then nitrosylate thiols and other biological molecules (Wade and Castro, 1990). In the process, the ferric iron is reduced to the ferrous state. The importance of this mechanism is probably limited *in vivo* by the availability of ferric iron.

While NO is unlikely to be toxic because of its overt chemical reactivity, it could be toxic by more subtle mechanisms. NO binds reversibly to many metal centres such as to cytochrome *c* oxidase. When produced in micromolar concentrations, NO may reversibly inhibit oxygen consumption by mitochondria, much like cyanide (Cleeter *et al.*, 1994). As the mitochondria become more reduced, they will be more likely to produce superoxide. The greater flux of superoxide within mitochondria will react with more NO to produce peroxynitrite, which is known to attack Complex I and II, resulting in irreversible inhibition of mitochondria (Radi *et al.*, 1994).

10.5 Diffusion of nitric oxide

NO has the highest diffusion coefficient of any biological molecule, being 1.4-fold higher than oxygen or carbon monoxide at 37°C. Because NO is hydrophobic, it crosses cell membranes as readily as oxygen or carbon dioxide. The distance a molecule can diffuse can be estimated from the following equation, which is readily derived from the Fick's laws of diffusion (Pryor, 1992).

$$x = \sqrt{2D \cdot t}$$

The distance x implies that 36% (one over the natural exponent e) of the initial NO concentration would have diffused past in time t. The diffusion coefficient (D) for NO is 4.8×10^{-5} cm^2 s^{-1} at 37°C (Wise and Houghton, 1968). For time (t), we use the half-lives of NO in normoxic physiological buffers about 30 s, and 1 s for NO in blood-free perfused guinea-pig heart (Kelm and Schrader, 1988). The substantially shorter half-life in heart is most likely due to reactions with myoglobin. The estimated mean diffusion distances of NO are then 540 μm in normoxic buffer and 100 μm in blood-free, isolated perfused heart in one biological half-life. These are large distances on the scale of a neuron in brain.

To understand the effects of NO diffusion in brain, compare NO to the spread of sound. Imagine that you are sitting in a lecture room, listening to a rather dull talk, and that each person in the audience is the equivalent of a neuron. Several people in the room can whisper to their neighbours, commenting upon the content of the lecture, and perhaps changing the view of others about the lecture. Several scattered discussions can occur simultaneously in the room without disturbing the general audience because sound, like NO, dampens radially from its source. In addition, the walls contain sound absorbent material to prevent echoes. However, if something should panic the audience, such as the announcement of a pop quiz, to cause everyone to yell as loudly as possible, the whole auditorium becomes chaotic and no one could make sense of what their neighbour is saying. That is what cerebral ischaemia can do to the brain by activating the NMDA receptor.

If all NOS-positive neurons are stimulated simultaneously, micromolar concentrations of NO can be produced. Lancaster (1994) has recently pointed out that the internal concentration of NO can be far greater in an NO-producing cell than that due only to the production of NO within that cell. When you are stuck in traffic, most of the NO you are breathing is not from your automobile but from the cumulative action of surrounding automobiles.

10.6 Haemoglobin is a major trap for nitric oxide *in vivo*

Not only can NO diffuse into and around other neurons, it can rapidly diffuse into capillaries. Both oxyhaemoglobin and oxymyoglobin react rapidly with NO to form nitrate and methaemoglobin.

$$Hb\text{–}Fe^{2+}\text{–}O_2 + \cdot NO \rightarrow Hb\text{-}Fe^{3+} + NO_3^-$$

The methaemoglobin is then reduced to deoxyhaemoglobin and thus can remove NO catalytically. The presence of myoglobin in heart makes the half-life of NO perfused through the heart rather short, with a half-life of less than 1 s (Kelm and Schrader, 1988). While haemoglobin is confined within red blood cells, the diffusion of NO is sufficiently great that it could diffuse from parenchymal cells into blood vessels within a few seconds. Clearly, oxygen is able to diffuse at a sufficiently fast rate from red blood cells to supply cellular needs for cells 100 μm from the blood vessel. Since NO has a slightly faster diffusion coefficient than oxygen and is hydrophobic enough to cross membrane, it should be able to diffuse 100 μm back to a red blood cell. From the diffusion calculations given above, 36% of nitric oxide could diffuse the 100 μm to a blood vessel in 1 s. Thus, diffusion to a blood vessel is even faster than the 6 s half-life of NO in an isolated organ perfusion cascade.

10.7 Superoxide in cerebral ischaemic injury

From 1% to 5% of all oxygen consumed is partially reduced to produce the oxygen radical superoxide ($O_2\cdot^-$). Activation of the NMDA receptor in many neurons can lead to the formation of superoxide through oxidation of fatty acids (Lafon-Cazal *et al.*, 1993). Normally, superoxide concentrations are kept at very low levels (10–100 pM) by the large amounts of endogenous superoxide dismutase, an enzyme that rapidly scavenges superoxide. However, intravenously administered polyethylene glycol conjugated superoxide dismutase (PEG-SOD) can reduce cerebral ischaemic injury, providing evidence that superoxide is involved in neural injury (Liu *et al.*, 1989). Transgenic mice overexpressing superoxide dismutase are also quite resistant to cerebral insults (Chan *et al.*, 1991; Kinouchi *et al.*, 1991; Yang *et al.*, 1994).

Superoxide is not a highly reactive molecule at neutral pH (Sawyer and Valentine, 1981), though there is evidence for direct toxicity (Fridovich, 1986; Gardner and Fridovich, 1991). It was given its name by Linus Pauling, who proposed its existence based upon Lewis dot structures. At neutral pH, superoxide exists as an anion (pK_a of 4.3). Small anions are not good oxidizing agents because that involves placing a second electron on a negatively charged molecule. At more acidic pH, superoxide becomes a better oxidizing agent because a larger fraction is protonated to the perhydroxyl or dioxydanyl radical ($HO_2\cdot$). To explain how strong oxidants are formed at neutral pH, the most commonly proposed mechanism is the iron-catalysed Haber–Weiss reaction (more appropriately called the superoxide-driven Fenton reaction). This is a complex cycle whereby superoxide reduces ferric iron and the resulting ferrous iron attacks hydrogen peroxide to form hydroxyl radical. The hydroxyl radical theory became widely accepted in part because no better explanation could be offered at the time for how superoxide dismutase and catalase could protect tissue.

However, there are several major problems with the Haber–Weiss reaction as an explanation for free radical toxicity. The iron-catalysed Haber–Weiss reaction requires the interaction of three molecules that are all at low concentrations *in vivo* due to efficient scavenging systems. The reaction rate of hydrogen peroxide with ferrous iron is slow ($\sim 10^3$ M^{-1} s^{-1}), so that endogenous catalases and glutathione peroxidases should be effective at out-competing the reaction. Hydroxyl radical reacts within a few Ångstroms of its site of formation which makes it unlikely to find a critical cellular target. To circumvent this problem, iron has been postulated to be bound in specific sites that are vulnerable to attack. Ascorbate also reduces ferric iron (Fe^{3+}) and is present in much higher concentrations than superoxide. Therefore, superoxide dismutase may not inhibit the Haber–Weiss reaction *in vivo*, since superoxide is only acting as a reductant that can be replaced by other cellular components.

These inconsistencies suggest that other reactions may be important for understanding superoxide toxicity and led to the investigation of the reaction of NO and superoxide. Both NO and superoxide are paramagnetic species that react by a diffusion-limited reaction (6.7×10^9 M^{-1} s^{-1}) to form the toxic oxidant peroxynitrite:

$$O_2{\cdot}^- + {\cdot}NO \rightarrow ONOO^-$$

NO is the only known biological molecule that can be produced in high enough concentrations and reacts fast enough to out-compete superoxide dismutase for superoxide, and therefore, can be an important target for superoxide. The rate constant for the reaction of superoxide with SOD is $2\text{–}3\times10^9$ M^{-1} s^{-1}, about 3-fold less than the rate of reacting with NO. Normally, SOD can out-compete NO for superoxide because SOD is present in 5–30 μM concentrations. The concentrations of NO used for signal transduction are about 0.1 μM. Pathological activation of NOS in brain or inflammatory cells may increase the concentrations of NO to levels as high as 2–4 μM (Malinski *et al.*, 1993). Under these conditions, a majority of superoxide would react with NO rather than superoxide dismutase, leading to the formation of peroxynitrite.

10.8 Peroxynitrite as an oxidant

Peroxynitrite is a powerful oxidant but not a free radical because the unpaired electrons on superoxide and NO have combined to form a new covalent bond. Peroxynitrite is remarkably stable in alkaline solution (Tsai *et al.*, 1994). At neutral to slightly acidic pH, peroxynitrous acid (ONOOH), decays homolytically to form a species with the reactivity of a hydroxyl radical (·OH) and nitrogen dioxide (NO_2)

as intermediates by the following series of reactions (Beckman *et al.*, 1990; Crow *et al.*, 1994):

$$ONOO^{-}+H^{+}\leftrightarrow HOONO\rightarrow \text{`}HO\cdot+NO_2\text{'} \quad (pK_a=6.8,\ t_{1/2}<1\ s)$$

However, peroxynitrite does not appear to separate completely into free hydroxyl radical and NO_2, but rather to react as an activated complex (Koppenol *et al.*, 1992). This energetic intermediate may be derived by bending of the N–O–O angle and stretching of the O–O bond of *trans* peroxynitrous acid, which forms a species resembling hydroxyl radical and NO_2 joined by a weak O–O bond.

The direct oxidizing reactions of peroxynitrite appear to be far more toxic than the hydroxyl radical-like reactivity of peroxynitrite, because these reactions are more rapid and specific for important cellular targets such as thiols and zinc fingers (Radi *et al.*, 1991; Crow *et al.*, 1995). Classical hydroxyl radical scavengers generally do not reduce injury by peroxynitrite and occasionally can increase injury (Zhu *et al.*, 1992). In the presence of hydroxyl radical scavengers, peroxynitrous acid will also form NO_2, which is also a potent cytotoxic oxidant.

At physiological pH, peroxynitrite is sufficiently stable to diffuse over several cell diameters to critical cellular targets before becoming protonated and decomposing. Peroxynitrite initiates lipid peroxidation and reacts directly with sulfhydryl groups at a thousand-fold greater rate than hydrogen peroxide at pH 7.4. In addition, peroxynitrite reacts with transition metals to form a powerful nitrating agent with reactivity suggestive of the nitronium ion (NO_2^+). Peroxynitrite is far more toxic to *Escherichia coli* than NO or superoxide generated by xanthine oxidase (Brunelli *et al.*, 1995). Whereas exposure to 1 mM NO either aerobically or anaerobically, xanthine oxidase or hydrogen peroxide was essentially non-toxic, peroxynitrite (1 mM) was completely lethal in a few seconds (Zhu *et al.*, 1992). Lipton *et al.* (1993) obtained similar results with cultured neurons, where the lethal dosage of peroxynitrite was only 20–50 μM.

10.9 The interaction of superoxide and nitric oxide in cerebral ischaemia

Many pathophysiological processes including reperfusion of ischaemic tissue, inflammation and sepsis initiate events that cause the simultaneous production of NO and superoxide. Ischaemia induces intracellular superoxide production by mitochondria, xanthine oxidase and other sources. NO, generated during reperfusion after the influx of calcium during ischaemia, can diffuse significant distances where it rapidly reacts with superoxide both intracellularly and in the vascular lumen to form peroxynitrite. Because the rate of peroxynitrite formation depends upon the product of superoxide and NO concentrations, each 10-fold increase in the concentrations of NO and superoxide results in 100-fold amplification in the rate of peroxynitrite formation. Because the formation of peroxynitrite requires the reaction of superoxide with NO, it seemed logical that increasing the amount of

SOD available to scavenge superoxide, or reducing the amount of NO produced by administering NOS inhibitors would reduce neurotoxicity.

10.10 Inhibition of nitric oxide synthesis in cerebral ischaemic

Simply inhibiting NO synthesis has been shown to enhance injury after cerebral ischaemia in some laboratories (Yamamoto *et al.*, 1992; Zhang *et al.*, 1994), and to provide substantial protection in others (Nowicki *et al.*, 1991; Nagafuji *et al.*, 1992). Using the middle cerebral artery (MCA)–carotid occlusion model we found that post-administration of 5 mg kg^{-1} nitroarginine-HC1 i.v. significantly reduced cerebral infarction by approximately 28% with no mortality (Chen *et al.*, 1994). However, pre-administration with nitroarginine before MCA–carotid occlusion caused substantial mortality and apparently made injury worse in surviving animals. The reasons for the toxicity with pre-administration are speculative, but several possibilities exist. Nitroarginine at the above dosage reduces cerebral blood flow by 40% without affecting systemic blood pressure (Beckman *et al.*, 1993a). Thus, nitroarginine by inhibiting even physiological concentrations of NO may be disrupting collateral circulation, which is important in limiting the extent of infarct. Delayed administration of nitroarginine until the time of reperfusion may be protective by reducing the excessive generation of NO. Recently, Huang *et al.* (1994) have shown that mice with the NOS gene knocked out are more resistant to cerebral ischaemia induced in the MCA territory, but treatment with nitroarginine still worsened the outcome.

10.11 Superoxide dismutase and cerebral ischaemia

Native superoxide dismutase has poor pharmacological characteristics which can be greatly improved by conjugation to polyethylene glycol (PEG-SOD) for administration intravenously (Beckman *et al.*, 1988). PEG-SOD can reduce cerebral ischaemic injury in gerbils subjected to 40 min bilateral carotid occlusion (Beckman *et al.*, 1986). PEG-SOD can also reduce infarct volume by 25% in an MCA stroke model (Liu *et al.*, 1989). Liposome-entrapped superoxide dismutase is also highly protective against cold oedema and ischaemia (Chan *et al.*, 1987). Transgenic mice over-expressing superoxide dismutase are also protected against ischaemia and trauma. PEG-SOD has recently been found to be protective in severe head trauma in humans (Muizelaar *et al.*, 1993). In a recent randomized, blinded phase II study, treatment of severe head trauma patients (Glasgow coma scale >3) with 5000 and 10 000 U kg^{-1} PEG-SOD resulted in a 50% decrease in morbidity and mortality compared to placebo-treated controls. The most striking effect was on the shorter duration of elevated intracranial pressure observed in patients receiving PEG-SOD.

Most animal studies of superoxide dismutase have used a single dose because of the cost. In myocardial ischaemia, 5-fold higher dosages of superoxide dismutase reduce efficacy and may make injury worse. The effect is not due to toxicity of superoxide dismutase itself, because much higher dosages of superoxide dismutase are well tolerated in control animals. Only a 3-fold increase in PEG-SOD above the protective dosage (10 000 U kg^{-1}) resulted in no protection (He *et al.*, 1993). Similar findings in ischaemic heart have been reported by Bernier *et al.* (1989), Omar *et al.* (1990) and Omar and McCord (1990). Transgenic mice over-expressing native Cu,Zn superoxide dismutase have been shown to exhibit abnormal proliferation of neuromuscular junctions, similar to what is observed in amyotrophic lateral sclerosis (ALS). These mice were originally developed as potential models of Down's syndrome, which exhibit a trisomy for chromosome 21 containing the superoxide dismutase gene. Over-expression of the extracellular superoxide dismutase gene resulted in greater susceptibility of mice exposed to hyperbaric oxygen (Oury *et al.*, 1992), while inhibition of NO was protective.

10.12 Superoxide dismutase-catalysed nitration

Peroxynitrite reacts directly with Cu,Zn superoxide dismutase via a nitronium ion (NO_2^+) intermediate that attacks tyrosine residues and converts them to nitrotyrosine. Peroxynitrite is attracted to the active site of superoxide dismutase like superoxide by a lining of positively charged amino acids around the copper atom. Once in the active site, peroxynitrite forms a transient cuprous adduct:

$$\text{SOD-Cu}^{2+}\ {}^{-}\text{OO–N=O} \rightarrow \text{SOD–Cu}^{+}\text{O}^{-}\cdots\text{NO}_2^{+}$$

This intermediate complex can donate a nitronium ion (NO_2^+) to phenolics to form nitrophenols. After releasing a hydroxyl ion, native superoxide dismutase is regenerated and therefore acts catalytically.

$$\text{SOD–Cu}^{+}\text{O}^{-}\cdots\text{NO}_2^{+}\text{+tyrosine} \rightarrow \text{SOD–Cu}^{2+}\text{+OH}^{-}\text{+NO}_2\text{-tyrosine}$$

Cu,Zn superoxide dismutase is not inactivated as a result of its reaction with peroxynitrite. Superoxide dismutase catalyses the nitration of a wide range of phenolics, including tyrosines in lysozyme and histone, at a reaction rate of $10^5\ \text{M}^{-1}\ \text{s}^{-1}$. Nitration of tyrosines can disrupt their ability to be phosphorylated, which may have significant effects on signal transduction.

Nitration is a permanent footprint and detection of nitrotyrosine in proteins is a marker for oxidants derived from NO such as peroxynitrite. Specific monoclonal and polyclonal antibodies to nitrotyrosine in proteins have been raised. Immunohistochemistry in human atherosclerotic lesions, particularly in the atheroma and around foam cells derived from macrophages, has demonstrated the presence of nitrotyrosine, indicating oxidants derived from NO such as peroxynitrite are generated in human atherosclerosis and may be involved in its pathogenesis (Beckman *et al.*, 1994). Additionally, nitrotyrosine and its metabolites, 3-nitro-4-hydroxyphenylacetic acid and 3-nitro-4-hydroxyphenylpropionic acid, have been detected in human urine. The presence of nitrotyrosine suggests that nitration is produced in humans and most likely eliminated by proteolysis (Ohshima *et al.*, 1990).

10.13 Motor neuron disease, superoxide dismutase and peroxynitrite

Motor neuron disease or ALS is a tragic disease of unknown aetiology that typically strikes people in their 40s to 70s. It is characterized by a relentless loss of large motor neurons over a 1–5 year period, with the remainder of the central nervous system remaining relatively intact. The disease initially presents as a weakness in one limb but then progressively spreads over several years to cause complete paralysis. While motor neurons are clearly dying, the disease is not simply due to neurodegeneration. There is considerable sprouting of neuromuscular junctions and synaptic remodelling as motor neurons attempt to compensate for the loss of neighbouring motor neurons. During the progression of the disease, single motor neurons may have innervated large numbers of muscle fibres that were originally controlled by dozens of motor neurons.

Most cases of ALS are sporadic, with no known cause, but less than 5% are due to an autosomal dominant mutation mapped to chromosome 21. Recently, the affected gene has been shown to encode Cu,Zn superoxide dismutase (Rosen *et al.*, 1993). A series of 25 independently arising mutations at 13 different amino acid positions have been identified at present (Deng *et al.*, 1993; Ogasawara *et al.*, 1993; Rosen *et al.*, 1993). All of the mutations are missense with some mutations changing only a single methyl group in the whole protein. None of the mutations occur in the active site, though one mutation recently identified in Japan changes one of the four histidines that binds copper in the active site (Ogasawara *et al.*, 1993).

The principal effect of the mutations appears to be slightly destabilizing the superoxide dismutase protein. Superoxide dismutase activities in people carrying the mutation appear to be decreased by 20–50% (Bowling *et al.*, 1993; Deng *et al.*, 1993). We have found that homozygous mutant superoxide dismutases expressed in *E. coli* have substantial activity, but are less stable and may be more susceptible to losing their copper. Recently, over-expression of one of the ALS mutations in transgenic mice has been shown to cause lower motor neuron degeneration at 4–6 months after birth (Gurney *et al.*, 1994). Mice expressing equivalent activities of the native human superoxide dismutase do not develop symptoms.

The effects of the mutations may be to increase nitration by reaction with peroxynitrite rather than decrease superoxide scavenging (Beckman *et al.*, 1993b). All individuals identified with a mutation are heterozygous with one normal allele for superoxide dismutase. Superoxide dismutase is normally present in a cell in a million-fold higher concentration than its substrate, superoxide. Thus, a 50% decrease in activity would at most double the steady-state concentration of superoxide. However, a doubling of the steady-state superoxide concentration will double the rate of peroxynitrite formation due to the reaction with NO. Peroxynitrite in turn reacts with superoxide dismutase at about 10^5 M^{-1} s^{-1} to nitrate tyrosine residues on proteins (Beckman *et al.*, 1992; Ischiropoulos *et al.*, 1992b).

Because of the massive axonal volume of motor neurons, they contain more neurofilaments than any other cell and abnormal assemblies of neurofilaments are commonly observed in ALS (Troost *et al.*, 1992; Brady, 1993). Some ALS motor neurons contain large swellings in the axon called spheroids that are swollen with neurofilaments. The lowest molecular weight of the three proteins making up neurofilaments, neurofilament L, consists of about 4% tyrosine and may also be the key target of nitration in motor neurons. Neurofilament L is also crucial to the proper assembly of neurofilaments and small deletions in the tyrosine-rich amino terminus can cause dominant effects on the assembly of normal neurofilaments. Over-expression of neurofilament L has been linked to the development of an ALS-like syndrome in mice (Xu *et al.*, 1993). We have recently found that neurofilaments may be particularly susceptible to superoxide dismutase-catalysed nitration.

A possible objection to the nitration hypothesis is that motor neurons do not normally make NO (Valtschanoff *et al.*, 1992). However, a wide range of injuries can induce NOS activity in neurons (Solodkin *et al.*, 1992; Wu, 1993; Wu and Scott, 1993). Avulsion of the ventral root in spinal cord induces NOS activity in affected motor neurons. Furthermore, motor neuron degeneration over the subsequent 6 weeks is largely prevented by chronic inhibition with low dosages of nitroarginine (Wu and Li, 1993). We propose that the propagation of the disease is due to the diffusion of NO from one motor neuron to the next. In ALS, motor neurons appear to be attempting to compensate by increasing their motor units. This requires synaptic remodelling, which may involve the induction of NOS. Since NO will diffuse to its neighbours, they will be subject to slightly greater nitration mediated by peroxynitrite. Neurofilaments are the most abundant proteins and are the most likely targets for nitration. If they begin to assemble abnormally, axonal transport will be disrupted and the motor neuron will be deprived of growth factors. It will then attempt to remodel and also induce NO synthesis. The effect of the superoxide dismutase mutants is to enhance the propagation of the disease. Interestingly, there appears to be no correlation with the age of onset in the familial superoxide dismutase mutations, but the rate of progression can be influenced by the type of mutation. In particular, people with the mutation alanine-4 to valine, have a life span of only a year compared to 3–5 years in other individuals.

In the present review, we have considered that the toxicity of NO must be more subtle than due to its direct reactivity. As a free radical, NO is not reactive. However, the reaction of NO and superoxide to form peroxynitrite is sufficiently rapid to outcompete endogenous defences. NO has diverse functions in brain, affecting neuronal activity, regulation of cerebral blood flow and systemic blood pressure, and adhesion of activated inflammatory cells. Inhibition of NO synthesis has yielded conflicting effects upon the outcome of cerebral and spinal cord insults, making injury worse under certain conditions. A better understanding of the balance between the physiological functions of nitric oxide and the mechanisms of toxicity will determine the therapeutic potential of modulating nitric oxide production.

Acknowledgements

This work was supported by grants HL46407, NS24338 and HL48676 from the National Institutes of Health and from the American Heart Association. J.S. Beckman is supported as an Established Investigator of the American Heart Association.

References

Beckman, J.S., Campbell, G.A., Hannan, J., Karfias, C.S. & Freeman, B.A. (1986). Involvement of superoxide and xanthine oxidase with death due to cerebral ischemia-induced seizures in gerbils. *Superoxide & Superoxide Dismutase in Chemistry, Biology and Medicine.* (ed. G. Rotillio) Elsevier Sci. Publ., Amsterdam, pp. 602–607.

Beckman, J.S., Minor Jr., R.M., White, C.J., Repine, J., Rosen, G.M. & Freeman, B.A. (1988) *J. Biol. Chem.* **263**, 6584–6802.

Beckman, J.S., Beckman, T.W., Chen, J., Marshall, P.M. & Freeman, B.A. (1990) *Proc. Natl. Acad. Sci. USA* **87**, 1620–1624.

Beckman, J.S., Ischiropoulos, H., Zhu, L., van der Woerd, M., Smith, C., Chen, J., Harrison, J., Martin, J.C. & Tsai, M. (1992) *Arch. Biochem. Biophys.* **298**, 438–445.

Beckman, J.S., Chen, J., Ischiropoulos, H., Zhu, L., Conger, K.A. & Halsey, Jr, J.H. (1993a) In *Oxygen Radicals and Tissue Injury* (eds Samson, F. & Tarr, M.), pp. 165–185. Boston, Birkheuser.

Beckman, J.S., Carson, M., Smith, C.D. & Koppenol, W.H. (1993b) *Nature* **364**, 584.

Beckman, J.S., Ye, Y.Z., Anderson, P., Chen, J., Accavetti, M.A., Tarpey, M.M. & White, C.R. (1994) *Biol. Chem. Hoppe-Seyler* **375**, 81–88.

Bernier, M., Manning, A.S. & Hearse, D.J. (1989) *Am. J. Physiol.* **256**, H1344–H1352.

Bowling, A.C., Schulz, J.B., Brown Jr., R.H. & Beal, M.F. (1993) *J. Neurochem.* **61**, 2322–2325.

Brady, S.T. (1993) *Cell* **73**, 1–3.

Bredt, D.S., Hwang, P.M. & Snyder, S.H. (1990) *Nature (Lond.)* **347**, 768–770.

Brunelli, L., Crow, J.P. & Beckman, J.S. (1995) *Arch. Biochem. Biophys.* **316**: in press.

Chan, P.H., Longar, S. & Fishman, R.A. (1987) *Ann. Neurol.* **21**, 540–547.

Chan, P.H., Yang, G.Y., Chen, S.F., Carlson, E. & Epstein, C.J. (1991) *Ann. Neurol.* **29**, 482–486.

Chen, J., Conger, K.A., Tan, M.J. & Beckman, J.S. (1994) In *Basic Mechanisms of Cerebral Ischemia* (eds. Hartman, A. & Yatsu, F.). Springer-Verlag, Berlin.

Cleeter, M.W.J., Cooper, J.M., Darley-Usmar, V.M., Moncada, S. & Schapira, A.H.V. (1994) *FEBS Lett.* **345**, 50–54.

Crow, J.P., Spruell, C., Chen, J., Gunn, C., Ischiropoulos, H., Tsai, M., Smith, C.D., Radi, R., Koppenol, W.H. & Beckman, J.S. (1994a) *Free Rad. Biol. Med.* **16**, 331–338.

Crow, J.P., McCord, J.M. & Beckman, J.S. (1995) *Biochemistry*, in press.

Dawson, V.L., Dawson, T.M., London, E.D., Bredt, D.S. & Snyder, S.H. (1991) *Proc. Natl. Acad. Sci. USA* **88**, 6368–6371.

Dawson, V., Dawson, T., Uhl, G. & Snyder, S. (1993a) *Proc. Natl. Acad. Sci. USA* **90**, 3256–3259.
Dawson, V., Dawson, T., Bartley, D., Uhl, G. & Snyder, S. (1993b) *J. Neurosci.* **13**, 2651–2661.
Deng, H.-X., Hentati, A., Tainer, J., Igbal, Z., Cayabyab, A., Hung, W.-Y., Getzoff, E., Hu, P., Herzfeldt, B., Roos, R., Warner, C., Deng, G., Soriano, E., Smyth, C., Parage, H., Ahmed, A., Roses. A., Hallewell, R., Pericak-Vance, M. & Siddique, T. (1993) *Science* **261** 1047–1051.
East, S.J. & Garthwaite, J. (1991) *Neurosci. Lett.* **123**, 17–19.
Ford, P.C., Wink, D.A. & Stanbury, D.M. (1993) *FEBS Lett.* **326**, 1–3.
Föstermann, U., Gorsky, L.D., Pollock, J.S., Schmidt, H.H.H.W., Heller, M. & Murad, F. (1990) *Biochem. Biophys. Res. Commun.* **168**, 727–732.
Fridovich, I. (1986) *Arch. Biochem. Biophys.* **247**, 1–11.
Furchgott, R.F. & Vanhoutte, P.M. (1989) *FASEB J.* **3**, 2007–2018.
Gardner, P.R. & Fridovich, I. (1991) *J. Biol. Chem.* **266**, 1478–1483.
Garthwaite, J., Charles, S.L. & Chess-Williams, R. (1988) *Nature (Lond.)* **336**, 385–388.
Gurney, M.E., Pu, H., Chiu, A.Y., Dal Corto, M.C., Polchow, C.Y., Alexander, D.D., Caliendo, J., Hentati, A., Kwon, Y.W., Deng, H.-X., Chen, W., Zhai, P., Sufit, R.L. & Siddique, T. (1994) *Science*; **264**, 1772–1774.
He, Y.Y., Hsu, C.Y., Ezrin, A.M., & Miller, M.B. (1993) *Am. J. Physiol.* **265**, H252–H256.
Hogg, N., Darley-Usmar, V.M., Wilson, M.T. & Moncada, S. (1992) *Biochem. J.* **281**, 419–424.
Huang, Z., Huang, P.L., Pahahian, Dalkara, T., Fishman, M.C. & Moskowitz, M.A. (1994) *Science* **265**, 1883–1885.
Iadecola, C., Pelligrino, D.A., Moscowitz, M.A. & Lassen, N.A. (1994) *J. Cereb. Blood Flow Metab.* **14**, 175–192.
Ignarro, L.J. (1991) *Biochem. Pharmacol.* **41**, 485–490.
Ischiropoulos, H., Zhu, L. & Beckman, J.S. (1992a) *Arch. Biochem. Biophys.* **298**, 446–451.
Ischiropoulos, H., Zhu, L., Chen, J., Tsai, H.M., Martin, J.C., Smith, C.D. & Beckman, J.S. (1992b) *Arch. Biochem. Biophys.* **298**, 431–437.
Janzen, E.G. (1994) *Chem. Eng. News* (March 14), 4.
Kelm, M. & Schrader, J. (1988) *Eur. J. Pharmacol.* **155**, 317–321.
Kinouchi, H., Epstein, C.J., Mizui, T., Carlson, E., Chen, S.F. & Chan, P.H. (1991) *Proc. Natl. Acad. Sci. USA* **88**, 11158–11162.
Koppenol, W.H., Moreno, J.J., Pryor, W.A., Ischiropoulos, H. & Beckman, J.S. (1992) *Chem. Res. Toxicol.* **5**, 834–842.
Lafon-Cazal, M., Pietri, S., Culcasi, M. & Bockaert, J. (1993) *Nature* **364**, 535–537.
Lancaster, J.R. (1994) *Proc. Natl. Acad. Sci. USA* **91**, 8137–8141.
Lipton, S.A., Choi, Y.B., Pan, Z.H., Lei, S.Z., Chen, H.S.V., Sucher, N.J., Loscalzo, J., Singel, D.J. & Stamler, J.S. (1993) *Nature* **364**, 626–631.
Liu, T.H., Beckman, J.S., Freeman, B.A., Hogan, E.L. & Hsu, C.Y. (1989) *Am. J. Physiol.* **256**, H589–H593.
Malinski, T., Bailey, F., Zhang, Z.G. & Chopp, M. (1993) *J. Cereb. Blood Flow Metab.* **13**, 355–358.
Moncada, S., Herman, A.G. & Vanhoutte, P.M. (1987) *TIPS* **8**, 365–368.
Muizelaar, J., Marmarou, A., Young, H., Schoi, S., Wolf, A., Schneider, R. & Kontos, H. (1993) *J. Neurosurg.* **78**, 375–382.
Nagafuji, T., Matsui, T., Koide, T. & Asano, T. (1992) *Neurosci. Lett.* **147**, 159–162.
Nowicki, J.P., Duval, D., Poignet, H. & Scatton, B. (1991) *Eur. J. Pharmacol.* **204**, 339–340.
Ogasawara, M., Matsubara, Y., Narisawa, K., Aoki, M., Nakamura, S., Itoyama, Y. & Abe, K. (1993) *Nat. Gen.* **5**, 323–324.
Ohshima, H., Friesen, M., Brouet, I. & Bartsch, H. (1990) *Food Chem. Tox.* **28**, 647–652.

Omar, B.A. & McCord, J.M. (1990) *Free Rad. Biol. Med.* **9**, 473–478.
Omar, B.A., Gad, N.M., Jordan, M.C., Striplin, S.P., Russell, W.J., Downey, J.M. & McCord, J.M. (1990) *Free Rad. Biol. Med.* **9**, 465–471.
Oury, T.D., Ho, Y.-S., Piantadosi, C.A. & Crapo, J.D. (1992) *Proc. Natl. Acad. Sci. USA* **89**, 9715–9719.
Palmer, R.M.J., Ferrige, A.G. & Moncada, S. (1987) *Nature (Lond.)* **327**, 523–526.
Peunova, N. & Enlkolopov, G. (1993) *Nature (Lond.)* **364**, 450–453.
Pryor, W. (1992) *Free Rad. Biol. Med.* **12**, 83–88.
Pryor, W.A., Church, D.F., Govindan, C.K. & Crank, G. (1982) *J. Org. Chem.* **147**, 156–158.
Radi, R., Beckman, J.S., Bush, K.M. & Freeman, B.A. (1991) *J. Biol. Chem.* **266**, 4244–4250.
Radi, R., Rodriguez, M., Castro, L. & Telleri, R. (1994) *Arch. Biochem. Biophys.* **308**, 89–95.
Rosen, D.R., Siddique, T., Patterson, D., Figlewicz, D.A., Sapp, P., Hentati, A., Donaldson, D., Goto, J., O'Regan, J.P., Deng, H.-X., Rahmani, Z., Krizus, A., McKenna-Yasek, D., Cayabyab, A., Gaston, S.M., Berger, R., Tanszi, R.E., Halperin, J.J., Herzfeldt, B., Van den Bergh, R., Hung, W.-Y., Bird, T., Deng, G., Mulder, D.W., Smyth, C., Lang, N.G., Soriana, E., Pericak-Vance, M.A., Haines, J., Rouleau, G.A., Gusella, J.S., Horvitz, H.R. & Brown, R.H.J. (1993). *Nature*; **362**, 59–62.
Rubanyi, G.M. & Vanhoutte, P.M. (1986) *Am. J. Physiol.* **250**, H822–H827.
Sawyer, D.T. & Valentine, J. (1981) *Acct. Chem. Res.* **14**, 393–400.
Seidell, A. (1919) *Solubilities of Inorganic and Organic Compounds*, Vol. 1, p. 461. New York, D. Van Nostrand Co.
Shibuki, K. & Okada, D. (1991) *Nature (Lond.)* **349**, 326–329.
Solodkin, A., Traub, R.J. & Gebhart, G.F. (1992) *Neuroscience* **51**, 495–499.
Southam, E. & Garthwaite, J. (1993) *Neuropharmacology* **32**, 1267–1277.
Stamler, J.S., Simon, D.I., Osborne, J.A., Mullins, M.E., Jaraki, O., Michel, T., Singel, D.J. & Loscalzo, J. (1992a) *Proc. Natl. Acad. Sci. USA* **89**, 444–448.
Stamler, J.S., Jaraki, O., Osborne, J., Simon, D.I., Keaney, J., Vita, J., Singel, D., Valeri, C.R. & Loscalzo, J. (1992b) *Proc. Natl. Acad. Sci. USA* **89**, 7674–7677.
Stamler, J.S., Singel, D.J. & Loscalzo, J. (1992c) *Science* **258**, 1898–1902.
Traylor, T.G. & Sharma, V.S. (1992) *Biochemistry* **31**, 2847–2849.
Troost, D., Sillevis-Smitt, P.A.E., de Jong, J.M.B.V. & Swaab, D.F. (1992) *Acta Neuropathol.* **84**, 664–673.
Tsai, J.-H.M., Hamilton, T.P., Harrison, J.G., Jablowski, M., Woerd, M.v.d., Martin, J.C. & Beckman, J.S. (1994) *J. Am. Chem. Soc.* **116**, 4115–4116.
Valtschanoff, J.G., Weinberg, R.J. & Rustioni, A. (1992) *J. Comp. Neurol.* **321**, 209–222.
Wade, R. & Castro, C. (1990) *Chem. Res. Toxicol.* **3**, 289–291.
Wise, D.L. & Houghton, G. (1968) *Chem. Eng. Sci.* **23**, 1211–1216.
Wu, W. (1993) *Exp. Neurol.* **120**, 153–159.
Wu, W. & Li, L. (1993) *Neurosci. Lett.* **153**, 121–124.
Wu, W. & Scott, D.E. (1993) *Exp. Neurol.* **121**, 279–283.
Xu, Z., Cork, L, Griffin, J. & Cleveland, D. (1993) *Cell* **73**, 23–33.
Yamamoto, S., Golanov, E.V., Berger, S.B. & Reis, D.J. (1992) *J. Cereb. Blood Flow Metab.* **12**, 717–726.
Yang, G., Chan, P.H., Chen, J., Carlson, E., Chen, S.F., Weinstein, P., Epstein, C.J. & Kamii, H. (1994) *Stroke* **25**, 165–170.
Zafiriou, O. & McFarland, M. (1980) *Anal. Chem.* **52**, 1662–1667.
Zhang, F., White, J.G. & Iadecola, C. (1994) *J. Cereb. Blood Flow Metab.* **14**, 217–226.
Zhu, L., Gunn, C. & Beckman, J.S. (1992) *Arch. Biochem. Biophys.* **298**, 452–457.

CHAPTER 11

NITRIC OXIDE AND THE REGULATION OF CEREBRAL ARTERIAL TONE

Noboru Toda

Department of Pharmacology, Shiga University of Medical Sciences, Seta, Ohtsu 520-21, Japan

Table of Contents

11.1 Introduction 207
11.2 NO derived from vasodilator nerve 208
11.2.1 Response to vasodilator nerve stimulation 209
11.2.2 Variations in the response to vasodilator nerve stimulation 212
11.2.3 Histological study on NOS-containing nerves 213
11.2.4 Evidence for and against the hypothesis that NO is a neurotransmitter 214
11.2.5 Prejunctional regulation of nitroxidergic nerve function 216
11.3 NO derived from the endothelium 217
11.3.1 EDRF 217
11.3.2 EDNO 218
11.3.3 Ageing 220
11.3.4 Shear stress, stretch and blood flow change 220
11.3.5 Basal release of EDNO 220
11.3.6 EDNO-mediated response *in vivo* 222
11.4 Conclusion 222
References 223

11.1 Introduction

Studies on the physiological and pathophysiological roles of nitric oxide (NO), initiated in the vascular endothelium and macrophage, have now been expanded to the central and peripheral nervous systems, cardiovascular system and immunology. Following the discovery of the unique gas NO radical, basic and clinical researchers have found clues to solve many of the problems known for decades. The neurohumoral transmitter mechanism in non-adrenergic, non-cholinergic nerves is an example.

Nitric Oxide in the Nervous System
ISBN 0-12-721985-4

Cerebral circulation is regulated by neural and humoral factors, and differently from circulation in peripheral organs and tissues. A unique feature of cerebral circulatory regulation is the weak functioning of adrenergic vasoconstrictor nerves and the dominant functioning of non-adrenergic, non-cholinergic vasodilator nerves. They are also characterized by their response to endogenous vasoactive substances which modulate the vascular tone by mechanisms dependent upon or independent of the endothelium. These features act to maintain necessary blood flow in the brain even in circulatory emergency.

The purposes of this contribution are to describe the mechanisms underlying the response to vasodilator nerve stimulation in cerebral arteries, to discuss the new concept of neurotransmission via NO, to interpret factors influencing the vasodilator nerve function and to review the cerebroarterial relaxation mediated by endothelium-derived relaxing factor (EDRF) or NO, as compared to the response associated with NO derived from nerves.

11.2 NO derived from vasodilator nerve

The discovery of the non-adrenergic, non-cholinergic nature of nicotine-induced relaxation of dog cerebral arteries (Toda, 1975) triggered the investigation of the mechanisms underlying neurally induced cerebrovascular dilatation. Nicotine and electrical square pulses applied transmurally via an electronic stimulator (transmural electrical stimulation) have been used to activate perivascular nerves; the selective neural stimulation is verified by abolition of the induced responses by hexamethonium and tetrodotoxin, respectively, in each preparation. Relaxant responses to these chemical and electrical stimulations are not affected by many pharmacological antagonists, including cyclo-oxygenase inhibitors, aminophylline, cimetidine, chlorpheniramine and ouabain (Toda, 1982). Although cerebroarterial relaxations caused by prostacyclin (Toda, 1980), adenosine, ATP (Toda *et al.*, 1982), histamine (Toda, 1990a) and activation of membrane ATPase (Toda, 1974) have been demonstrated, the involvement of prostaglandins, purinoceptors, histamine receptors and the electrogenic Na^+ pump in the response are thus excluded in this preparation. The other endogenous vasodilator substances include polypeptides, such as substance P, vasoactive intestinal polypeptide (VIP; Toda, 1982; Owman, 1990), calcitonin gene-related peptide (CGRP; Toda and Okamura, 1991a) and atrial natriuretic peptide (ANP; Okamura *et al.*, 1989). These peptides, except for ANP, have histochemically been demonstrated to be present in perivascular nerves innervating the cerebral arterial wall (Owman, 1990), leading us to speculate that the response to vasodilator nerve stimulation is associated with the release of these peptides. The neurally induced relaxation is endothelium independent, whereas the relaxation caused by substance P depends on the endothelium (Onoue *et al.*, 1988); thus, the involvement of substance P in neurally induced relaxation is ruled out. The other peptides, though able to profoundly relax cerebral arteries, easily lose

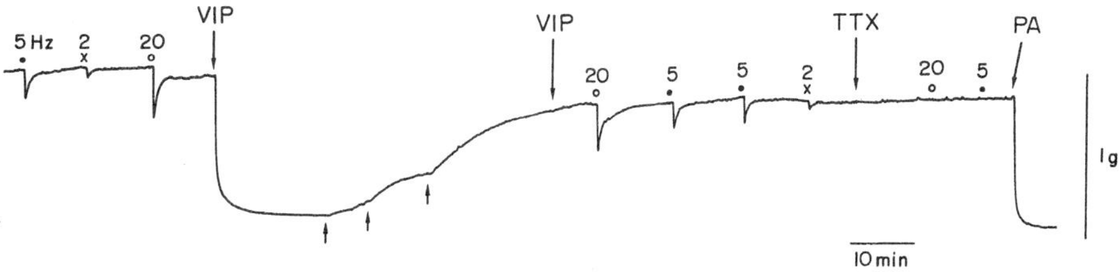

Figure 1 Relaxant responses to transmural electrical stimulation (2, 5 and 20 Hz) in a monkey middle cerebral arterial strip made unresponsive to VIP by prior treatment with a high concentration (10^{-7} M) of VIP. The strip was partially contracted with $PGF_{2\alpha}$. Upward arrows indicate supplemental additions of $PGF_{2\alpha}$ to restore the active tone. The responses did not differ before and after the addition of VIP. TTX=3×10^{-7} M tetrodotoxin; PA=10^{-4} M papaverine, which produced the maximal relaxation.

their vasodilator ability following treatment with high concentrations of the same peptides. In arteries made insensitive to these peptides, nerve stimulation still produces a relaxation that is comparable to that seen before peptide treatment (Figure 1). Therefore, VIP, CGRP and ANP are also excluded as candidates for neurotransmitter-induced relaxation (Toda, 1982; Okamura *et al.*, 1989; Toda and Okamura, 1991a). This conclusion is supported by the fact that treatment with capsaicin in concentrations sufficient to abolish the response mediated by neurogenic CGRP and substance P does not suppress the response to nerve stimulation in dog cerebral arteries (Okamura and Toda, 1994).

11.2.1 Response to vasodilator nerve stimulation

11.2.1.1 Mechanical response

As already mentioned, relaxations caused by nicotine, unlike those induced by electrical stimulation, are abolished by hexamethonium but not by tetrodotoxin. Generation of action potentials does not seem to be evoked by nicotine, which liberates neurotransmitters by acting on nicotinic receptors on nerve terminals and increasing the inward movement of Ca^{2+} across neural membranes (Nedergaard, 1988). Relaxations associated with vasodilator nerve stimulation by nicotine or electrical pulses are abolished by treatment with methylene blue or oxyhaemoglobin (Toda, 1987; Linnik and Lee, 1989). Methylene blue is known to be an inhibitor of soluble guanylyl cyclase (Gruetter *et al.*, 1981), the enzyme being activated by nitro- or nitroso-compounds and EDRF (Furchgott, 1983). Oxyhaemoglobin scavenges NO (Martin *et al.*, 1985). These data allowed us to speculate that the neurogenic vasodilatation is mediated by cyclic GMP (Toda, 1987) which might be produced by nitroso-compounds. Until the discovery of NO synthase (NOS), the involvement of nitroso-compounds was not clarified.

On the basis of accumulated data, EDRF was suggested to be NO by Furchgott (1988), Palmer *et al.* (1987) and Ignarro *et al.* (1987a) or an NO analogue, like

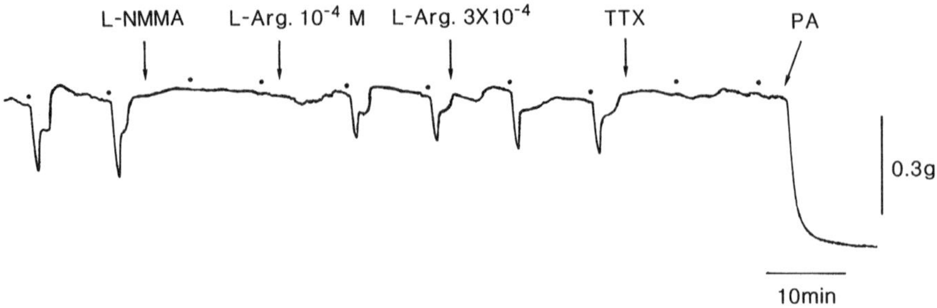

Figure 2 Modification by N^G-monomethyl-L-arginine (L-NMMA, 10^{-4} M) and L-arginine (L-Arg) of the response to transmural electrical stimulation at 5 Hz in a dog middle cerebral arterial strip partially contracted with $PGF_{2\alpha}$. TTX=3×10^{-7} M tetrodotoxin; PA=10^{-4} M papaverine. (Reproduced with permission from Toda and Okamura, 1990a.)

S-nitrosocysteine (Myers *et al.*, 1990). This concept is now widely accepted. NO is synthesized from L-arginine, and the synthesis is inhibited by L-arginine analogues (Palmer *et al.*, 1988), such as N^G-monomethyl-L-arginine (L-NMMA), N^G-nitro-L-arginine (L-NNA) and L-NNA methyl ester. The development of these drugs allowed the first demonstration of a suppression by L-NMMA of the response to transmural electrical stimulation in a dog cerebral arterial strip (Figure 2; Toda and Okamura, 1990a). D-NMMA was without the effect. The inhibition by L-NMMA was reversed by L-arginine, a substrate of NOS, but not by D-arginine. Similar inhibition was seen with L-NNA, which is approximately 100 times more effective than L-NMMA (Toda *et al.*, 1990). Nicotine-induced relaxations were also abolished by L-NMMA and L-NNA, and the response restored by L-arginine (Toda and Okamura, 1990b,c, 1991b,c). We had previously demonstrated that acetylcholine in high concentrations (10^{-4} M or higher) elicits relaxations of dog cerebral artery, which are not influenced by atropine but are abolished by hexamethonium (Toda, 1979). Although the mechanism was not determined at that time, this response is now considered to derive from actions on nicotinic receptors on vasodilator nerve terminals. The hexamethonium-sensitive relaxation elicited by high concentrations of acetylcholine is abolished by treatment with NOS inhibitors in dog ophthalmic arteries (Wang *et al.*, 1993).

Nicotine injected into the vertebral artery dilates the basilar artery in anaesthetized dogs in which the arterial diameter is angiographically measured. Treatment with L-NNA or hexamethonium abolishes the effect of nicotine (Toda *et al.*, 1993c). Nicotine may therefore produce this basilar artery dilatation by a neurogenic mechanism indentical to that seen above in isolated cerebral arteries.

11.2.1.2 Release of NO and content of cGMP

In superfused endothelium-denuded dog cerebral arterial strips, transmural electrical stimulation or nicotine increases the release of nitroxy compounds (NO_x), the increments being depressed by treatment with NOS inhibitors (Toda and Okamura, 1990b). The effects of electrical stimulation and nicotine are abolished by tetrodotoxin and hexamethonium, respectively. The electrical and chemical stimulations also increase the content of cGMP in this preparation (Toda and Okamura, 1990b, 1991), as do nitro- or nitroso-compounds (Ignarro *et al.*, 1987b). NO liberated from stimulated nerves appears to activate soluble guanylyl cyclase in smooth muscle cells and increase the production of cGMP which decreases the concentration of intracellular Ca^{2+} responsible for myofibrillar contraction and/or the sensitivity to Ca^{2+} of contractile machinery, resulting in muscular relaxation.

11.2.1.3 Comparison of the response to NO derived from nerve and endothelium

Characteristics in the production and the action of NO are similar in cerebral vasodilator nerves and endothelial cells. In addition, histological study has demonstrated that antiserum raised against NOS purified from the rat cerebellum is effective in staining the enzyme in the perivascular nerves and endothelial cells (Bredt *et al.*, 1990). However, we have had an impression that the neurally induced response is more susceptible to NOS inhibitors than the response mediated by endothelium-derived NO. It is hypothesized that there are two distinct NO synthesizing enzymes of constitutive type in the cerebellum, having low and high affinities for L-NNA (East and Garthwaite, 1990). Therefore, quantitative comparisons were made of the effectiveness of NOS inhibitors in inhibiting the response to nerve stimulation and EDRF-releasing peptides (Toda *et al.*, 1993a). Per cent inhibitions by 10^{-6} M L-NNA of the responses to nicotine, substance P and arginine vasopressin at concentrations producing similar magnitudes of relaxation in dog cerebral arteries were 66%, 58% and 79%, respectively. The greater inhibitory effect on vasopressin-induced relaxation would be due to the ability of this peptide to elicit appreciable contraction by acting directly on smooth muscle (Onoue *et al.*, 1988). If the inhibitory potency reflects the susceptibility of NOS to L-NNA, the different natures of this enzyme in the nerve and endothelium cannot be detected.

Removal of Ca^{2+} from the bathing media abolishes the responses to nerve stimulation (N. Toda *et al.*, unpublished data) and those mediated by EDRF in blood vessels (Griffith *et al.*, 1986), suggesting the necessity of Ca^{2+} in the production and release of NO in the nerve and endothelium. Dihydropyridine-type Ca^{2+} entry blockers, selective inhibitors of the L-type Ca^{2+} channel, do not interfere with the release of neurotransmitter from nerves (Godfraind *et al.*, 1986) nor with the release of EDRF from the endothelium (Miller *et al.*, 1985). Nicardipine, a dihydropyridine-type blocker, does not inhibit the responses to vasodilator nerve stimulation or substance P, an EDRF-releasing peptide, in dog cerebral arteries (Toda and Okamura, 1992a). On the other hand, non-selective Ca^{2+} entry blockers, such as Ni^{2+}, Co^{2+}, Cd^{2+}, Mn^{2+} and La^{3+}, depress the sustained increase by EDRF-releasing substances in intracellular Ca^{2+} and the production and release of NO in cultured endothelial cells (Robertson and Gallin, 1986; Colden-Stanfield *et al.*, 1987).

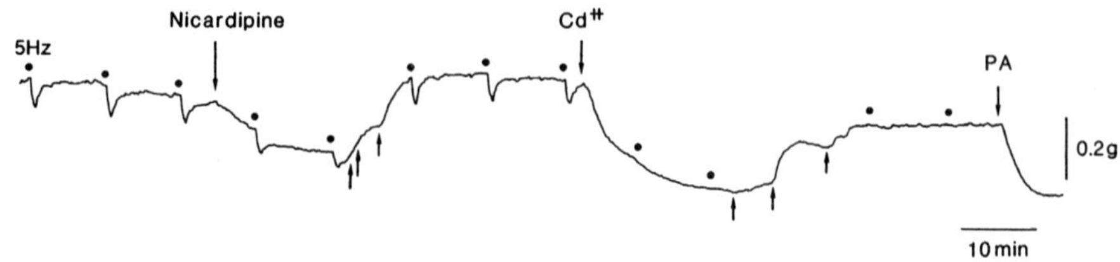

Figure 3 Modification by Ca^{2+} entry blockers, cadmium (Cd^{2+}) and nicardipine, of the response to transmural electrical stimulation at 5 Hz in a dog middle cerebral arterial strip denuded of the endothelium. The strip was contracted with $PGF_{2\alpha}$. Upward arrows indicate supplemental additions of $PGF_{2\alpha}$ to restore the active tone. PA=10^{-4} M papaverine. (Reproduced with permission from Toda and Okamura, 1992a.)

Cd^{2+} also inhibits the mechanical response (Figure 3) and the increased production of cGMP by vasodilator nerve stimulation or substance P in dog cerebral arteries (Toda and Okamura, 1992a). The activity of constitutive NOS present in nerve cells and fibres and in endothelial cells is related directly to concentrations of Ca^{2+} (Bredt and Snyder, 1990; Janssens *et al.*, 1992). These findings suggest that Ca^{2+} introduced from the external fluid via Ca^{2+} channels of a non-L-type is quite important or a prerequisite for activating the enzyme and synthesizing NO in these tissues. There is some evidence that Ca^{2+} liberated from intracellularly stored sites is also available for the synthesis of NO in the endothelial cells.

11.2.2 Variations in the response to vasodilator nerve stimulation

11.2.2.1 Species variation

Relaxant responses to nerve stimulation by electrical pulses and nicotine are also obtained in human (Figure 4; Toda, 1993), Japanese monkey (Toda and Okamura, 1990c), bovine (Gonzalez and Estrada, 1991; Ayajiki *et al.*, 1993), porcine (Lee *et al.*, 1991) and feline cerebral arteries. Modifications of the responses by NOS inhibitors, arginine,oxyhaemoglobin and methylene blue are principally identical in these species. It is difficult to quantitatively compare the response in these species. However, a trend toward more responsiveness of the monkey arteries than those of other mammals can be seen. There have been histochemical demonstrations of nerve fibres containing NOS in these cerebral arteries, except for the porcine artery. Human and feline cerebral arterial preparations are more responsive to nicotine than to electrical stimulation. In contrast, bovine basilar arterial strips respond exclusively to electrical stimulation with a relaxation. Although the unresponsiveness to nicotine may be explained by a lack of nicotinic receptors in nerve terminals innervating that artery, the reason for the lower responsiveness to electrical stimulation is unclear. Leakage of electrical current, inadequate clearance between the tissue and stimulating electrode, tissue thickness, uneven distribution of nerve fibres, etc. may be considered.

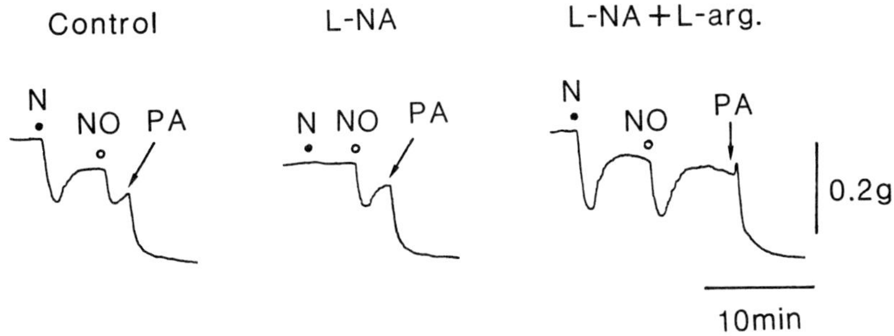

Figure 4 Modification by N^G-nitro-L-arginine (L-NA, 10^{-6} M) and L-arginine (L-arg, 3×10^{-4} M) of the responses to nicotine (10^{-4} M) and NO (2×10^{-7} M) in a human cerebral arterial strip contracted with $PGF_{2\alpha}$. PA=10^{-4} M papaverine. (Reproduced with permission from Toda, 1993.)

11.2.2.2 Ageing

Nicotine-induced relaxations do not differ between the proximal middle cerebral and the basilar arteries isolated from beagles of 30 days, 3 months, 1 year and 3 years of age (Toda *et al.*, 1986). The relaxant response is greater in proximal middle cerebral arteries from the adult dogs than in the arteries of the distal portion (Toda and Miyazaki, 1984). In contrast to the dog arteries, responses of monkey basilar and middle cerebral arteries to electrical stimulation and nicotine vary with age (Toda, 1991). In the young (3–4 weeks of age) monkey arteries, nerve stimulation elicits a contraction, which is abolished or reversed to a slight relaxation by treatment with phentolamine. The arteries from adult monkeys (4 years or older) respond to the stimulation only with a relaxation. The relaxant response seen in the presence of α-adrenoceptor blockade is directly related to ageing from 3 weeks to 4 years. Similar results are also obtained with nicotine-induced relaxation. In these primate arteries, the vasodilator nerve function develops with age, although it has not been determined whether this is due to the age-dependent maturation of the perivascular nerve or to the increased sensitivity with age of the smooth muscle to nitro- and nitroso-vasodilators.

11.2.3 Histological study on NOS-containing nerves

Bredt *et al.* (1990) first demonstrated the presence of perivascular nerves containing NOS immunoreactivity in large cerebral arteries of the rat brain. By the use of the same NOS antiserum, we determined the dense network of nerve bundles and fibres in the wall of dog cerebral arteries (Figure 5; Yoshida *et al.*, 1993). Similar innervation is also seen in monkey cerebral arteries. The fibres mainly distribute to the adventitia, and fine fibres are also present in the outer layer of the media. Such an innervation in the medial layer seems to be important in allowing a labile substance

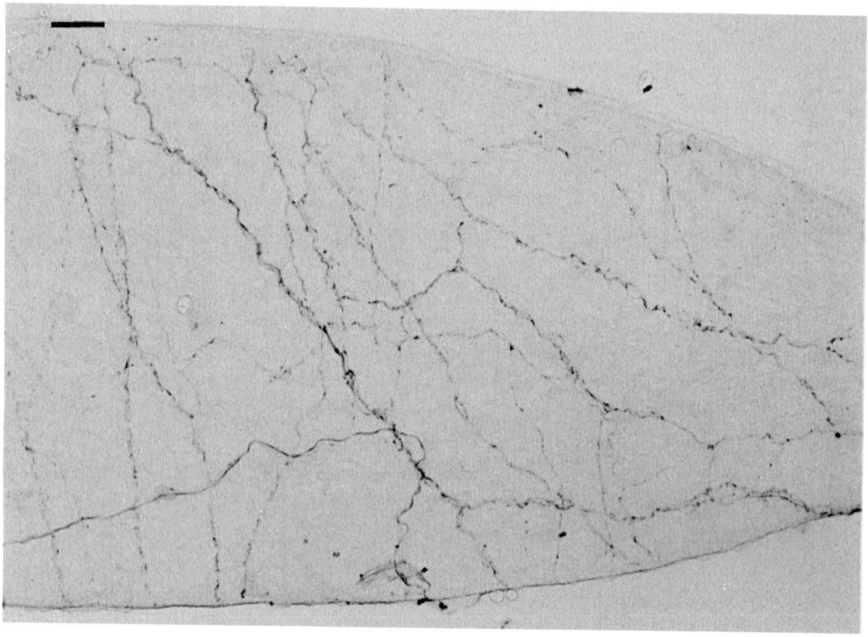

Figure 5 Histochemical demonstration of perivascular nerve fibres containing NOS immunoreactivity in a whole mount preparation of a dog middle cerebral artery.

like NO to transmit information from nerve to muscle. In contrast to the data on rat cerebral arteries, NOS-positive fibres are also seen in distal cerebral arteries from the dog. Recent studies by Nozaki *et al.* (1993) have shown the dense network of perivascular nerve fibres containing NOS immunoreactivity in human cerebral arteries.

Distribution of nerve cells, bundles and fibres stained by the NADPH diaphorase method is reportedly identical to that of nerves containing NOS immunoreactivity (Dawson *et al.*, 1991). Our findings on dog cerebral arteries, in which the nerve fibres are stained by the NADPH diaphorase method (Toda *et al.*, 1993c) and NOS antiserum (Yoshida *et al.*, 1993), support their conclusion. We obtained the positive staining of NADPH diaphorase in perivascular nerves also in the bovine basilar artery.

11.2.4 Evidence for and against the hypothesis that NO is a neurotransmitter

The findings so far presented strongly support the hypothesis that NO acts as a neurotransmitter in the vasodilator nerves in the cerebral arteries from primate and subprimate mammals. Therefore, we call these nerves 'nitroxidergic' (Toda *et*

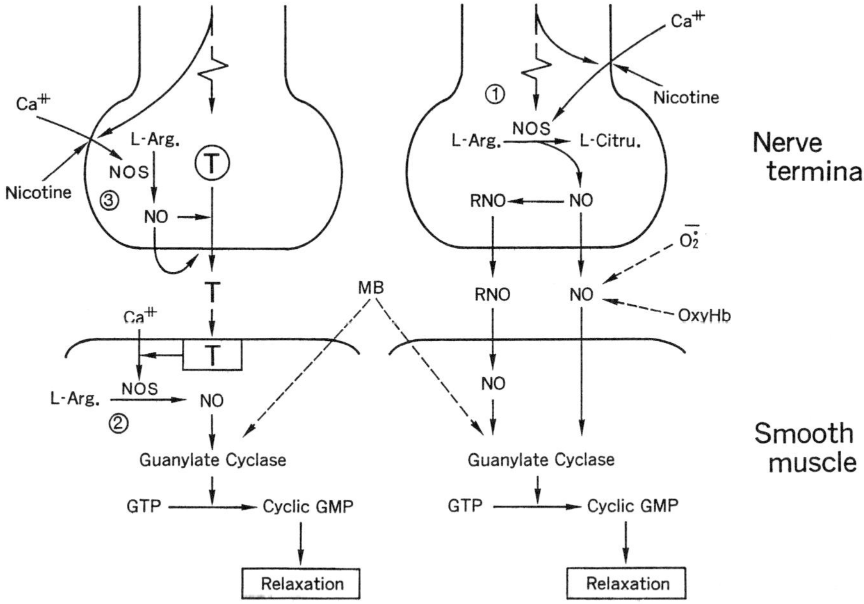

Figure 6 Possible mechanisms underlying the relaxation induced by stimulation of vasodilator nerve with electrical pulses or nicotine in cerebral arteries. L-Arg.=L-arginine; L-Citru.=L-citrulline; RNO=NO analogue, such as *S*-nitrosothiole; T=unknown vasodilator transmitter; O_2^-=superoxide anion; MB=methylene blue. (Reproduced with permission from Toda and Okamura, 1990c, with some modifications.)

al., 1991c; Toda and Okamura, 1992b). NOS inhibitors impair the response to nerve stimulation but do not affect the relaxation caused by NO. Oxyhaemoglobin and methylene blue abolish these responses. NO_x is released during nerve stimulation from the arteries denuded of the endothelium. Nerve stimulation and NO increase the content of cGMP in the tissue. There are NOS-containing nerve fibres in the arterial wall. These data are sufficient to support the concept of neurohumoral transmission with NO from vasodilator nerve to cerebroarterial smooth muscle. However, one of the classical criteria for neurotransmitter status is not met; NO itself is not stored in nerve terminals. Our hypothesis on this problem is that such an extremely labile substance is not necessarily present in a stored form but is synthesized from L-arginine by NOS that is activated immediately upon increased cellular Ca^{2+} associated with nerve activation ('1' in Figure 6). Although there is no direct evidence supporting such a way of synthesis and release of NO, it is perhaps not necessary to adhere to the classical concept of neurotransmission, regarding the storage of transmitter before nerve action potentials reach the site of release.

In addition to the mechanisms of NO production, there are some comments raised against the hypothesis of NO as a neurotransmitter. One of these is the possibility of producing NO in extraneuronal tissues. Smooth muscle may be a probable site ('2' in Figure 6). After we reported the possible mechanism of action of NO in neurotransmission, Wood *et al.* (1990) and Moritoki *et al.* (1991) provided evidence for a muscle-derived relaxing factor that was characterized to be functionally identical to NO. However, in our experimental system with cerebral arteries, we did not obtain evidence indicating the production of NO, even though the arteries were incubated for 24 h in the bathing media and stimulated repeatedly by L-arginine (Toda and Okamura, 1990c), as were the blood vessels used in their experiments. Recent studies on cultured smooth muscle cells indicate that the NO production is due to expression of an NOS produced following cytokine or lipopolysaccharide stimulation (Busse and Mülsch, 1990), which is not activated by Ca^{2+}. Glial cells (mainly microglia) also synthesize NO by an NOS of the inducible type (Galea *et al.*, 1992; Simmons and Murphy, 1992). Astrocytes also appear to possess constitutive NOS, although the level of expression is lower than in neurons (Murphy *et al.*, 1993). NOS is largely induced in astrocytes, as in microglia. We have reported evidence showing nitroxidergic innervation also in extracranial arteries, such as dog and monkey mesenteric and temporal arteries (Toda and Okamura, 1992b), which are not surrounded by glial and astrocytic cells. These data exclude the possible involvement of smooth muscle and glial cells in the neurally induced response. One other comment is on a stimulating action of neurogenic NO on the prejunctional site ('3' in Figure 6) which participates in the synthesis of an unknown substance (T) as a vasodilator neurotransmitter. This possibility can be ruled out by the fact that in the arteries in which the neurally induced relaxation is abolished by treatment with NOS inhibitors, the addition of NO or NO donors, such as sodium nitroprusside and SIN-1, does not restore the response (Toda and Okamura, 1990c).

11.2.5 Prejunctional regulation of nitroxidergic nerve function

In dog and monkey cerebral arteries, the relaxation elicited by nerve stimulation is not inhibited or potentiated by treatment with atropine. On the other hand, bovine basilar artery relaxations thus stimulated are significantly potentiated by the muscarinic receptor antagonist (Toda and Ayajiki, 1990). The relaxation is mediated by NO, as already explained. Histochemical examination of the rat brain detected cholinergic and VIPergic innervation in cerebral arteries that arises from the pterygopalatine ganglion (Hara *et al.*, 1985). A recent study adds further information indicating that nerve cells and fibres containing NOS or NADPH diaphorase are also present in the ganglion (Nozaki *et al.*, 1993). Therefore, we investigated the implication of cholinergic and VIPergic nerves in the nitroxidergic nerve function in bovine basilar arteries (Ayajiki *et al.*, 1993). The relaxation induced by electrical stimulation is attenuated by acetylcholine in a concentration-dependent manner, the inhibition being reversed by atropine

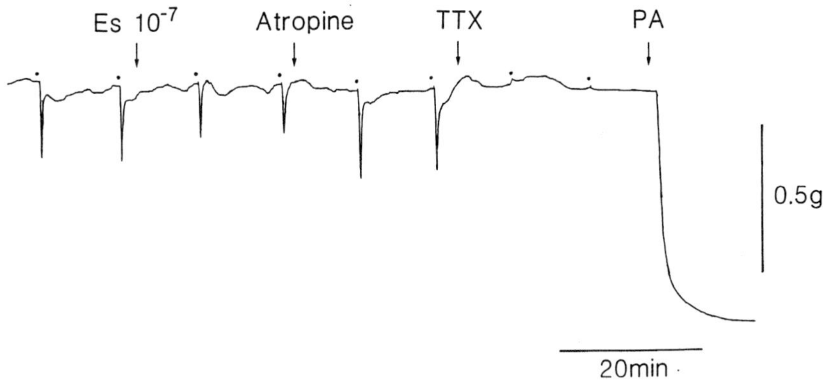

Figure 7 Modifications by physostigmine (Es, 10^{-7} M) and atropine (10^{-7} M) of the response to transmural electrical stimulation (10 Hz) of a bovine basilar arterial strip contracted with $PGF_{2\alpha}$. TTX=3×10^{-7} M tetrodotoxin; PA=10^{-4} M papaverine. (Reproduced partially with permission from Toda and Ayajiki, 1990.)

(Figure 7). In addition, the response is potentiated by atropine and inhibited by physostigmine, a cholinesterase inhibitor, that does not stimulate muscarinic receptors directly. There are abundant nerve fibres containing cholinesterase in the cerebroarterial wall, demonstrated by histochemical study (Toda and Ayajiki, 1990). Acetylcholine liberated from cholinergic nerves by electrical stimulation is expected to act on muscarinic receptors on nitroxidergic nerve terminals, resulting in a suppression of the synthesis and release of NO. This is the first demonstration of cholinergic prejunctional inhibition in perivascular nerves. There is no evidence like this in cerebral arteries from other mammals so far tested.

The other substance that may influence the neurogenic response is VIP. Treatment with this peptide in different concentrations does not significantly alter the neurally induced relaxation, suggesting that neurogenic VIP even though liberated fails to participate in the regulation of cerebrovascular tone by acting on pre- and postjunctional sites (Ayajiki *et al.*, 1993).

11.3 NO derived from the endothelium

11.3.1 EDRF

Furchgott and Zawadzki (1980) discovered that the endothelium when stimulated by acetylcholine liberated a vasodilator substance, thereby producing smooth muscle relaxation. The substance was therefore named 'endothelium-derived relaxing factor' (Furchgott, 1983). Other substances including bradykinin, substance P, arginine vasopressin, histamine, serotonin, noradrenaline, ADP, ATP, Ca^{2+} ionophores, etc. also have the ability to liberate EDRF. The vascular effect of EDRF, a quite labile substance, is abolished by oxyhaemoglobin and antioxidants or by methylene blue, an inhibitor of soluble guanylyl cyclase (Gruetter *et al.*, 1981). After a long effort by many investigators, EDRF was determined to be NO or an NO analogue, as described in the previous section. The agonist-induced, endothelium-dependent relaxation that is abolished by NOS inhibitors, such as L-NMMA, L-NNA and L-NNA methyl ester, can be regarded as a response mediated by NO derived from the endothelium (EDNO). Recent studies on a variety of blood vessels indicate that EDRF may not necessarily be identical to EDNO, since high concentrations of NOS inhibitors do not always abolish the endothelium-dependent relaxation (Vanhoutte, 1993; Enokibori *et al.*, 1994).

The endothelium of cerebral arteries from various mammals also has the ability to liberate EDRF. The relaxation of cerebral arteries caused by agonists, such as acetylcholine, substance P, vasopressin, histamine, ADP, thrombin and Ca^{2+} ionophore, is endothelium dependent, and is suppressed by treatment with oxyhaemoglobin, antioxidants and methylene blue but not by cyclo-oxygenase inhibitors. Responses to acetylcholine vary between animal species. This compound produces an endothelium-dependent relaxation in human, cat and rabbit cerebral arteries that is smaller in magnitude than that seen in peripheral arteries from the same species. On the other hand, acetylcholine does not produce measurable endothelium-dependent relaxations in dog and monkey cerebral arteries. Summarized in Table 1 are the agonists and the preparations that participate in the release of EDRF. Release of EDRF from isolated basilar arterial segments (Kim *et al.*, 1989) and perfused cat pial arterioles *in vivo* (Kontos *et al.*, 1988) can be detected by bioassay methods. Responses mediated by EDRF in the cerebral microcirculation *in vivo* are summarized in a review article by Marshall and Kontos (1990). Among the substances listed in the table, only a few agonists and preparations have been re-evaluated to determine whether EDNO is actually involved in the cerebroarterial relaxation.

11.3.2 EDNO

Endothelium-dependent relaxations of dog cerebral arteries induced by substance P or arginine vasopressin and those produced by acetylcholine in cat cerebral arteries are abolished or reversed to contractions by treatment with NOS inhibitors (Alonso *et al.*, 1992; Katusic, 1992; Toda *et al.*, 1993a). This inhibition is reversed by L-arginine. Endothelial cells containing NOS immunoreactivity have been demonstrated in rat cerebral arteries (Bredt *et al.*, 1990). As far as we know, there is no evidence indicating that an EDRF other than NO is involved in the endothelium-dependent relaxation, except in the rabbit middle cerebral arterial segments, which respond to acetylcholine with a hyperpolarization and vasodilatation (Brayden, 1990). The substance(s) derived from the endothelium that hyperpolarizes

Table 1 Endothelium-dependent relaxation in cerebral arteries

Species	Artery	Vasodilator	Reference
Dog	BA	Vasopressin	Katusic *et al.* (1984)
		Oxytocin	Katusic *et al.* (1986)
		Bradykinin	
		A23187	Kanamaru *et al.* (1987)
		Thrombin	Kim *et al.* (1988)
	MCA	Substance P	Onoue *et al.* (1988)
		A23187	
	PComA	Vasopressin	
		A23187	
	BA	Bradykinin	Katusic *et al.* (1989)*
	BA, MCA	Substance P	Toda *et al.* (1993a)*
		ATP, ADP	
Rabbit	BA	Acetylcholine	Fujiwara *et al.* (1986)
		ATP	
Human	BA	Acetylcholine	Whalley *et al.* (1987)
		Bradykinin	
		A23187	Kanamaru *et al.* (1989)
		Thrombin	Hatake *et al.* (1990)
		Bradykinin	
	MCA	Histamine	Toda (1990a)
		Acetylcholine	
		Substance P	
Monkey	BA, MCA	Histamine	Toda (1990a)
			Ayajiki *et al.* (1992)*
		ATP, ADP	Toda *et al.* (1991a)
		A23187	
Pig	BA	ADP	Shimokawa *et al.* (1988)
Cat	MCA	Acetylcholine	Dauphin and Hamel (1990)
	MCA	Acetylcholine	Alonso *et al.* (1992)*

BA, basilar artery; MCA, middle cerebral artery; PComA, posterior communicating artery.
* Mediation by endothelium-derived NO was determined.

vascular muscle cell membranes is called endothelium-derived hyperpolarizing factor (EDHF) and is distinct from EDRF (Suzuki and Chen, 1990).

The histamine-induced phasic relaxation of monkey cerebral arteries is dependent on the endothelium and is abolished by chlorpheniramine or L-NNA, suggesting that EDNO released by stimulation of the histamine H_1 receptor subtype in the endothelium mediates the response (Figure 8; Toda, 1990a; Ayajiki *et al.*, 1992).

11.3.3 Ageing

The endothelium-dependent relaxations of isolated human basilar arteries by bradykinin and the Ca^{2+} ionophore A23187 are not altered with increasing age from 15 to 91 years (Hatake *et al.*, 1990). In contrast, the dilatation of rat cerebral arterioles *in vivo* by acetylcholine and bradykinin that is presumably mediated by EDRF is reduced with advancing age from 6–8 months to 22–24 months, whereas nitroglycerin-induced vasodilatation does not differ between the adult and aged rats (Mayhan *et al.*, 1990). We have determined that the responsiveness of human coronary arteries mediated by EDRF is not reduced with ageing (Toda and Okamura, 1989), whereas relaxations of dog coronary and mesenteric arteries mediated by EDRF are related inversely to increasing age (Shimizu and Toda, 1986; Toda *et al.*, 1987). Whether such a discrepancy is due to different species or to the different size of the arteries (basilar artery vs. cerebral arteriole) remains to be ascertained.

11.3.4 Shear stress, stretch and blood flow change

Shear stress applied to cultured aortic endothelial cells opens an ion channel that is permeable to Ca^{2+} (Lansman *et al.*, 1987) and activates a K^+ current (Olesen *et al.*, 1988), possibly triggering the production of EDRF. Flow-dependent dilatation in dog conduit coronary and femoral arteries (Holtz *et al.*, 1984; Rubanyi *et al.*, 1986) and rabbit and rat peripheral resistance vessels (Griffith *et al.*, 1987; Koller and Kaley, 1990) is mediated by EDRF. Elevation of transmural pressure in isolated, perfused cat middle cerebral arterial segments produces membrane depolarization, generates action potentials and reduces internal diameter, only when the endothelium is intact (Harder, 1987). On the other hand, the myogenic response to increasing intraluminal pressure in isolated perfused rat posterior cerebral arteries and to stretch in dog cerebral arterial strips is preserved after the endothelium is removed (McCarron *et al.*, 1989; Nakayama *et al.*, 1989). Further study is required to determine whether cerebral vascular endothelial cells are resistant to physical stimuli or whether the functions of the endothelial mechanosensor differ in a variety of blood vessels from different mammals.

11.3.5 Basal release of EDNO

Contractions caused by NOS inhibitors or oxyhaemoglobin of dog and monkey cerebral arteries are dependent partially on the endothelium, suggesting that

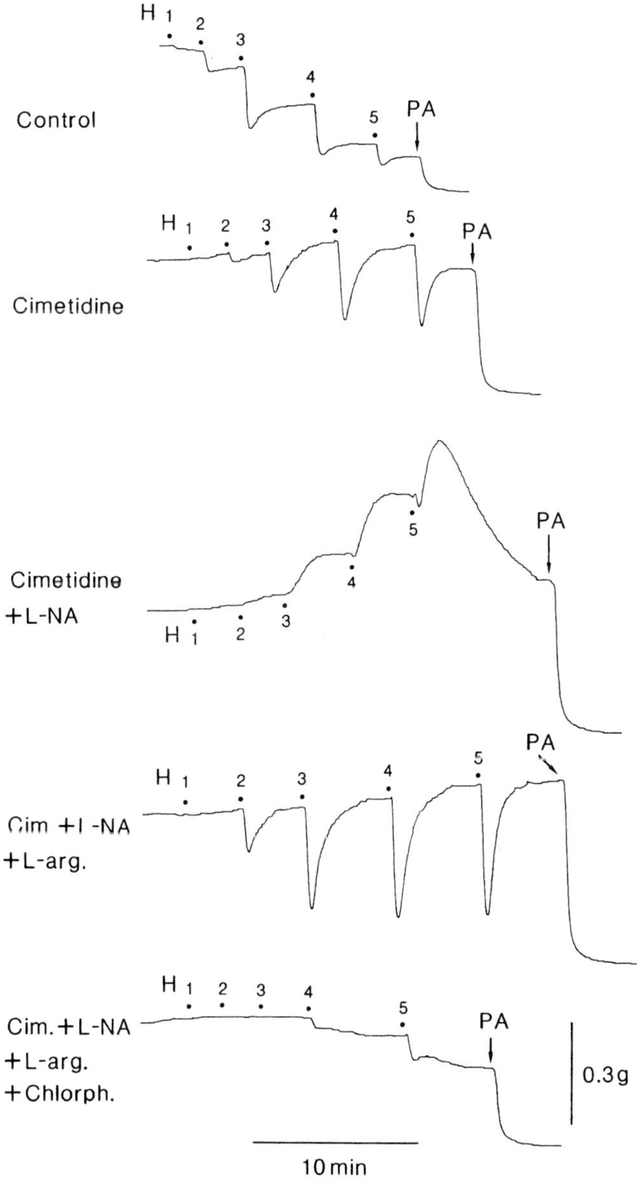

Figure 8 Responses to histamine of a monkey basilar arterial strip with intact endothelium before and after treatment with various blocking agents. The strip was contracted with $PGF_{2\alpha}$. Concentrations of histamine from 1 to 5=2×10^{-8}, 10^{-7}, 5×10^{-7}, 2×10^{-6} and 10^{-5} M, respectively. PA=10^{-4} M papaverine; Cim.=10^{-5} cimetidine; Chlorph.=10^{-6} M chlorphenir-amine; L-NA=10^{-6} M N^{G}-nitro-L-arginine; L-arg.=3×10^{-4} M L-arginine. (Reproduced with permission from Ayajiki *et al.*, 1992.)

spontaneously released EDNO significantly dilates the arteries (Toda, 1990b; Toda *et al.*, 1993b; Brian and Kennedy, 1993). The L-NNA-induced contraction was reversed by L-arginine but not by D-arginine (Toda *et al.*, 1993b). The basal release of NO may be associated with stretch of the preparation that is required to allow agonist-induced contraction (Dainty *et al.*, 1990). Intracisternal application of NOS inhibitors or oxyhaemoglobin (Toda *et al.*, 1991b) constricts basilar arteries in anaesthetized dogs. The induced vasoconstriction by NOS inhibitors is reversed by L-arginine (Toda *et al.*, 1993d). Therefore, basal release of NO is considered to play an important role in the regulation of cerebral vascular tone *in vivo.*

11.3.6 EDNO-mediated response *in vivo*

Vasodilator responses to arginine vasopressin and substance P angiographically measured are markedly attenuated by L-NMMA (Suzuki *et al.*, 1993) and L-NNA (K. Ayajiki *et al.*, unpublished data) in anaesthetized dogs, whereas those to VIP and nitroglycerin are not influenced. L-Arginine prevents or reverses the inhibitory effect of the NOS inhibitors. The responses to vasopressin differ between regions of cerebral arteries *in vivo* (Suzuki *et al.*, 1993), as seen in isolated dog cerebral arteries (Onoue *et al.*, 1988). Although endothelium dependency of the response cannot be determined *in vivo*, the data on NOS inhibitors and L-arginine, together with those obtained with isolated cerebral arteries, suggest that EDNO is involved in cerebral arterial dilatation *in vivo.*

11.4 Conclusion

The findings so far presented indicate that cerebral arterial relaxation associated with perivascular nerve stimulation is mediated by NO liberated from nerve terminals which activates soluble guanylyl cyclase in smooth muscle and increases the synthesis of cGMP. NO liberated from the endothelium also mediates the cerebroarterial dilatation in response to chemical (vasodilator substances) or physical (shear stress, blood flow change, etc.) stimuli. Neurogenic responses via NO are similarly observed in primate and subprimate mammals, whereas the responses mediated via EDNO following activation of drug receptors frequently vary between animal species. Nevertheless, NO derived from both nerve and endothelium undoubtedly plays an important role in the regulation of cerebral vascular tone *in vitro* and *in vivo.* Changes in vascular functions caused by NOS inhibitors *in vivo* reflect the physiological role of NO liberated from the nerve and/or endothelium under resting conditions. The effectiveness of ganglionic blocking agents would appear to be the only way to determine the involvement of neurogenic NO (Toda *et al.*, 1993e) at this moment, since we do not yet have a reliable method to examine the participation of endothelial cells *in vivo.*

References

Alonso, M.J., Salaices, M., Sanchez-Ferrer, C.F. & Marin, J. (1992) *J. Pharmacol. Exp. Ther.* **261**, 12–20.

Ayajiki, K., Okamura, T. & Toda, N. (1992) *Jpn. J. Pharmacol.* **60**, 357–362.

Ayajiki, K., Okamura, T. & Toda, N. (1993) *Neuroscience* **54**, 819–825.

Brayden, J.E. (1990) *Am.J. Physiol.* **259**, H668–H673.

Bredt, D.S. & Snyder, S.H. (1990) *Proc. Natl. Acad. Sci. USA* **87**, 682–685.

Bredt, D.S., Hwang, P.M. & Snyder, S.H. (1990) *Nature* **347**, 768–770.

Brian, J.E. & Kennedy, R.H. (1993) *Am. J. Physiol.* **264**, H1245–H1250.

Busse, R. & Mülsch, A. (1990) *FEBS Lett.* **275**, 87–90.

Colden-Stanfield, M., Schilling, W.P., Ritchie, A.K., Eskin, S.G., Nararro, L.T. & Kunze, D.L. (1987) *Circ. Res.* **61**, 632–640.

Dainty, I.A., McGrath, J.C., Spedding, M. & Templeton, A.G.B. (1990) *Br. J. Pharmacol.* **100**, 767–773.

Dauphin, F. & Hamel, E. (1990) *Eur. J. Pharmacol.* **178**, 203–213.

Dawson, T.M., Bredt, D.S., Fotuhi, M., Hwang, P.M. & Snyder, S. (1991) *Proc. Natl. Acad. Sci. USA* **88**, 7797–7801.

East, S.J. & Garthwaite, J. (1990) *Eur. J. Pharmacol.* **184**, 311–313.

Enokibori, M., Okamura, T. & Toda, N. (1994) *Br. J. Pharmacol.* **111**, 77–82.

Fujiwara, S., Kassell, N.F., Sasaki, T., Nakagomi, T. & Lehman, R.M. (1986) *J. Neurosurg.* **64**, 445–452.

Furchgott, R.F. (1983) *Circ. Res.* **53**, 557–573.

Furchgott, R.F. & Zawadzki, J.V. (1980) *Nature* **288**, 373–376.

Furchgott, R.F. (1988) *Vasodilatation* (ed. Vanhoutte P.M.) Raven Press, New York, pp. 401–414.

Galea, E., Feinstein, D.L. & Reis, D.J. (1992) *Proc. Natl. Acad. Sci. USA* **89**, 10945–10949.

Godfraind, T., Miller, R. & Wibo, M. (1986) *Pharmacol. Rev.* **38**, 321–416.

Gonzalez, C. & Estrada, C. (1991) *J. Cereb. Blood Flow Metab.* **11**, 366–370.

Griffith, T.M., Edwards, D.H., Newby, A.C., Lewis, M.J. & Henderson, A.H. (1986) *Cardiovasc. Res.* **20**, 7–12.

Griffith, T.M., Edwards, D.H., Davies, R.L.L., Harrison, T.J. & Evans, K.T. (1987) *Nature* **329**, 442–445.

Gruetter, C.A., Kadowitz, P.J. & Ignarro, L.J. (1981) *Can. J. Physiol. Pharmacol.* **59**, 150–156.

Hara, H., Hamill, G.S. & Jacobowitz, D.M. (1985) *Brain Res. Bull.* **14**, 179–188.

Harder, D.R. (1987) *Circ. Res.* **60**, 102–107.

Hatake, K., Kakishita, E., Wakabayashi, I., Sakiyama, N. & Hashida, S. (1990) *Stroke* **21**, 1039–1043.

Holtz, J., Forstermann, U., Pohl, U., Giesler, M. & Bassenge, E. (1984) *J. Cardiovasc. Pharmacol.* **6**, 1161–1169.

Ignarro, L.J., Byrns, R.E., Buga, G.M. & Wood, K.S. (1987a) *Circ. Res.* **60**, 82–92.

Ignarro, L.J., Byrns, R.E., Buga, G.M., Wood, K.S. & Chaudhuri, G. (1987b) *Proc. Natl. Acad. Sci. USA* **84**, 9265–9269.

Janssens, S.P., Shimouchi, A., Quertermous, T., Bloch, D.B. & Bloch, K.D. (1992) *J. Biol. Chem.* **267**, 14519–14522.

Kanamaru, K., Waga, S., Kojima, T., Fujimoto, K. & Itoh, H. (1987) *Stroke* **18**, 932–937.

Kanamaru, K., Waga, S., Fujimoto, K., Itoh, H. & Kubo, Y. (1989) *Stroke* **20**, 1208–1211.

Katusic, Z.S. (1992) *Am. J. Physiol.* **262**, H1557–H1562.

Katusic, Z.S., Shepherd, J.T. & Vanhoutte, P.M. (1984) *Circ. Res.* **55**, 575–579.

Katusic, Z.S., Shepherd, J.T. & Vanhoutte, P.M. (1986) *J. Pharmacol. Exp. Ther.* **236**, 166–170.

Katusic, Z.S., Marshall, J.J., Kontos, H.A. & Vanhoutte, P.M. (1989) *Am. J. Physiol.* **257**, H1235–H1239.

Kim, P., Sundt, T.M. & Vanhoutte, P.M. (1988) *J. Neurosurg.* **69**, 239–246.
Kim, P., Lorenz, R.R., Sundt, T.M. & Vanhoutte, P.M. (1989) *J. Neurosurg.* **70**, 108–114.
Koller, A. & Kaley, G. (1990) *Am. J. Physiol.* **258**, H916–H920.
Kontos, H.A., Wei, E.P. & Marshall, J.J. (1988) *Am. J. Physiol.* **255**, H1259–H1262.
Lansman, J.B., Hallan, T.J. & Rink, T.J. (1987) *Nature* **325**, 811–813.
Lee, T.J.F. & Sarwinski, S. (1991) *Blood Vessels* **28**, 407–412.
Linnik, M.D. & Lee, T.J.F. (1989) *J. Cereb. Blood Flow Metab.* **9**, 219–225.
Marshall, J.J. & Kontos, H.A. (1990) *Hypertension* **16**, 371–386.
Martin, W., Villani, G.M., Jothianandan, D. & Furchgott, R.F. (1985) *J. Pharmacol. Exp. Ther.* **232**, 708–716.
Mayhan, W.G., Faraci, F., Baumbach, G.L. & Heistad, D.D. (1990) *Am. J. Physiol.* **258**, H1138–H1143.
McCarron, J.G., Osol, G. & Halpern, W. (1989) *Blood Vessels* **26**, 315–319.
Miller, R.C., Schoeffter, P. & Stocklet, J.C. (1985) *Br. J. Pharmacol.* **85**, 481–487.
Moritoki, H., Ueda, H., Yamamoto, T., Hisayama, T. & Takeuchi, S. (1991) *Br. J. Pharmacol.* **102**, 841–846.
Murphy, S., Simmons, M.L., Agullo, L., Garcia, A., Feinstein, D., Galea, E., Reis, D.J., Minc-Golomg, D. & Schwartz, J.P. (1993) *Trends Neurosci.* **16**, 323–328.
Myers, P.R., Minor, R.L., Guerra, R., Bates, J.N. & Harrison, D.G. (1990) *Nature* **345**, 161–163.
Nakayama, K., Tanaka, Y. & Fujishima, K. (1989) *Eur. J. Pharmacol.* **169**, 33–42.
Nedergaard, O.A. (1988) In *The Pharmacology of Nicotine* (eds Rand, M. & Thurau, K.), pp. 143–162. Oxford, IRL Press.
Nozaki, K., Moskowitz, M.A., Maynard, K.I., Koketsu, N., Dawson, T.M., Bredt, D.S. & Snyder, H.S. (1993) *J. Cereb. Blood Flow Metab.* **13**, 70–79.
Okamura, T. and Toda, N. (1994) *J. Auton. Nerv. System* **49**, S55–S58.
Okamura, T., Inoue, S. & Toda, N. (1989) *Br. J. Pharmacol.* **97**, 1258–1264.
Olesen, S.-P., Clapham, D.E. & Davies, P.F. (1988) *Nature* **331**, 168–170.
Onoue, H., Nakamura, N. & Toda, N. (1988) *Stroke* **19**, 1388–1394.
Owman, C. (1990) *Blood Vessels* **27**, 73–93.
Palmer, R.M.J., Ferrige, A.G. & Moncada, S. (1987) *Nature* **327**, 524–526.
Palmer, R.M.J., Rees, D.D., Ashton, D.S. & Moncada, S. (1988) *Biochem. Biophys. Res. Commun.* **153**, 1251–1256.
Robertson, D. & Gallin, J.I. (1986) *J. Cell Biol.* **103**, 2379–2387.
Rubanyi, G.M., Romero, J.C. & Vanhoutte, P.M. (1986) *Am. J. Physiol.* **250**, H1145–H1149.
Shimizu, I. & Toda, N. (1986) *Br. J. Pharmacol.* **89**, 769–778.
Shimokawa, H., Kim, P. & Vanhoutte, P.M. (1988) *Circ. Res.* **63**, 604–612.
Simmons, M.L. & Murphy, S. (1992) *J. Neurochem.* **59**, 897–905.
Suzuki, H. & Chen, G. (1990) *News Physiol. Sci.* **5**, 212–215.
Suzuki, Y., Satoh, S., Oyama, H., Takayasu, M. & Shibuya, M. (1993) *Stroke* **24**, 1049–1054.
Toda, N. (1974) *Am.J. Physiol.* **227**, 1206–1211.
Toda, N. (1975) *J. Pharmacol. Exp. Ther.* **193**, 376–384.
Toda, N. (1979) *J. Pharmacol. Exp. Ther.* **209**, 352–358.
Toda, N. (1980) *Am. J. Physiol.* **238**, H111–H117.
Toda, N. (1982) *Am. J. Physiol.* **243**, H145–H153.
Toda, N. (1987) *J. Cereb. Blood Flow Metab.* **8**, 46–53.
Toda, N. (1990a) *Am. J. Physiol.* **258**, H311–H317.
Toda, N. (1990b) *Am. J. Physiol.* **258**, H57–H63.
Toda, N. (1991) *Am. J. Physiol.* **260**, H1443–H1448.
Toda, N. (1993) *Experientia* **49**, 51–53.
Toda, N. & Ayajiki, K. (1990) *Am. J. Physiol.* **258**, H983–H986.
Toda, N. & Miyazaki, M. (1984) *J. Cardiovasc. Pharmacol.* **6**, 1230–1237.

Toda, N. & Okamura, T. (1989) *Am. J. Physiol.* **257**, H988–H995.
Toda, N. & Okamura, T. (1990a) *Jpn. J. Pharmacol.* **52**, 170–173.
Toda, N. & Okamura, T. (1990b) *Biochem. Biophys. Res. Commun.* **170**, 308–313.
Toda, N. & Okamura, T. (1990c) *Am. J. Physiol.* **259**, H1511–H1517.
Toda, N. & Okamura, T. (1991a) *J. Cardiovasc. Pharmacol.* **17** (Suppl. 3), S234–S237.
Toda, N. & Okamura, T. (1991b) *J. Pharmacol. Exp. Ther.* **258**, 1027–1032.
Toda, N. & Okamura, T. (1992a) *J. Pharmacol. Exp. Ther.* **261**, 234–239.
Toda, N. & Okamura, T. (1992b) *News Physiol. Sci.* **7**, 148–152.
Toda, N., Okunishi, H., Taniyama, K. & Miyazaki, M. (1982) *Blood Vessels* **19**, 226–236.
Toda, N., Shimizu, I., Okamura, T. & Miyazaki, M. (1986) *J. Cardiovasc. Pharmacol.* **8**, 681–688.
Toda, N., Bian, K. & Inoue, S. (1987) *Naunyn-Schmiedebergs Arch. Pharmacol.* **336**, 359–364.
Toda, N., Minami, Y. & Okamura, T. (1990) *Life Sci.* **47**, 345–351.
Toda, N., Kawakami, M., Yamazaki, M. & Okamura, T. (1991a) *Br. J. Pharmacol.* **102**, 805–810.
Toda, N., Kawakami, M. & Yoshida, K. (1991b) *Am. J. Physiol.* **260**, H420–H425.
Toda, N., Kitamura, Y. & Okamura, T. (1991c) *J. Vasc. Med. Biol.* **3**, 235–241.
Toda, N., Ayajiki, K. & Okamura, T. (1993a) *J. Vasc. Res.* **30**, 61–67.
Toda, N., Ayajiki, K. & Okamura, T. (1993b) *Stroke* **24**, 1584–1589.
Toda, N., Ayajiki, K. & Okamura, T. (1993c) *Am. J. Physiol.* **265**, H103–H107.
Toda, N., Ayajiki, K., Yoshida, K., Kimura, H. & Okamura, T. (1993d) *Circ. Res.* **72**, 206–213.
Toda, N., Kitamura, Y. & Okamura, T. (1993e) *Hypertension* **21**, 3–8.
Vanhoutte, P.M. (1993) *Circulation* **87** (Suppl. V), V-9-V-17.
Wang, Y., Okamura, T. & Toda, N. (1993) *Exp. Eye Res.* **57**, 275–281.
Whalley, E.T., Amure, Y.O. & Lye, R.H. (1987) *Naunyn-Schmiedebergs Arch. Pharmacol.* **335**, 433–437.
Wood, K.S., Buga, G.M., Byrns, R.E. & Ignarro, L.J. (1990) *Biochem. Biophys. Res. Commun.* **170**, 80–88.
Yoshida, K., Okamura, T., Kimura, H., Bredt, D.S., Snyder, S.H. & Toda, N. (1993) *Brain Res.* **629**, 67–72.

CHAPTER 12

NITRIC OXIDE IN THE AUTONOMIC AND ENTERIC NERVOUS SYSTEMS*

M. J. Rand and C. G. Li

Pharmacology Research Laboratory, Department of Medical Laboratory Science, Royal Melbourne Institute of Technology, Victoria 3001, Australia

Table of Contents

12.1	Discovery of NO-mediated neuroeffector transmission	228
12.2	Criteria for establishing nitrergic transmission	230
	12.2.1 Synthesis of the nitrergic transmitter	230
	12.2.2 Inhibition of NOS	233
	12.2.3 Storage of the nitrergic transmitter	235
	12.2.4 Release of the nitrergic transmitter	236
	12.2.5 Mimicry of the nitrergic transmitter	237
	12.2.6 Modifying agents	240
	12.2.7 Effector mechanisms in nitrergic transmission	243
12.3	Tissues innervated by autonomic nitrergic nerves	245
	12.3.1 Anococcygeus and retractor penis muscles	245
	12.3.2 Urinary tract	247
	12.3.3 Respiratory tract	248
	12.3.4 Cardiovascular system	249
	12.3.5 Other organs	250
12.4	Nitrergic transmission in the gastrointestinal tract	250
	12.4.1 Extrinsic nitrergic autonomic nerves	250
	12.4.2 Electrophysiological observations	251
	12.4.3 Sphincters	252
	12.4.4 Segments between sphincters	254
	12.4.5 Nitrergic involvement in enteric reflexes	259
12.5	Mechanisms of nitrergic transmission	260
	12.5.1 Nature of the nitrergic transmitter	260
	12.5.2 Is there exocytosis in nitrergic transmission?	261
	12.5.3 Spontaneous release of the nitrergic transmitter	263

*NO-mediated transmission to erectile tissue and cerebral blood vessels are dealt with separately in Chapters 11 and 13 and are not referred to in the following. We have surveyed the bulk of the relevant literature available to us up to the end of 1993 (bearing in mind the delay in arrival of journals in Australia), but the limitation on space has forced us to be highly selective in citing much of it. We would be grateful if readers would inform us about notable omissions.

Nitric Oxide in the Nervous System
ISBN 0-12-721985-4

12.5.4 Modulation of nitrergic transmission 263
12.5.5 Co-transmitters, independent transmitter or tandem transmitters 265
12.5.6 Excitation of nitrergic transmission by agonists 266
12.6 Interactions between NO and other autonomic transmitters 267
12.6.1 Noradrenaline 267
12.6.2 Acetylcholine 268
12.7 Concluding remarks 268

Acknowledgements 269
References 269

12.1 Discovery of NO-mediated neuroeffector transmission

Much of the research leading to the initial recognition of the involvement of nitric oxide (NO) in neuroeffector transmission was carried out with rat anococcygeus and bovine retractor penis muscles (for reviews, see Gillespie, 1987; Gillespie *et al.*, 1990; Martin and Gillespie, 1991). In these muscles, field stimulation produces a noradrenaline-mediated contraction, but when the contractile response is blocked and the tone is raised, there is a neurogenic non-adrenergic, non-cholinergic (NANC) relaxation that was not attributable to any of the other mediators recognized before the discovery of NO-mediated transmission. The key to the nature of the mediator was provided by the discovery that the endothelium-derived relaxing factor (EDRF) was NO (or an NO-like substance) derived from L-arginine, and that the production of this factor, which could now be termed endothelium-derived NO (EDNO), was selectively blocked by certain N^{G}-substituted analogues of L-arginine (for review, see Moncada *et al.*, 1991). It was already known that there were many similarities between EDRF and the inhibitory transmitter of the anococcygeus and retractor penis muscles (Gillespie, 1987; Furchgott, 1988). The obvious next step, the use of the newly developed inhibitors of NO synthesis, provided the first definitive evidence for NO-mediated transmission in anococcygeus muscles. This was done almost simultaneously by four independent groups in Glasgow (Gillespie *et al.*, 1989), London (Gibson and Mirzazadeh, 1989; Gibson *et al.*, 1989), Melbourne (Li and Rand, 1989b) and North Carolina (Ramagopal and Leighton, 1989). Details of the methods and findings have been reviewed by Rand (1992a).

We suggest that the process of NO-mediated transmission be termed 'nitrergic' (Li and Rand, 1989b; Rand and Li, 1990) following the precedent set by Dale (1933) when he coined the terms cholinergic and adrenergic. The term nitrergic is not meant to imply a commitment to NO as such, but that the transmitter is NO-like. Other terms that have been suggested include *nitrinergic* (Schmidt *et al.*, 1992b), *nitroxergic* (Desai *et al.*, 1991b) and *nitroxidergic* (Toda and Okamura, 1992a,b). In this review, we have used the term nitrergic in relation to the transmission process and

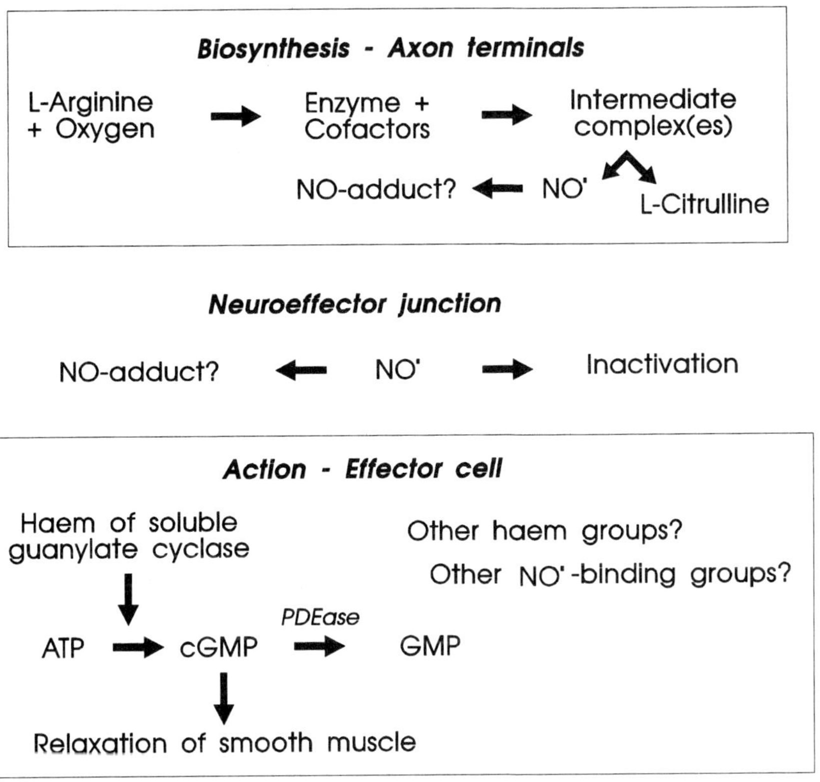

Figure 1 Schematic representation of the mechanisms of nitrergic transmission. The key enzyme in the biosynthesis of the nitrergic transmitter is a NOS which, with its cofactors, converts the substrates L-arginine and O_2 via an intermediate into L-citrulline and NO (see text and Figure 2). In the scheme, these processes are shown in the nerve terminals; however, for another possibility, see Section 12.5.2. The transmitter is more likely to be an NO-yielding adduct than free NO (see Section 12.5.1). In the neuroeffector junction, the transmitter may be inactivated by superoxide anions or the formation of NO adducts that do not yield NO. The main target in the smooth muscle of effector cells is the haem prosthetic group of soluble guanylyl cyclase, and relaxation is mediated by its product, cGMP (see Section 12.2.7).

the nature of the transmitter whenever, in our opinion, the balance of the evidence suggested a neurogenic action was mediated by a NO-like substance, even though the authors whose work we cite may have been more circumspect.

In the 3 years that have elapsed since the first demonstration, substantial evidence has been accumulated for nitrergic neuroeffector transmission in many organs from various species (for reviews, see Rand, 1992a,b; Sanders and Ward, 1992; Sneddon and Graham, 1992; Stark and Szurszewski, 1992). The general nature of the mechanisms of nitrergic transmission that emerged is shown in Figure 1. But before describing the nitrergic innervation of individual organs, it is instructive to consider the extent to which NO, or an NO-like substance, fulfils the criteria for acceptance as a transmitter.

12.2 Criteria for establishing nitrergic transmission

The evidence for the neurotransmitter roles of the so-called classical transmitters acetylcholine and noradrenaline in the autonomic nervous system is established textbook material (see, for example, Bowman and Rand, 1980). Identification of the nature of a transmitter requires that possible candidates fulfil a number of criteria that have been elaborated *post hoc* on the basis of known transmitter mechanisms. The main features of hitherto generally accepted criteria are as follows:

(1) A system for synthesis of the putative transmitter must be present.
(2) There must be a store of the putative transmitter in the axon terminals.
(3) The putative transmitter must be released by nerve stimulation.
(4) Administration of the putative transmitter must produce a response that mimics that produced by nerve stimulation.
(5) Drugs that modify the responses to the putative transmitter should have corresponding effects on the responses to nerve stimulation.

The criteria derived from established modes of transmission may not necessarily apply in all respects to nitrergic transmission, or indeed to other yet to be discovered modes of transmission. Consequently, the acceptance of nitrergic transmission should be on assessment of the totality of the evidence rather than on insistence that an arbitrary set of criteria be met rigorously. In due course, a revised set of criteria will be established that will accommodate nitrergic with other modes of transmission.

12.2.1 Synthesis of the nitrergic transmitter

The key element in nitrergic transmission is the enzyme NO synthase (NOS; see Chapters 1 and 2), which belongs to a family of enzymes that include cytochrome P_{450} reductase (Bredt *et al.*, 1991). These enzymes contain a haem group (White and Marletta, 1992) and require the cofactors NADPH, the flavines FAD and FMN (Moncada *et al.*, 1991), and tetrohydrobiopterin (Schmidt *et al.*, 1992a), and neuronal NOS is Ca^{2+}/calmodulin dependent (Bredt and Snyder, 1990). The nitrogen atom of the NO produced by NOS is derived from one of the two equivalent terminal guanidino groups of L-arginine and molecular oxygen is the source of the oxygen atoms incorporated into NO and L-citrulline (see Moncada *et al.*, 1991). In the process, an intermediate, N^{ω}-hydroxy-L-arginine, is formed and this can also be converted to citrulline and NO by NOS (Stuehr *et al.*, 1991b; Fukuto *et al.*, 1992; Nathan, 1992; Stuehr and Griffith, 1992) or by cytochrome $P_{450}3A$ (Renaud *et al.*, 1993). The reactions are shown in Figure 2.

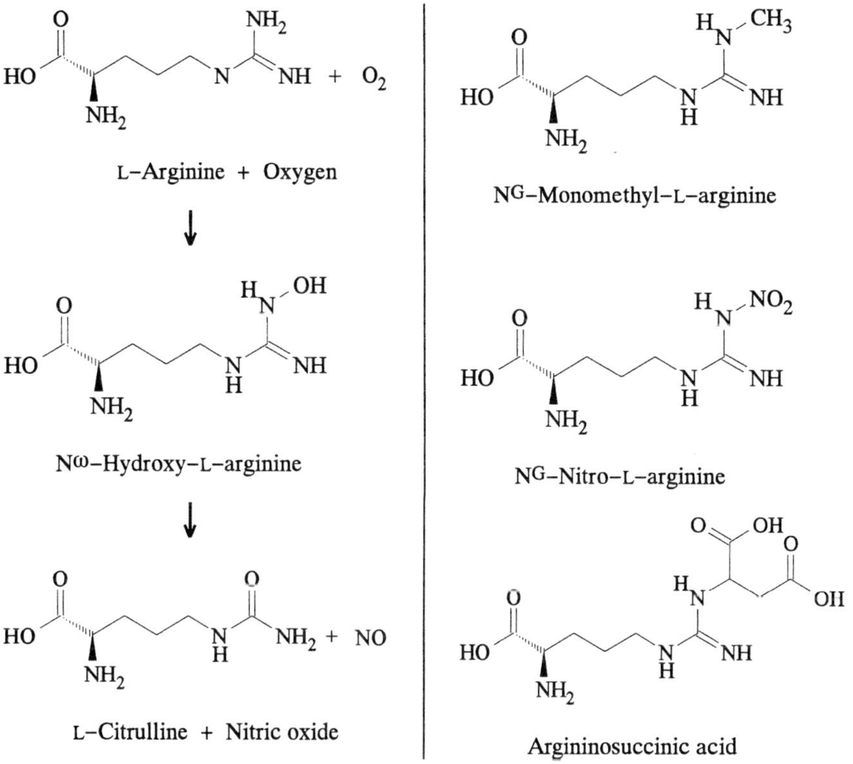

Figure 2 The left column shows the chemical steps in the conversion of L-arginine and O_2 to L-citrulline by NOS via the intermediate N^{ω} hydroxy-L-arginine (see text and Figure 1). The right column shows the structures of the NOS inhibitors N^{G}-monomethyl-L-arginine (L-NMMA) and N^{G}-nitro-L-arginine (L–NNA), and the compound argininosuccinic acid which is an intermediate in the resynthesis of L-arginine from L-citrulline, and resembles the NOS inhibitors in being an N^{G}-L-arginine derivative. Note that L-NMMA and the dimethyl analogue are possible endogenous inhibitors of NOS and that argininosuccinic acid has activity as a NOS inhibitor (see Section 12.2.2.1).

12.2.1.1 Localization of neuronal NOS

The detection of NOS in peripheral neurons by histochemical methods in many tissues has provided the morphological basis for nitrergic neuroeffector transmission. The first account of immunohistochemical localization of NOS by Bredt *et al.* (1990) revealed the presence of peripheral as well as central immunoreactive neurons. Another more simple method to detect NOS histochemically in various tissues is based on the finding that neuronal NOS possesses NADPH diaphorase activity (Hope *et al.*, 1991; Dawson *et al.*, 1991). These techniques have been widely used (see Table 1 for references and Chapter 5). NOS immunoreactivity and

Table 1 Localization of NOS immunoreactive and/or NADPH diaphorase-positive autonomic and enteric nerves in mammalian organs, except for the penis (see Chapter 13) and cerebral vasculature (see Chapter 11). This list is not exhaustive, being limited to those organs in which there is also functional evidence for nitrergic transmission, as discussed in the text (see Sections 12.3 and 12.4)

Organ	Species	References
Anococcygeus	Rat	Schmidt *et al.* (1992b), Dail *et al.* (1993), Song *et al.*, (1993)
	Mouse	Brave *et al.* (1993b)
Retractor penis	Rat	Dail *et al.* (1993)
Airways	Guinea-pig	Fischer *et al.* (1993), Hassall *et al.* (1993)
	Rat	Kobzik *et al.* (1993)
	Ferret	Dey *et al.* (1993)
	Human	Diaz de Rada *et al.* (1993), Kobzik *et al.* (1993)
GI tract	Rat	Bredt *et al.* (1990), Dawson *et al.* (1991), Belai *et al.* (1992), Schmidt *et al.* (1992b), Aimi *et al.* (1993), Alm *et al.* (1993), Forster and Southam (1993)
	Mouse	Grozdanovic *et al.* (1992)
	Guinea-pig	Costa *et al.* (1992), Furness *et al.* (1992a), Llewellyn-Smith *et al.* (1992), Nichols *et al.* (1992), McConalogue and Furness (1993), Young *et al.* (1992)
	Dog	Ward *et al.* (1992b)
	Pig	Barbiers *et al.* (1993), Krammer *et al.* (1993)
	Human	Springall *et al.* (1992), Faussone-Pellegrini *et al.* (1993)
Gall bladder	Guinea-pig	Talmage and Mawe (1993)
	Mouse	Grozdanovic *et al.* (1992)
Urinary tract	Rat	McNeill *et al.* (1992), Schmidt *et al.* (1992b), Alm *et al.* (1993), Vizzard *et al.* (1993)
	Pig	Larsson *et al.* (1992), Persson *et al.* (1993)
	Sheep	Triguero *et al.* (1993)
Uterus	Rat	Papka and McNeill (1992), Shew *et al.* (1993)
Heart	Rat	Klimaschewski *et al.* (1992), Schmidt *et al.* (1992b)
	Guinea-pig	Hassall *et al.* (1992), Klimaschewski *et al.* (1992), Tanaka *et al.* (1993)
Blood vessels	Guinea-pig	Kummer *et al.* (1992), Kummer and Mayer (1993), Klimaschewski *et al.* (1992)
	Rat	Klimaschewski *et al.* (1992)
	Dog	Yoshida *et al.* (1993)

NADPH diaphorase activity have a one-to-one correspondence in enteric neurons of dog colon (Ward *et al.*, 1992b), guinea-pig ileum and colon (Young *et al.*, 1992) and cultured myenteric neurons (Saffrey *et al.*, 1992). Such neurons are widespread: in the guinea-pig small intestine and caecum they comprise about 20% of nerve cell bodies and about 30% of the axons supplying smooth muscle (Costa *et al.*, 1992; Furness *et al.*, 1992a).

L-Citrulline, one of the products of NOS, has been detected by immunohistochemistry in enteric neurons and dorsal root ganglia, whereas its substrate, L-arginine, also detected by immunohistochemistry, was in glial cells (Aoki *et al.*, 1993).

12.2.2 Inhibition of NOS

It follows from the postulate of a mode of transmission that inhibition of the biosynthesis of the transmitter should result eventually in failure of the process without any corresponding reduction in the responses to mimicking agents. For nitrergic transmission, this is the case for all of the inhibitors of neuronal NOS so far tested.

Nathan (1992) proposed five classes of NOS inhibitors: substrate competitors; nucleotide cofactor competitors; tetrohydrobiopterin inhibitors; calmodulin binders; and haem binders. In addition, since molecular oxygen is required, hypoxia inhibits the activity of NOS.

12.2.2.1 N^G-substituted analogues of L-arginine

Moncada *et al.* (1991) described the development of inhibitors of NOS. Those most commonly used in studies on neuroeffector transmission are N^G-monomethyl-L-arginine (L-NMMA), N^G-nitro-L-arginine (L-NNA) and its methyl ester (L-NAME). The structures of L-NMMA and L-NNA are shown in Figure 2. They are stereospecific, the corresponding analogues of D-arginine being inactive. The inhibitory actions of L-NMMA and low concentrations of L-NNA and L-NAME are competitive since they can be overcome by raising the concentration of L-arginine, but D-arginine is ineffective. With high concentrations of the nitroarginine derivatives, the interaction with NOS is irreversible, but their inhibitory action is blocked or attenuated by prior exposure to a high concentration of L-arginine (Martin *et al.*, 1993). In the bovine retractor penis, L-NMMA was more potent than L-arginine in reversing submaximal blockade by L-NNA or L-NAME, and in the rat anococcygeus muscle, a low concentration of L-NMMA producing partial reductions of nitrergic relaxations inhibited the subsequent blocking effect of L-NNA, and submaximal blockage by L-NNA was partially reversed by L-NMMA to an even greater extent than by L-arginine (Martin *et al.*, 1993). Inhibition of nitrergic relaxations of the mouse anococcygeus muscle produced by L-NNA were reversed by N^{ω}-hydroxy-L-arginine (Gibson *et al.*, 1992). Despite these complications, stereospecific blockade of neurogenic relaxations by N^G-substituted analogues of L-arginine and stereospecific reversal or prevention of the blockade by L-arginine has provided the 'gold standard' for identification of nitrergic transmission.

Methylation of L-arginine residues in proteins forms L-NMMA and dimethyl L-arginine (L-NDMA) which are continually released during protein catabolism and then hydrolysed by a specific enzyme to form citrulline and monomethylamine or dimethylamine (Kimoto *et al.*, 1993). The relatively low potency of L-NMMA as an NOS inhibitor may be due to its metabolism to citrulline and thence to L-arginine. The amounts of endogenous N^G-methylated arginines in various tissues are about 3–20% of the content of free L-arginine, which may be sufficient for action as inhibitory modulators of NOS (Kimoto *et al.*, 1993). Levels of L-NDMA are markedly raised in patients with chronic renal failure, and this may account for part of the associated symptoms (Vallance *et al.*, 1992a,b). The finding that L-arginine enhances responses to nitrergic nerve stimulation in many tissue could be due to its overcoming the inhibitory effect of endogenous methylated arginines.

L-NMMA inhibits the uptake of L-arginine into endothelial cells (Bogle *et al.*, 1992) and cultured neurons (Beart *et al.*, 1993). L-NNA and L-NAME have only very weak uptake blocking activity.

Argininosuccinic acid is an intermediary in the biosynthesis of L-arginine from L-citrulline and is yet another N^G-substituted derivative of L-arginine (Figure 2). We showed that it had a definite but transient effect in inhibiting nitrergic transmission in the rat anococcygeus muscle (Rand and Li, 1992b). The transient nature of the effect was probably due its rapid metabolism to L-arginine.

12.2.2.2 Cofactor inhibitors

Nitro blue tetrazolium, the substrate used for detecting NADPH diaphorase reactivity, inhibits the NOS activity of the enzyme (Hope *et al.*, 1991), presumably by competing for the NADPH reducing equivalents, but a number of other mechanisms have been suggested (Schmidt *et al.*, 1993). We were unable to determine whether it inhibited nitrergic relaxations in the rat anococcygeus muscle since it produced a complete and long-lasting relaxation by a mechanism that we have not yet explored (C.G. Li and M.J. Rand, unpublished observations).

Diphenylene iodonium (DPI) inhibited macrophage NOS, apparently by blocking attachment of NADPH and perhaps other nucleotide cofactors (Stuehr *et al.*, 1991a) and, in accord with inhibition of other NOSs, it blocked endothelium-dependent relaxations of rabbit (Stuehr *et al.*, 1991a) and rat (Rand and Li, 1993a; Wang *et al.*, 1993) aortic smooth muscle and blocked nitrergic relaxations of rat anococcygeus muscle by a mechanism that differed from that of L-NAME since its effect was not countered by L-arginine (Rand and Li, 1993a).

We are not aware of any studies on peripheral nitrergic transmission in which tetrohydrobiopterin inhibitors of NOS have been used.

12.2.2.3 Calmodulin inhibitors

Various calmodulin inhibitors inhibit NOS activity (Bredt and Snyder, 1990; Förstermann *et al.*, 1991; Schmidt *et al.*, 1993), but most of them interfere with contractile mechanisms in effector cells and in our experience are not useful for functional studies on nitrergic neuroeffector function.

12.2.2.4 Haem inhibitors

Since NOS is a haem-containing enzyme with a similar structure to cytochrome P_{450} it is not surprising that substances that bind to haem groups such as CO inhibit NOS activity (White and Marletta, 1992). Phencyclidine is an irreversible inhibitor of rat brain NOS (Osawa and Davila, 1993) and it was postulated that the effect was similar to the inactivation of microsomal P_{450}, in which intermediates are formed that alter the haem group. To our knowledge, haem inhibitors have not been used so far in studies on nitrergic neuroeffector transmission.

12.2.2.5 Other NOS inhibitors

The P_{450} substrate and competitive inhibitor 7-ethoxyresorufin inhibited NOS activity in rat brain homogenates and human placental tissue (Li and Rand, 1993b; Di Iulio *et al.*, 1993). However, it is also a generator of superoxide anions (see Section 12.2.6.4) and this was responsible for its inhibitory effect on nitrergic relaxations in rat anococcygeus muscles (Rand and Li, 1993d).

A number of thiol-binding compounds or sulphydryl inhibitors inhibit rat brain NADPH diaphorase (Kemp *et al.*, 1988; Hope and Vincent, 1989). The diuretic drug ethacrynic acid is a sulphydryl-binding compound, and was specifically developed with this property in mind to replace the more toxic mercurial diuretics. We found that ethacrynic acid inhibited relaxations of rat anococcygeus muscle elicited by NO, the NO donors SNP and GTN and nitrergic nerve stimulation (Li *et al.*, 1994). Some effects of ethacrynic acid are shown in Figure 3. These effects were almost completely prevented if the thiol cysteine was present during the period of incubation with ethacrynic acid. The addition of cysteine after the effects of ethacrynic acid were established slightly but significantly restored the response to nitrergic stimulation, but not those to the NO donors.

12.2.2.6 Hypoxia

Molecular oxygen is the source of the oxygen atoms incorporated into NO and L-citrulline by NOS, therefore it is not surprising that hypoxia impairs NO synthesis (Rengasamy and Johns, 1991), thereby decreasing endothelium-dependent relaxations (see Lüscher and Vanhoutte, 1990) and nitrergic relaxations in the anococcygeus muscles of the rat (Bowman and McGrath, 1985) and mouse (Gibson and Mirzazadeh, 1989).

12.2.3 Storage of the nitrergic transmitter

It is highly improbable that such a readily diffusible and labile substance as NO could be stored in a biological compartment, but an NO-yielding substance (such

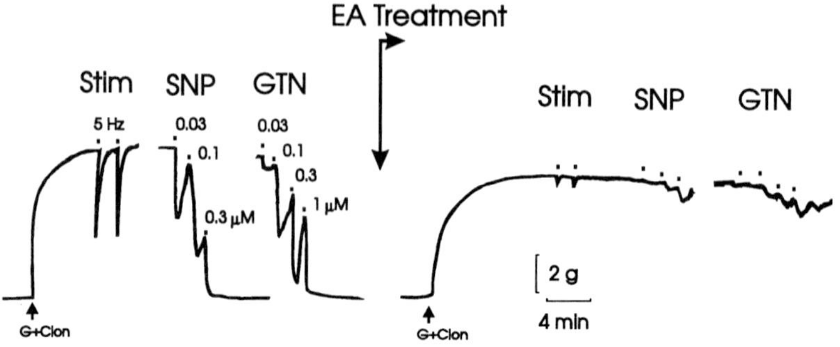

Figure 3 Inhibition by ethacrynic acid (EA) of relaxations of a rat isolated anococcygeus muscle elicited nitrergic nerve stimulation at 5 Hz for 10 s (Stim) and by the NO donors SNP (0.03–0.3 μM) and GTN (0.03–1 μM). Tone was raised by guanethidine (G, 30 μM) plus clonidine (Clon, 5 μM). After making the control observations, the muscle was exposed to EA (100 μM) for 45 s and repeatedly washed for 30 min with EA-free bathing solution; tone was restored but the relaxant responses were almost abolished.

as a nitrosothiol) could conceivably be contained in a transmitter storage vesicle (Ignarro, 1990). However, the rapid action of NOS inhibitors in blocking nitrergic transmission, as illustrated in Figure 4, suggests that there is little or no preformed store so the transmitter must be formed on demand (Li and Rand, 1989b; Stark *et al.*, 1991). This finding is in striking contrast to that obtaining after inhibition of biosynthesis of acetylcholine or noradrenaline when the reduction of stimulation-induced responses only starts after long-lasting repetitive stimulation has almost exhausted preformed stores.

12.2.4 Release of the nitrergic transmitter

Methods for the measurement of NO have recently been reviewed (Archer, 1993). Chemiluminescence assay has been used for measuring release of NO into head-space gas during nitrergic nerve stimulation of the opossum internal anal sphincter (Chakder and Rattan, 1993a). Bioassay, using the cascade superfusion technique, has been used to detect the stimulation-induced release of NO (or an NO-like substance) from isolated preparations of the canine ileocolonic junction and rat gastric fundus (Bult *et al.*, 1990; Boeckxstaens *et al.*, 1991b,c,d): the activity of the factor was increased by L-arginine and SOD and abolished by L-NNA and oxyhaemoglobin. The production of [^{3}H]-citrulline after loading with L-[^{3}H]-arginine, indicating NOS activation, resulted from stimulation of isolated myenteric ganglia (Grider and Jin, 1993) and strips of rat colon circular muscle (Grider, 1993).

Since in many tissues the nitrergic transmitter has different properties from those of NO in aqueous solution or various NO donors, as discussed below (Section 12.2.6), it may be an NO-donating adduct, formed after release of NO from NOS. The methods employed so far to detect stimulation-induced release do not

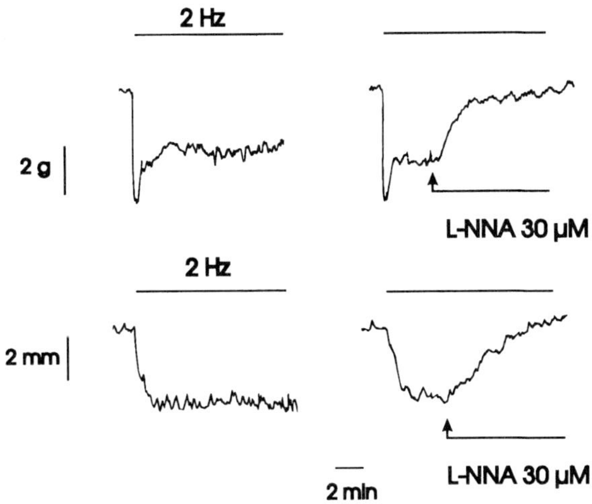

Figure 4 The effects of the NOS inhibitor L-NNA (30 μM) on sustained relaxations of a rat isolated anococcygeus muscle (isometric tension, upper records) and a gastric fundus strip (isotonic length, scale corrected for magnification, lower records) during low frequency (2 Hz) repetitive nitrergic nerve stimulation for the periods indicated by the horizontal lines above the records. Control records (left hand) show that the relaxations reached steady states that were maintained for at least 8 min. Addition of L-NNA during the period of stimulation (right hand records) resulted in almost immediate decreases in the relaxations and complete abolition within 8 min. In the anococcygeus muscle, tone was raised and the noradrenergic response blocked by guanethidine (30 μM). In the gastric fundus, noradrenergic and cholinergic effects were blocked by guanethidine (10 μM) and atropine (1 μM), respectively, and the VIP component of the response was blocked by chymotrypsin (1 unit/ml).

discriminate between free NO and NO-yielding adducts, so the question of what actually mediates nitrergic transmission has not yet been resolved.

12.2.5 Mimicry of the nitrergic transmitter

Mimicry of the response to nerve stimulation by the exogenous administration of the putative transmitter is an essential component of the evidence for any particular mediator; hence, we have considered the actions of NO and NO donors in some detail. Since the properties of the nitrergic transmitter do not correspond to those of free NO, one of the on-going tasks is in establishing which NO donor corresponds more closely.

12.2.5.1 NO

Many investigators have shown that aqueous solutions of NO, prepared by bubbling NO gas through deoxygenated water, mimic responses to nitrergic nerve stimulation, but its potency is less than that of many NO donors.

It is often assumed that the functionally active product made of NOS is the neutral free radical NO•, but Stamler *et al.* (1992) pointed out that there are three interconvertible redox forms of nitrogen monoxide with distinctly different properties and biological reactivities:

$$\underset{\text{Nitrosium ion}}{NO^+} \underset{-e}{\overset{+e}{\rightleftharpoons}} \underset{\text{Nitric oxide}}{NO\bullet} \underset{-e}{\overset{+e}{\rightleftharpoons}} \underset{\text{Nitroxyl anion}}{NO^-}$$

The form produced by NOS has not been determined with certainty, and probably depends on the prevailing conditions.

12.2.5.2 Inorganic and organic nitrites

Acidified solutions of sodium nitrite have been used as a source of NO (Furchgott, 1988). The main form so produced is the nitrosium ion, NO^+ (Stamler *et al.*, 1992).

Amyl nitrite is the first of the substances now known as nitrovasodilators to be used in therapeutics when it was shown to be effective in relieving the pain of angina pectoris (Brunton, 1867). A number of related compounds are known; they can be considered as adducts of NO with the corresponding alcohol (see Section 12.2.6.3).

12.2.5.3 Glyceryl trinitrate (GTN) and related organic nitrates

The NO-donating action of organic nitrates is different from that for organic nitrites (Kowaluk and Fung, 1991). It is unlikely that an organic nitrate serves a biological role as an NO-donating substance, but GTN is frequently used as an NO donor and it is worth considering briefly what is known about its mechanism of action. The NO released in biological systems from nitrate esters requires the presence of free sulphydryl groups (Needleman *et al.*, 1973; Ignarro *et al.*, 1981), and involves metabolic activation by, for example, a microsomal cytochrome P_{450} (Servent *et al.*, 1989; Marks *et al.*, 1991; Delaforge *et al.*, 1993).

12.2.5.4 Sodium nitroprusside (SNP)

The NO moiety in SNP has the characteristics of the nitrosium ion. It has been suggested that SNP releases NO spontaneously (Katsuki *et al.*, 1977), but this is disputed, and it may require reaction with a thiol before acting as an NO donor (Ignarro *et al.*, 1981) and/or enzymatic activation (Marks *et al.*, 1991), but the mechanism is different from that for GTN (Kowaluk *et al.*, 1992).

12.2.5.5 Nitrosothiols

Thiols readily form nitrosothiols by reaction with sodium nitrite under acid conditions (Kowaluk and Fung, 1990) and nitrosothiols decompose to yield NO and the corresponding disulphide (Williams, 1985). It has been pointed out that the reversibility of the reaction provides a convenient way to store and transport NO and may play a role in the interconversion of thiols and disulphides (Girard and Potier, 1993).

The nitrosothiols most commonly used in pharmacological experiments are *S*-nitroso-*N*-acetylpenicillamine (SNAP), *S*-nitroso-*N*-acetylcysteine (SNAC), *S*-nitrosocysteine (NOCys), *S*-nitrosoglutathione (NOGS), and *S*-nitrosocoenzyme A (NOCoA). Of these, the last three may be physiologically relevant.

It has been postulated that the endothelium-derived NO-like relaxing factor is a nitrosothiol (Ignarro, 1989; Myers *et al.*, 1990; Bates *et al.*, 1991; Rubanyi *et al.*, 1991) or a dinitrosyl iron cysteine complex (Vedernikov *et al.*, 1992), and that the nitrergic transmitter is a nitrosothiol (Thornbury *et al.*, 1991; Gibson *et al.*, 1992; Li and Rand, 1993a; Rand and Li, 1993b). The activity of acid-treated inhibitory factor extracted from the bovine retractor penis (see Section 12.3.1) has been attributed to formation of a nitrosothiol (Kerr *et al.*, 1992). NOCys is present in human bronchial mucus and was regarded as an endogenous bronchodilator (Barnes and Belvisi, 1993) and nitrosothiols have bronchodilator activity in guinea-pigs (Dupuy *et al.*, 1992) and on guinea-pig tracheal smooth muscle (Jansen *et al.*, 1992). NOCys produced hyperpolarization of smooth muscle resembling that produced by nitrergic nerve stimulation in the rat gastric fundus (Kitamura *et al.*, 1993) and dog colon (Thornbury *et al.*, 1991). Likewise, NOCys gave good mimicry of relaxant responses of the rat anococcygeus muscle to nitrergic nerve stimulation (Li and Rand, 1993a; Rand and Li, 1993b).

Nitrosothiols have very low lipophilicity (octanol/buffer at pH 7.4 partition coefficients of 0.02–0.05), therefore passive diffusion through cell membranes is highly unlikely (Kowaluk and Fung, 1990). The rates of spontaneous liberation of NO from a range of nitrosothiols did not correlate with their order of relaxant potencies on rabbit aortic smooth muscle (Kowaluk and Fung, 1990), or with their relaxant potencies on the mouse anococcygeus muscle (Gibson *et al.*, 1992). Therefore, there could be a membrane transport mechanism, or transfer of the NO moiety to an intracellular carrier in effector cells.

Nitrosothiols are considerably more potent than an equivalent concentration of NO. This was strikingly demonstrated by J.S. Gillespie, X. Lui and W. Martin (personal communication, 1992), who showed that the relaxation of the rat anococcygeus muscle produced by adding NO to a cysteine solution beforehand was much greater than that produced by an aqueous solution of NO, although cysteine alone had no effect on muscle tone (Gillespie *et al.*, 1994). Following this lead, we showed that in the presence of cysteine the relaxant action of NO was much greater than in its absence, as illustrated in Figure 5. However, cysteine did not potentiate nitrergic relaxations (Rand and Li, 1992c). Cysteine potentiated relaxations and cGMP production in rabbit mesenteric arteries elicited by acidic sodium nitrite or SIN-1

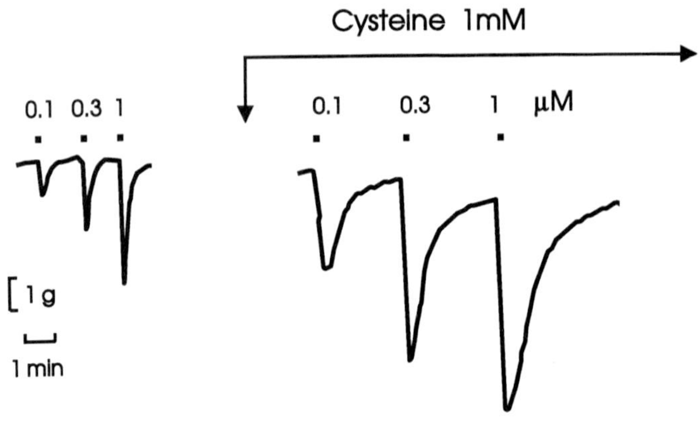

Figure 5 Comparison of the relaxant actions on a partly contracted rat isolated anococcygeus muscle of 0.1, 0.3 and 1 μM of NO alone (administered in an aqueous solution) and in the presence of L-cysteine (1 mM). Cysteine itself had no effect on the tone, but greatly enhanced and prolonged the responses to NO.

but not by EDNO (Arvola *et al.*, 1992). The L-isomer of cysteine has usually been used, but the D-isomer gives the same results (Thornbury *et al.*, 1991; A. McLaren, C.G. Li & M.J. Rand, unpublished observation).

12.2.5.6 Other NO donors

In studies on nitrergic neuroeffector transmission, only occasional use has been made of SIN-1 (linsidomine), which releases NO spontaneously (Feelisch, 1991) and also generates superoxide (Woolsey and Van der Loos, 1970), hydroxylamine, which requires enzymatic conversion to generate NO, and sodium azide, which requires O_2-dependent activation to yield NO (Waldman and Murad, 1987).

We are not aware of relevant studies in which the NO donors hydroxyguanidine- and hydroxyarginine-NO adducts (Zembowicz *et al.*, 1991, 1992) and NONOates (Morley and Keefer, 1993) have been used.

12.2.6 Modifying agents

The criterion that agents modifying responses to a putative transmitter should have corresponding effects on responses to nerve stimulation is critical. The use of such agents has eliminated a number of possible candidates as the nitrergic transmitter, but has not yet led to its identification.

12.2.6.1 Oxyhaemoglobin

This is one of the more important tools for establishing a role for NO in physiological processes. Oxyhaemoglobin inhibits endothelium-derived NO-mediated vasodilatation and relaxations elicited by NO and NO donors (for reviews, see Gillespie *et al.*, 1990; Lüscher and Vanhoutte, 1990; Martin and Gillespie, 1991; Moncada *et al.*, 1991). The effect of oxyhaemoglobin is due to binding of NO with high affinity (Gibson and Roughton, 1957). The binding is irreversible since oxyhaemoglobin and NO react rapidly (<0.1 s) to form methaemoglobin and NO_3^+ (Kelm and Schrader 1990). This is probably critical since, by mass action, oxyhaemoglobin could abstract NO from adducts with a higher affinity for it.

It was already known that oxyhaemoglobin, but not methaemoglobin, selectively blocked what was subsequently found to be nitrergic transmission in the retractor penis and anococcygeus muscles (Bowman *et al.*, 1982; Bowman and Gillespie, 1982; Gibson and Tucker, 1982; Byrne and Muir, 1984). More recently, it has been shown uniformly in many tissues that oxyhaemoglobin inhibits stimulation-induced relaxations in every case in which they have been shown by other means to be nitrergically mediated, for example in the rabbit anococcygeus muscle (Graham and Sneddon, 1993) and in many parts of the gastrointestinal tract of several species (11 examples are listed in the review by Stark and Szurszewski, 1992).

The haemoglobin within intact erythrocytes also blocks responses to NO and EDNO, but not nitrergic relaxations of the bovine retractor penis muscle (Gillespie and Sheng, 1988, 1989), indicating that EDNO, like the highly permeant NO, readily crosses the erythrocyte membrane, but the nitrergic transmitter does not.

12.2.6.2 Hydroxocobalamin

The cobalt atom of hydroxocobalamin can bind NO, being converted to nitrosocobalamin. We investigated the possibility that hydroxocobalamin might affect NO-mediated actions by sequestering NO (Li and Rand, 1993a; Rajanayagam *et al.*, 1993; Rand and Li, 1993b). In rat anococcygeus muscles, it inhibited relaxations induced by NO and SNP but had little or no effect on responses to GTN, NOCys, SNAP and nitrergic stimulation, and enhanced responses to NOGS and NOCoA, whereas oxyhaemoglobin blocked relaxations elicited by all these agents. The inhibition by hydroxocobalamin of relaxations elicited by NO, but not those elicited by nitrergic nerve stimulation, indicates that the nitrergic transmitter is unlikely to be free NO, and more closely resembles a nitrosothiol.

Since hydroxocobalamin did not affect nitrergic nerve stimulation-induced relaxations, but blocked the relaxant action of EDNO in rat aortic rings (Rajanayagam *et al.*, 1993), we suggest that it might be a useful agent in discriminating between primary nitrergic transmission and an indirect effect resulting from the action of other transmitters on endothelial cells, which may be the case, for example, in the salivary glands. NADPH diaphorase activity is present in neurons of mouse submandibular glands (Grozdanovic *et al.*, 1992) and nerve stimulation-induced NANC vasodilator responses of the cat submandibular gland *in situ* were blocked by L-NAME (Schachter *et al.*, 1992; Edwards and Garrett, 1993). However, this is not conclusive evidence for a direct nitrergic innervation since another non-nitrergic NANC transmitter (possibly VIP; see Section 12.5.5.1) could have released EDNO.

12.2.6.3 Alcohols

Relaxations of the rat and rabbit anococcygeus and the bovine retractor penis muscles, which are now known to be nitrergically mediated, were inhibited by ethanol (Gillespie *et al.*, 1982). We found that a range of alcohols inhibited relaxations elicited by nitrergic nerve stimulation, NO and SNP in rat anococcygeus muscles and gastric fundus strips (Rand and Li, 1994), and attributed this to sequestration of NO to form nitroso-alcohols. The reaction NO+R-OH→R-O-N=O proceeds readily (Taylor and Baker, 1945). Certain nitroso-alcohols (e.g. butyl and amyl nitrite, or nitrosobutanol and nitrosopentanol, respectively, as they may also be termed) are well known as NO donors (nitrovasodilators); therefore, we suggest that the reaction is reversible, and NO is sequestered when the alcohol is in excess, but NO is released when the nitroso-alcohol is in excess.

12.2.6.4 Superoxide and superoxide generators

Superoxide rapidly inactivates EDNO, and elimination of superoxide with superoxide dismutase (SOD) prolongs the half-life of NO in biological systems (Gryglewski *et al.*, 1986; Rubanyi and Vanhoutte, 1990). The presence of SOD was necessary for detection of an NO-like factor released by nerve stimulation (see Section 12.2.4). SOD enhanced and prolonged nitrergically mediated descending inhibition in the rat colon (Hata *et al.*, 1990) and ileum (Kanada *et al.*, 1992). However, SOD did not enhance responses to nitrergic stimulation in some tissues, including inhibitory junction potentials (IJPs) in the circular muscle of the dog jejunum (Stark *et al.*, 1991), and relaxations of the rat anococcygeus muscle (Rand and Li, 1993a) and opossum internal anal sphincter (Chakder and Rattan, 1992), although responses to NO were enhanced. The superoxide-generating substances pyrogallol and hydroquinone reduced response to nitrergic stimulation in the rat anococcygeus muscle but not in the bovine retractor penis (Gillespie and Sheng, 1990), the mouse anococcygeus muscle (Hobbs *et al.*, 1991), or the opossum internal anal sphincter (Chakder and Rattan, 1992). SOD overcame the effect of pyrogallol in the rat anococcygeus muscle (Gillespie and Sheng, 1990) and hydroquinone on the opossum anal sphincter (Chakder and Rattan, 1992). The differences between tissues and species in the effects of superoxide generators and SOD presumably reflect differences in intrinsic SOD activity.

The substances considered under the heading of guanylyl cyclase inhibitors (see Section 12.2.7.1) may owe part or all of their actions to generation of superoxide.

12.2.6.5 Free radical scavengers

Hydroquinone has free radical scavenging activity as well as being a superoxide generator. It inhibited relaxations in response to nitrergic stimulation in rat anococcygeus muscles but not in the bovine retractor penis muscle (Gillespie and Sheng, 1990), mouse anococcygeus muscle or guinea-pig trachea (Hobbs *et al.*, 1991).

Responses to NO were greatly reduced, but those to SNP, sodium azide or hydroxylamine were not affected (Hobbs *et al.*, 1991). Hydroquinone did not affect the actions of nitrosothiols, with the exception of NOCys, on the mouse anococcygeus muscle (Gibson *et al.*, 1992). However, the hyperpolarization of rat gastric fundus smooth muscle produced by NOCys was blocked by hydroquinone whereas that produced by SNP was not (Kitamura *et al.*, 1993).

12.2.7 Effector mechanisms in nitrergic transmission

The target for the smooth muscle relaxing action of the NO, NO donors, EDNO and the nitrergic transmitter is the haem group of soluble guanylyl cyclase and the activated enzyme produces the second messenger cGMP from GTP (see Chapter 3). This signal transduction system has been comprehensively reviewed recently by Schmidt *et al.* (1993). The mechanisms for the link between the rise in the cGMP content and the relaxation of smooth muscle has not been fully elucidated, but several have been proposed (see Goy, 1991; Lincoln and Cornwell, 1993; Chapter 4).

The localization of guanylyl cyclase (Schmidt *et al.*, 1992b) and its product cGMP (Young *et al.*, 1993), determined by immunohistochemical techniques, showed that they have a complementary distribution to that of NOS. In guinea-pig small intestine and colon, cGMP immunoreactivity in myenteric neurons and smooth muscle was seen only after exposure to SNP (Young *et al.*, 1993). In many neurons on which NOS-containing terminals impinged, cGMP was not detected; this may have been because the level was too low, or perhaps the target in these neurons was not guanylate cyclase.

Nitrergic nerve stimulation increased levels of cGMP but not cAMP in the rat and mouse anococcygeus (Bowman and Drummond, 1984; Mirzazadeh *et al.*, 1991), rat gastric fundus (Ito *et al.*, 1990), ileum (Kanada *et al.*, 1992) and proximal colon (Suthammatpong *et al.*, 1993), opossum internal anal sphincter (Chakder and Rattan, 1993b), dog proximal colon (Ward *et al.*, 1992c) and internal anal sphincter (Grous *et al.*, 1991), the lower oesophageal sphincter of the opossum (Torphy *et al.*, 1986; Barnette *et al.*, 1989; Murray *et al.*, 1992) and human (Barnette *et al.*, 1991), and the rabbit urethra (Dokita *et al.*, 1991; Andersson and Persson, 1993).

The cell-permeable analogue 8-bromo-cGMP mimicked the effect of NO in hyperpolarizing muscle cells in the opossum oesophagus (Du *et al.*, 1991) and the dog pyloric sphincter (Bayguinov and Sanders, 1993a), in relaxing the rat gastric fundus (Barbier and Lefebvre, 1992a) and ileum (Kanada *et al.*, 1993) and the guinea-pig ileum (Osthaus and Galligan, 1992), and in inhibiting spontaneous contractions in the dog proximal colon (Ward *et al.*, 1992c). It is a useful tool for determining the site of disruption of the NOS/guanylate cyclase pathway in neuroeffector transmission by various agents.

12.2.7.1 Guanylyl cyclase inhibitors

Methylene blue is generally regarded as a selective inhibitor of soluble guanylyl cyclase (Ignarro, 1989). However, methylene blue is also known to generate superoxide and the consequent inactivation of NO or an NO adduct could explain the apparent enzyme inhibiting action (Wolin *et al.*, 1990; Gryglewski *et al.*, 1992; Marczin *et al.*, 1992). Furthermore, methylene blue has also been reported to inhibit NOS (Mayer *et al.*, 1993). Thus, despite the fact that it has been frequently used, methylene blue is somewhat flawed as a tool.

Methylene blue does not block uniformly responses to various agents that owe their relaxant action to activation of guanylyl cyclase (literature reviewed by Gryglewski *et al.*, 1992); in fact, the relaxant actions of SNP and sodium nitrite on vascular smooth muscle were potentiated by methylene blue (Gryglewski *et al.*, 1992; Matsumoto *et al.*, 1993). It was reported that methylene blue had no effect on IJPs and relaxations elicited by nitrergic nerve stimulation in smooth muscle from the dog proximal colon (Ward *et al.*, 1992c) but this was not confirmed (Huizinga *et al.*, 1992) and it blocked similar responses in opossum oesophageal smooth muscle (Conklin and Du, 1992). Methylene blue did not inhibit hyperpolarization of rat gastric fundus smooth muscle produced by NOCys, SNP or nitrergic stimulation (Kitamura *et al.*, 1993).

LY-83583 (6-analino-5,8-quinolinedione) was reported to be an inhibitor of soluble guanylyl cyclase (Mülsch *et al.*, 1988), but is also a generator of superoxide anions (Furchgott and Jothianandan, 1991). It did not affect nitrergic nerve stimulation-induced relaxations in the rat gastric fundus although it inhibited responses to exogenous NO and this was partially prevented by SOD (Barbier and Lefebvre, 1992b).

Other substances that are said to inhibit guanylyl cyclase include the following. *N*-Methylhydroxylamine selectively reduced NO-induced relaxations of mouse anococcygeus muscle, nitrosothiol-induced relaxations being unaffected (Gibson *et al.*, 1992). Cystamine (Rapoport and Murad, 1983) blocked the production of IJPs elicited by nitrergic nerve stimulation in the opossum oesophagus (Du *et al.*, 1991). Pyocyanin is chemically related to methylene blue and had similar properties (Gryglewski *et al.*, 1992). Riboflavin (Murad *et al.*, 1979) inactivated NO (Warren *et al.*, 1990). We are not aware of the use of pyocyanin or riboflavine so far in studies on peripheral nitrergic transmission.

12.2.7.2 cGMP phosphodiesterase inhibitors

The specific cGMP phosphodiesterase (PDE) inhibitor M&B 22948 (zaprinast) produced hyperpolarization of the smooth muscle cells and prolonged the duration of IJPs elicited by nitrergic nerve stimulation in the dog pyloric sphincter (Bayguinov and Sanders, 1993a), ileocolonic sphincter and proximal colon (Ward *et al.*, 1992a,c), and increased nitrergic relaxations in the rat anococcygeus muscle (Mirzazadeh *et al.*, 1991). The cGMP-specific PDE inhibitor zaprinast potentiated SNP-induced relaxations in the rat gastric fundus, but surprisingly did not affect relaxations induced by NO or the nitrergic transmitter (Barbier and Lefebvre, 1992a). We have found that cGMP PDE inhibitors are difficult to use in functional studies with the rat anococcygeus muscle because of their relaxant action, which is presumably due to potentiation of the effect of the nitrergic transmitter released spontaneously (see Section 12.5.4).

12.2.7.3 Other targets for the nitrergic transmitter

NO has the potential for interaction with metalloproteins other than guanylyl cyclase and with free sulphydryl groups that may be concerned in regulating the activity of enzymes, receptors or other functional proteins (Stamler *et al.*, 1992; Henry *et al.*, 1993; Schmidt *et al.*, 1993). It has been suggested that NO can act directly on contractile proteins in smooth muscle to reduce their sensitivity to Ca^{2+} (Ozaki *et al.*, 1992). In the central nervous system, nitrosylation of thiol groups in the NMDA receptor channel complex results in loss of responsiveness to agonists (Lei *et al.*, 1992). NO donors increase the activity of a cytosolic ADP ribosyltransferase that catalyses ribosylation of a 39 kDa protein in various rat tissues (Brüne and Lapetina, 1990; Dimmeler and Brüne, 1991; Molina y Vedia *et al.*, 1992; Zhang and Snyder, 1992). The implications of these effects, if they occur in effector tissues with a nitrergic innervation, is obscure.

12.3 Tissues innervated by autonomic nitrergic nerves

The sites at which nitrergic autonomic neuroeffector transmission has been demonstrated are summarized in Table 2. NOS has been detected in nerves supplying these tissues (cf. Table 1), and NOS is also present in autonomic ganglia (Kummer *et al.*, 1992; Sheng *et al.*, 1993; Santer and Symons, 1993).

12.3.1 Anococcygeus and retractor penis muscles

The main evidence for a nitrergic innervation of these muscles has already been given in the foregoing material. In hindsight it can be seen that convincing evidence for the identification of an NO-like substance as the inhibitory transmitter had been accumulated before the introduction of NOS inhibitors. Thus, stimulation-induced relaxations were blocked by oxyhaemoglobin (Bowman *et al.*, 1982; Bowman and Gillespie, 1982; Byrne and Muir, 1984; Gillespie and Sheng, 1989) and hypoxia (Bowman and McGrath, 1985), were reduced by ethanol (Gillespie *et al.*, 1982), and were mediated by cGMP production (Bowman and Drummond, 1984).

In both muscles there is a dense network of varicose terminal axons comprising a typical autonomic ground plexus in these tissues and the ultrastructure is like that of other terminal autonomic axons in containing a variety of vesicles within the varicosities (see Martin and Gillespie, 1991). The presence of nitrergic nerves has been demonstrated histochemically (Table 1).

The retractor penis muscle was used in attempts to isolate the inhibitory transmitter. An acid extract of bovine retractor penis mimicked stimulation-induced

Table 2 Tissues in which relaxations and other responses elicited by stimulation of autonomic nerves have been demonstrated to be nitrergically mediated by blockade of the response with NOS inhibitors and counteraction of the blockade by L-arginine, and in most cases by others means. The table does not include the gastrointestinal tract (see Tables 3 and 4), penile tissue (see Chapter 13) or cerebral blood vessels (see Chapter 11)

Tissues	Species	References
Anococcygeus muscle	Rat	Gillespie *et al.* (1989), Li and Rand (1989b), Ramagopal and Leighton (1989), Hobbs and Gibson (1990), Liu *et al.* (1991), Rand (1992b), Martin *et al.* (1993)
	Mouse	Gibson *et al.* (1990)
	Rabbit	Graham and Sneddon (1993)
Retractor penis muscle	Bovine	Liu *et al.* (1991), Martin *et al.* (1993)
Urinary tract	Rat	Persson *et al.* (1992), Parlani *et al.* (1993)
	Rabbit	Dokita *et al.* (1991), Andersson *et al.* (1991, 1992)
	Dog	Hashimoto *et al.* (1993)
	Pig	Persson and Andersson (1992), Persson *et al.* (1993)
	Sheep	Garcia-Pascual *et al.* (1991), Thornbury *et al.* (1992)
	Human	James *et al.* (1993)
Respiratory tract	Guinea-pig	Tucker *et al.* (1990), Hobbs *et al.* (1991), Li and Rand (1991)
	Cat	Fisher *et al.* (1993)
	Pig	Kannan *et al.* (1992a,b)
	Human	Belvis *et al.* (1992a,b), Ellis and Undem (1992), Bai and Bramley (1993)
Mesenteric arteries	Dog	Toda and Okamura (1990)
	Monkey	Toda and Okamura (1992a)
	Bovine	Leckström *et al.* (1993)
Portal vein	Rabbit	Brizzolara *et al.* (1993)
Pulmonary artery	Guinea-pig	Liu *et al.* (1992)
Uterine artery	Guinea-pig	Morris (1993)

relaxations (Ambache *et al.*, 1975), and this extract, termed inhibitory factor (IF), has been extensively studied (see Gillespie *et al.*, 1990; Martin and Gillespie, 1991). IF lacked biological activity unless exposed to acid and the actions of an acid-activated extract were attributed to nitric oxide (Martin *et al.*, 1988). More recently, it has been suggested that activated IF is actually a nitrosothiol (Kerr *et al.*, 1992). A major hindrance to regarding IF as the inhibitory transmitter is that it can be extracted from many tissues that lack a nitrergic innervation; nevertheless, the properties of IF are similar to those of the inhibitory transmitter.

Stimulation-induced relaxations of the bovine retractor penis were not affected by L-NMMA but the more potent NOS inhibitor L-NNA was effective (Liu *et al.*, 1991; Martin *et al.*, 1993). The reversal of the inhibitory effect of L-NNA required a 30-fold excess of L-arginine, whereas a 3-fold excess was sufficient to reverse the inhibitory effect of L-NMMA in the rat anococcygeus. In fact, in the retractor penis L-NMMA was more potent than L-arginine in counteracting the blocking actions of L-NNA and L-NAME.

These findings suggest that the NOS differs between the rat and the bull, or between the anococcygeus and retractor penis muscles despite the fact that they are regarded as homologous tissues. Curiously enough, there has not been a comparative study of the two muscles from the same species.

12.3.2 Urinary tract

The nitrergic innervation of the lower urinary tract has recently been comprehensively reviewed (Andersson and Persson, 1993). NOS-immunoreactive and NADPH diaphorase-positive nerve fibres have been detected in the lower urinary tract of several species (Table 1), particularly in the trigone and urethra, and in some studies the enzyme was also detected in the urothelium (Andersson and Persson, 1993) Functional studies are listed in Table 2.

In keeping with the relatively sparse nitrergic innervation of the detrusor muscle, there was only a small contribution of nitrergic transmission in relaxations of isolated preparations or in the bladder *in situ* (Andersson and Persson, 1993; James *et al.*, 1993). However, there was evidence for greater nitrergic relaxation of the precontracted trigone of the pig bladder in that L-NNA blocked them and this effect was prevented by pretreatment with L-arginine (Andersson and Persson, 1992).

The evidence for nitrergic relaxation is stronger for the bladder neck and urethra. In the rabbit urethra, L-NNA and L-arginine produced their expected effects on NANC relaxations and acidified sodium nitrite mimicked the relaxation (Andersson *et al.*, 1991, 1992; Dockita *et al.*, 1991). In addition, methylene blue reduced and inhibition of cGMP PDE increased NANC relaxations (Dokita *et al.*, 1991), and the relaxations were accompanied by a rise in the cGMP content of the urethra which was greater in the presence of the selective cGMP PDE inhibitor zaprinast (Persson and Andersson, 1993). In rat, rabbit, pig, sheep and human isolated urethras, stimulation-induced relaxations were almost entirely abolished by NOS inhibitors (Andersson *et al.*, 1991, 1992; Dokita *et al.*, 1991; Garcia-Pascual *et al.*, 1991; Persson *et al.*, 1992, 1993; Persson and Andersson, 1992; Thornbury *et al.*, 1992; Parlani *et al.*, 1993), indicating that the inhibitory transmision was predominantly nitrergic. Yet in the dog urethra a slowly developing component of the relaxations elicited by stimulation at high frequencies (5 and 10 Hz) was not affected by L-NNA suggesting release of an additional inhibitory transmitter, but the slow relaxation was not affected by a VIP antagonist despite the fact that VIP-containing nerves are present (Hashimoto *et al.*, 1993), so the identity of the non-nitrergic component is obscure.

12.3.3 Respiratory tract

The physiological and pathophysiological roles of NO-mediated mechanisms in the lung have recently been comprehensively reviewed by Barnes and Belvisi (1993). In addition to the well-established cholinergic and noradrenergic innervations there are also NANC excitatory and inhibitory nerves that subserve functional roles (Stretton, 1991). Excitatory NANC responses are probably mediated by sensory neuropeptides such as substance P and neurokinin A, while inhibitory NANC responses were believed for some time to be mediated by VIP and related peptides (Stretton, 1991; Lammers *et al.*, 1992). Recently, nitrergic transmission has been demonstrated to mediate at least part of inhibitory NANC responses (Table 2). NOS-containing nerves are present in the airways of several species (Table 1).

Functional evidence for nitrergic transmission to airways smooth muscle was first demonstrated in the guinea-pig trachea (Tucker *et al.*, 1990; Li and Rand, 1990b, 1991), in which the peptidase-resistant, non-VIP NANC relaxation was reduced by NOS inhibitors and the response was partially restored by L-arginine. Relaxations elicited by stimulation at relatively low frequencies are mediated largely by the nitrergic transmitter, whereas responses to stimulation at higher frequencies are predominantly VIP mediated. The extent of NO involvement in NANC responses varies between species and tissue segments, since it has been demonstrated that L-NAME partly (about 50%) inhibited stimulation-induced relaxations in the guinea-pig trachea (Li and Rand, 1991) and human bronchi (Bai and Bramley, 1993) but virtually abolished them in human, pig and cat tracheal segments (Belvisi *et al.*, 1992a,b; Ellis and Undem, 1992; Kannan and Johnson, 1992a,b; Fisher *et al.*, 1993).

In addition to mediating relaxation, NO and NO donors relax or reduce while NOS inhibitors enhance cholinergic or non-cholinergic contractions of airways smooth muscle in the guinea-pig, rabbit, dog and human *in vitro* or *in vivo*, suggesting a modulating role of endogenous NO on excitatory transmission (Brave *et al.*, 1991; Jansen *et al.*, 1992; Belvisi *et al.*, 1991, 1993; Högman *et al.*, 1993; Lei *et al.*, 1993; Nijkamo *et al.*, 1993; Gao and Vanhoutte, 1993; Ward *et al.*, 1993).

12.3.4 Cardiovascular system

12.3.4.1 Heart

Cardiac function is regulated by both sympathetic noradrenergic and vagal cholinergic nerves, but there is also an extensive NANC motor and sensorimotor innervation the functions of which are poorly understood. Recently, NOS was detected in fibres innervating conducting and contractile cardiocytes, cardiac ganglion cells and coronary arteries of the rat and guinea-pig heart (Table 1) and both constitutive and inducible NOS activities are present in the human myocardium (de Belder *et al.*, 1993). There is evidence for modulation of cardiac function by NO donors; however, the extent to which the nitrergic innervation specifically affects cardiac function has not yet been determined.

12.3.4.2 Blood vessels

Dense networks of NOS-containing nerve fibres are detectable histochemically in a number of blood vessels (Table 1), including the pharyngeal artery, and arteries supplying the trachea, lower lip and tongue (Kummer *et al.*, 1992; Kummer and Mayer, 1993), uterine and colonic arteries and pulmonary veins (Kummer *et al.*, 1992; Klimaschewski *et al.*, 1992). In many cases the NOS was co-localized with VIP (Kummer *et al.*, 1992).

The functional study of the nitrergic innervation of blood vessels is complicated by the contribution of EDNO, which counteracts vasoconstrictor responses to perivascular nerve stimulation (Vo *et al.*, 1992). Furthermore, acetylcholine released from cholinergic neurons of submucosal arterioles in the guinea-pig ileum acted on muscarinic cholinoceptors of endothelial cells, releasing EDNO and producing relaxation (Andriantsitohaina and Surprenant, 1992). In such cases, blockade of neurogenic vasodilatation by an NOS inhibitor could be wrongly interpreted as evidence for a nitrergic innervation. Nevertheless, there is clear evidence for nitrergic transmission to endothelium-denuded vascular smooth muscle in blood vessels from a number of regions. The most comprehensively studied are in the cerebral vasculature (see Chapter 11) and the penis (see Chapter 13). In addition, NOS inhibitors and other agents that disrupt nitrergic transmission reduced stimulation-induced vasodilator responses of endothelium-denuded dog, bovine and monkey mesenteric arteries (Toda and Okamura, 1990, 1992a; Ahlner *et al.*, 1991; Leckström *et al.*, 1993). In the guinea-pig pulmonary artery there is a component of neuronal nitrergic relaxation since after removal of the endothelium L-NAME further enhanced stimulation-induced contractions (Liu *et al.*, 1992). Both VIP and NO mediate neurogenic vasodilation in the guinea-pig uterine artery (Morris, 1993).

12.3.5 Other organs

There is morphological evidence of a nitrergic innervation in several organs and tissues, such as the adrenal gland and pancreas, but its functional role is often more a matter of surmise than experimentally demonstrated and discussion of such matters is beyond the scope of this review.

12.4 Nitrergic transmission in the gastrointestinal tract

The gastrointestinal tract has an *extrinsic* innervation consisting of autonomic parasympathetic supplies from the cranial and pelvic regions to the upper and lower segments, respectively, of the tract, an autonomic sympathetic supply that, for the most part, accompanies blood vessels supplying the tract, and sensory neurons arising from dorsal root ganglion cells, and an *intrinsic* innervation, which is regarded in its own right as the enteric nervous system, consisting of a complex assembly of sensory and motor neurons subserving excitatory and inhibitory functions in the myenteric and submucous plexuses (Hoyle and Burnstock, 1989; Bywater and Taylor, 1991; McKirdy, 1991; Furness *et al.*, 1992b). The main candidates as inhibitory transmitters in the enteric nervous system are ATP, VIP, and more recently NO (Sanders and Ward, 1992; Stark and Szurszewski, 1992). Studies on the histochemical localization of NOS in enteric nerves are summarized in Table 1. NOS is present in the synaptosomes of dog ileal myenteric plexus (Kostka *et al.*, 1993). However, there are other sources of NO-like mediators, such as interstitial cells (Publicover *et al.*, 1993), gastric smooth muscle cells (Grider *et al.*, 1992; Grider, 1993; Murphy *et al.*, 1993), macrophages and other cells of the reticuloendothelial defence system and vascular endothelial cells (see Moncada *et al.*, 1991) and gastric mucosa (Brown *et al.*, 1992).

12.4.1 Extrinsic nitrergic autonomic nerves

Nitrergic nerves mediate vagally induced relaxation in the isolated stomach of the guinea-pig (Desai *et al.*, 1991b; Meulemans *et al.*, 1993) and in the stomach *in situ* in the rat and ferret (Lefebvre *et al.*, 1992a; Grundy *et al.*, 1993), in the dog pylorus (Allescher *et al.*, 1992a), and in the opossum lower oesophageal sphincter (Tøttrup *et al.*, 1991a). In these cases, it appears that the postganglionic parasympathetic neurons are nitrergic. However, there is no corresponding evidence for postganglionic nitrergic sympathetic neurons. The histochemically detected neuronal NOS in the guinea-pig colon was not affected by sympathectomy (McConalogue and Furness, 1993), and in accord with this, NOS inhibitors had no effect on relaxations of the colon elicited by stimulation of the extrinsic sympathetic nerves (P. Thaina, C.G. Li and M.J. Rand, unpublished observations).

12.4.2 Electrophysiological observations

The study of junction potentials evoked by nerve stimulation in effector cells has been of value in attempts to elucidate modes and mechanisms of transmission (Hoyle and Burnstock, 1989). The technique is particularly apt when the transmitter is a ligand for an ion-gating polymeric receptor complex (e.g. nicotinic cholinoceptors, $GABA_A$ receptors, NMDA receptors) since an ion flux and associated change in membrane potential is the primary event. However, the only functional target so far identified for the nitrergic transmitter is the haem group of soluble guanylyl cyclase, and the consequence of activating this enzyme, a rise in the intracellular cGMP, certainly mediates relaxation of smooth muscle, but changes in membrane potential may be secondary to the intracellular ion changes involved in inhibiting the contractile mechanisms; if so, they can be considered as epiphenomenona, as previously suggested (Rand, 1992b). Nevertheless, it has been established by several investigators that stimulation-induced relaxation is preceded by hyperpolarization and NO and NO donors produced a similar hyperpolarization (see Sanders and Ward, 1992). However, there are discrepant findings; for example, NOCys but not SNP elicited apamin-sensitive hyperpolarization in the rat gastric fundus (Kitamura *et al.*, 1993).

Stimulation with single pulses in tissues that exhibit nitrergic relaxations elicits IJPs in the smooth muscle cells. Such IJPs have been studied in opossum oesophagus (see Tables 3 and 4 for references), guinea-pig ileum (Lyster *et al.*, 1992; He and Goyal, 1993), dog ileum (Christinck *et al.*, 1991) and ileocolonic sphincter (Ward *et al.*, 1992a), dog and human jejunum (Stark *et al.*, 1991, 1993) and dog colon (Dalziel *et al.*, 1991; Thornbury *et al.*, 1991; Huizinga *et al.*, 1992; Ward *et al.*, 1992d). These IJPs often contain two components. In circular muscle from guinea-pig ileum there was a fast component that was blocked by apamin (Lyster *et al.*, 1992) and mediated by ATP (He and Goyal, 1993). The slow apamin-resistant IJP was greatly reduced by L-NNA and partially restored by L-arginine (Lyster *et al.*, 1992). However, He and Goyal (1993) also showed that it was mimicked by NO and VIP, and was mediated by VIP and NO acting in series. In strips of circular muscle from the dog ileum, IJPs produced by field stimulation were greatly reduced by L-NAME and excitatory junction potentials were revealed (Christinck *et al.*, 1991).

The ionic basis of NO-mediated IJPs and hyperpolarization is an increase in K^+ conductance (Stark *et al.*, 1991) due to an increase in the open probability of Ca^{2+}-activated K^+ channels (Thornbury *et al.*, 1991), but it is not clear how cGMP contributes to these membrane effects. The relaxation accompanying hyperpolarization could be due to inhibition of excitatory action potentials or inactivation of voltage-dependent Ca^{2+} channels. In accord with the latter, nitrergic stimulation of the rat anococcygeus muscle decreased the intracellular Ca^{2+} concentration (detected by the fura-2 method) which could explain the relaxation (Ramagopal and Leighton, 1989), but the cause of the decreased Ca^{2+} concentration is not clear (see Section 12.2.7).

12.4.3 Sphincters

In terms of the structural and functional organization of the gastrointestinal tract, it is convenient to consider two types of segments: sphincters and segments between sphincters (Papasova, 1989), and it is convenient to deal first with sphincters since their functions are simpler and the role of their nitrergic innervation is clear.

In oral to aboral order the sphincters are the (1) lower oesophageal sphincter, (2) pyloric sphincter, (3) ileocolonic/ileocaecal sphincter, or junction, and (4) internal sphincter (noting that the external anal sphincter that surrounds it and extends more distally to the anal orifice is composed of striated muscle); in addition, the sphincter of Oddi at the junction of the bile duct with the duodenum can be included. The characteristics of sphincters are that the musculature is predominantly circularly arranged, the resting tone is high compared to that of neighbouring segments of the tract, a rise in pressure in the adjacent proximal segment causes reflex relaxation, and a rise in pressure in the adjacent distal segment causes reflex contraction (Papasova, 1989). The predominant relaxant transmitter is nitrergic, as discussed below. The main findings are briefly summarized in Table 3.

12.4.3.1 Lower oesophageal sphincter

The sphincter of the opossum has been most frequently used. Field stimulation produced frequency-dependent relaxations accompanied by slow long-lasting hyperpolarizations which were inhibited by L-NNA and restored by L-arginine, and similar results were obtained in other species (see Table 3). In addition, the relaxations induced by vagal stimulation and swallowing were reduced by NOS inhibitors (Tøttrup *et al.*, 1991b; Yamato *et al.*, 1992a,b; Paterson *et al.*, 1992). This evidence for nitrergic transmission was strengthened by the observations that relaxation of the sphincter produced by nerve stimulation and NO donors was associated with an increase in the cGMP content of the muscle that was blocked by putative guanylyl cyclase inhibitors (Torphy *et al.*, 1986; Conklin and Du, 1992). However, it was suggested that cGMP may not be the only intracellular mediator of nerve-induced relaxations (Murray *et al.*, 1992).

12.4.3.2 Pyloric sphincter

In the dog pyloric sphincter *in situ*, L-NAME increased tone and phasic motor activity (Allescher *et al.*, 1992a). Field stimulation of the duodenum increased motor activity, and this was inhibited by superimposing stimulation of the vagus nerve or the distal antrum: these inhibitory effects were blocked by L-NAME. In dog isolated pyloric sphincter muscle, field stimulation produced relaxations that were blocked by L-NAME in an L-arginine-reversible manner, and by oxyhaemoglobin and apamin (Allescher *et al.*, 1992a; Bayguinov and Sanders, 1993a). Other references are given in Table 3.

Table 3 Evidence for a role of nitrergic transmission in functional control of gastrointestinal sphincters

Sphincter	Species	Function studied	References
Oesophageal	Opossum	Relaxation and/or IJP	Murray *et al.* (1991), Tøttrup *et al.* (1991a,b), Jury *et al.* (1992), Knudsen *et al.* (1992), Conklin *et al.* (1993)
		Vagal and reflex relaxation	Tøttrup *et al.* (1991b), Paterson *et al.* (1992)
		Swallow relaxation	Yamato *et al.* (1992a)
	Dog	Relaxation and/or IJP	Barnette *et al.* (1989), De Man *et al.* (1991), Jury *et al.* (1992)
	Human	Relaxation	McKirdy *et al.* (1992)
Pyloric	Dog	Relaxation and/or IJP	Allescher *et al.* (1992a), Bayguinov and Sanders (1993a,b)
	Ferret	Vagal relaxation	Grundy *et al.* (1993)
Oddi's	Possum	Relaxation	Baker *et al.* (1991)
	Opossum	Relaxation	Allescher *et al.* (1993)
	Guinea-pig	Relaxation	Pauletzki *et al.* (1993)
	Dog	Increased motility	Kaufman *et al.* (1993)
Ileocolonic	Dog	Relaxation	Boeckxstaens *et al.* (1990a,b,c, 1991,a,c), Bult *et al.* (1990), Bogers *et al.* (1991)
		IJP	Ward *et al.* (1992a)
Internal anal	Opossum	Relaxation	Chakder and Rattan (1992, 1993a,b), Tøttrup *et al.* (1992), Rattan and Chakder (1992)
		Reflex relaxation	Rattan *et al.* (1992)
	Human	Relaxation	Burleigh (1992), Okelly *et al.* (1993)

12.4.3.3 Sphincter of Oddi

The sphincter of Oddi is regulated by inhibitory neural mechanisms that include a nitrergic component since NOS-containing nerves are present (Table 1) and L-NAME significantly inhibited, in an L-arginine reversible manner, stimulation-induced relaxations of the sphincter from the Australian possum (Baker *et al.*, 1991) and the guinea-pig (Pauletzki *et al.*, 1993). In the anaesthetized dog, L-NAME increased motility of the sphincter of Oddi and duodenum, and its effect was blocked by L-arginine (Kaufman *et al.*, 1993).

12.4.3.4 Ileocolonic sphincter

The evidence for nitrergic transmission at this site is particularly strong and most of it has been obtained by a group of Belgian investigators using the tissue from the dog (Bult *et al.*, 1990; Boeckxstaens *et al.*, 1990a,b,c, 1991a,c; Bogers *et al.*, 1991). They showed that NANC-induced relaxations were inhibited by L-NMMA and L-NNA, and their effects were overcome by L- but not D-arginine. Oxyhaemoglobin abolished responses to NO and reduced those to NANC stimulation. Nerve stimulation released a NO-like factor (see Section 12.2.4). There is also electrophysiological evidence for a nitrergic innervation since stimulation-induced IJPs were blocked by NOS inhibitors and oxyhaemoglobin, increased by cGMP PDE inhibition, and mimicked by NO (Ward *et al.*, 1992a).

12.4.3.5 Anal sphincter

Nitrergic relaxations of the internal anal sphincter have been demonstrated in isolated preparations from opossum and human (see Table 3 for references). The evidence consists of mimicry by NO or NO donors, reduction by NOS inhibitors that was counteracted by L-arginine, and reduction by oxyhaemoglobin and hydroquinone. The release of nitric oxide was detected by chemiluminescence assay (Chakder and Rattan, 1993a). The rectoanal inhibitory reflex relaxation *in situ* was inhibitied by L-NNA (Rattan *et al.*, 1992).

12.4.4 Segments between sphincters

The functions of tracts between sphincters are more complex. Apart from those concerned with digestive processes (addition of enzymes to the contents and absorption) there are coordinated movements for mixing and propulsion of the contents. Those regions in which nitrergic transmission from enteric nerves has been demonstrated are summarized in Table 4.

12.4.4.1 Oesophagus

In the circular smooth muscle of the oesophagus, tone is low (in contrast to the sphincter) and field stimulation produces little in the way of relaxation but there is

Table 4 Regions of the gastrointestinal tract between sphincters in which relaxations and/or IJPs elicited by stimulation of enteric nitrergic nerves have been demonstrated

Region	Species	References
Oesophagus	Opossum	Christinck *et al.* (1991), Du *et al.* (1991), Knudsen *et al.* (1991), Murray *et al.* (1991)
	Rat	Will *et al.* (1990)
Gastric fundus	Rat	Li and Rand (1990a), Barbier and Lefebvre (1992b), Boeckxstaens *et al.* (1992), D'Amato *et al.* (1992), Lefebvre *et al.* (1992a), Kitamura *et al.* (1993), Shimamura *et al.* (1993)
	Guinea-pig	Lefebvre *et al.* (1992b), Meulemans *et al.* (1993)
	Cat	Barbier and Lefebvre (1993)
	Dog	Ozaki *et al.* (1992)
Duodenum	Dog	Toda *et al.* (1990a, 1991), Bayguinov *et al.* (1992)
Jejunum	Dog	Stark *et al.* (1991, 1993)
	Human	Stark *et al.* (1993)
Ileum	Guinea-pig	Osthaus and Galligan (1992), He and Goyal (1993)
	Rat	Kanada *et al.* (1992)
	Dog	Boeckxstaens *et al.* (1990a,c, 1991a), Christinck *et al.* (1991)
	Human	Maggi *et al.* (1991)
Caecum	Guinea-pig	Knudsen and Tøttrup (1992b), Shuttleworth *et al.* (1991)
Colon	Rat	Hata *et al.* (1990), Grider (1993), Middleton *et al.* (1993), Suthamnatpong *et al.* (1993)
	Guinea-pig	Maggi and Giuliani (1993)
	Dog	Dalziel *et al.* (1991), Thornbury *et al.* (1991), Huizinga *et al.* (1992), Ward *et al.* (1992c,d)
	Human	Burleigh (1992), Boeckxstaens *et al.* (1993a), Keef *et al.* (1993)
Taenia coli	Human	Tam and Hillier (1992)

nitrergically mediated hyperpolarization followed by depolarization, and an 'off-contraction' after the cessation of stimulation (see Table 4 for references). There does not appear to be a nitrergic innervation of the longitudinal smooth muscle (Murray *et al.*, 1991). The physiological role of the nitrergic innervation was demonstrated by the finding that L-NNA reduced oesophageal peristalsis (Yamato *et al.*, 1992b; Tøttrup, 1993).

12.4.4.2 Stomach

The roles played by the main NANC mediators of neurotransmission in the body of the stomach, ATP, VIP and the nitrergic transmitter, have been recently reviewed by Lefebvre (1993). The first evidence for a nitrergic innervation was obtained by Li and Rand (1990a), using strips of rat gastric fundus, in which NANC relaxations were reduced by L- but not D-NMMA and restored by L- but not D-arginine. When the component of the NANC relaxation that was mediated by VIP was blocked with a specific VIP antibody or by enzymatic destruction of VIP, the residual NANC relaxation was further reduced or abolished by L-NMMA. Similar findings were obtained by others (Table 4). In addition, nerve stimulation released an NO-like substance (see Section 12.2.4). Compared to the nitrergic component, the VIP component was slower to develop, lasted longer, and required higher frequencies of stimulation to elicit it (Li and Rand, 1990a), and was greater with long than with short train lengths of stimulation (D'Amato *et al.*, 1992; Boeckxstaens *et al.*, 1992). The finding that there is a nitrergic as well as a VIP-mediated contribution to the NANC relaxations correlated well with earlier observations that the initial part of the NANC relaxations could not be blocked by VIP antiserum or peptidases. However, in guinea-pig gastric fundus strips, relaxations elicited by both long and short trains of stimulation over a range of frequencies were inhibited by L-NNA (Lefebvre *et al.*, 1992b). It was also observed that VIP-induced relaxation was not affected, but there are contrary findings (see Section 12.5.5.1). In dog gastric smooth muscle, the force of spontaneous contractions was increased by L-NMMA, oxyhaemoglobin and methylene blue and decreased by L-arginine (Ozaki *et al.*, 1992).

12.4.4.3 Duodenum

The findings of nitrergic stimulation-induced relaxations of longitudinal and circular muscle are listed in Table 4. In addition, the amount of the bicarbonate-rich alkaline fluid secreted by the duodenal mucosa was increased by L-NAME and this effect was antagonized by L-arginine; after vagotomy, the effect of L-NAME was greatly reduced (Takeuchi *et al.*, 1993). These observations suggest that the cholinergic drive that is responsible for duodenal mucosal secretion is opposed by an NO-mediated mechanism.

12.4.4.4 Gall bladder

This organ is considered here because of its connection with duodenum. NOS has been detected histochemically in neurons of the gall bladder (Table 1) and is colocalized with VIP. L-NAME significantly enhanced CCK- and bethanechol-induced increases in intraluminal pressure of the gall bladder in anaesthetized guinea-pig and contractions of isolated strips of gall bladder (Mourelle *et al.*, 1993), indicating a modulatory role of NO on motility. Nitrergic nerves could contribute to such modulation since L-NAME enhanced field stimulation-induced contractions of the guinea-pig isolated gall bladder (C.G. Li and M.J. Rand, unpublished observations). However, field stimulation produced only a small relaxation when the tone was raised, suggesting there may not be a direct nitrergic innervation of the smooth muscle.

12.4.4.5 Jejunum

In human and dog jejunal circular muscle, NANC relaxations were inhibited by NOS inhibitors (Stark *et al.*, 1993). Stimulation-induced IJPs in human jejunum consisted of an initial fast hyperpolarization followed by a late sustained hyperpolarization; the latter was mimicked by exogenous NO, while in the dog stimulation-induced IJPs consisted only of a fast hyperpolarization which was blocked by NOS inhibitors (Stark *et al.*, 1991), but hyperpolarization was only delayed in the human tissue, suggesting that another transmitter may be involved (Stark *et al.*, 1993). L-NAME increased pressure and phasic contractions in the jejunum *in situ* in the rat (Calignano *et al.*, 1992) and the cat (Gustafsson and Delbro, 1993), suggesting a tonic inhibitory nitrergic tone.

12.4.4.6 Ileum

Stimulation-induced nitrergic responses of the ileum from various species consist of relaxation of the smooth muscle and production of IJPs (see Table 4 and Section 12.4.2), and also modulation of contractions and changes in the pattern of reflex peristalsis.

Nitrergic transmission explained only part of the field stimulation-induced relaxations of human ileal circular smooth muscle (Maggi *et al.*, 1991). In rat ileum, the stimulation-induced initial contraction was inhibited by L-NNA and mimicked by SNP (Bartho *et al.*, 1992), indicating a nitrergic mechanism may mediate contraction or, more likely, change the release of other mediators. Further evidence for NO-mediated inhibitory transmission in circular muscle from the dog terminal ileum was provided by Boeckxstaens *et al.* (1990a) in studies in which the adjacent ileocolonic junction was also used (Section 12.4.3.4).

Several studies have been concerned with a nitrergic modulation of responses to other mediators. Gustafsson *et al.* (1990) demonstrated that L-NMMA selectively enhanced stimulation-induced contractions of segments of guinea-pig ileum and suggested that the nitrergic transmitter acted prejunctionally to inhibit release of substance P. An alternative mechanism suggested by Osthaus and Galligan (1992) was that a nitrergic mechanism unmasked the action of the excitatory transmitter. It has also been suggested that an acetylcholine receptor-activated nitrergic mechanism may be important in modulation of excitatory responses (Wiklund *et al.*, 1993).

12.4.4.7 Caecum

The longitudinally arranged smooth muscle of this organ is arranged in three bands (taenia) on the surface of the caecum. They are sometimes, erroneously, referred to as taenia coli (see Campbell, 1991). The taenia of the guinea-pig caecum was the first tissue in which a distinct claim for an NANC inhibitory innervation was made (Burnstock *et al.*, 1966). ATP and VIP have been proposed as inhibitory transmitters: the evidence favours ATP as the principal one (for review, see White, 1988). The possibility of a nitrergic component was first tested by Li and Rand (1989b), who found that L-NMMA had virtually no effect on the neurogenic NANC relaxation. Knudsen and Tøttrup (1992b), using L-NNA, obtained the same result; however, L-NNA inhibited the relaxation phase of the biphasic response to field stimulation in the absence of atropine, and concluded that the nitrergic transmitter had a prejunctional inhibitory effect on the cholinergic nerves.

Using a preparation of the circular muscle between the taenia of the caecum, Shuttleworth *et al.* (1991) showed that stimulation-induced relaxations were reduced by L-NNA and oxyhaemoglobin by up to 50%, indicating a nitrergic component. This is in accord with the histochemical evidence that many NOS-immunoreactive neurons were located in the myenteric plexus of the guinea-pig caecum, and the fibres run to circular smooth muscle and longitudinal smooth muscle between the taenia (Furness *et al.*, 1992a).

12.4.4.8 Colon

In the longitudinal muscle of the rat colon, the mediators of neurogenic relaxation depend on the segment: in the proximal segment inhibitory transmission is mainly nitrergic; in the middle segment the nature of the mediator is uncertain; in the distal segment VIP is the most likely inhibitory transmitter (Suthamnatpong *et al.*, 1993). In human colon (Burleigh, 1992) and taenia coli (Tam and Hillier, 1992), L-NNA partially inhibited NANC relaxations.

In colonic circular smooth muscle from guinea-pigs (Maggi and Giuliani, 1993) and humans (Boeckxstaens *et al.*, 1993a), NANC relaxations are mediated by an apamin-sensitive transmitter, probably ATP, and the apamin-resistant transmitter, the effect of which was reduced or abolished by NOS inhibitors. In the circularly arranged smooth muscle of the rat distal colon, spontaneous rhythmic contractions were increased by NOS inhibition (Niklasson *et al.*, 1992; Middleton *et al.*, 1993) and also by tetrodotoxin, suggesting that tonic activity of nitrergic nerves has a modulatory effect in dampening the contractions (Middleton *et al.*, 1993).

The stimulation-induced hyperpolarization and inhibition of slow waves of depolarization of smooth muscle emanating from pacemaker cells in the colon were nitrergically mediated, being mimicked by NO and NOCys and blocked by L-NAME, L-NNA or oxyhaemoglobin in segments from dogs (Dalziel *et al.*, 1991; Ward *et al.*, 1992d; Huizinga *et al.*, 1992) and humans (Keef *et al.*, 1993). However, in the human segments, the fast hyperpolarization elicited by a single stimulus was not affected by L-NAME or oxyhaemoglobin but was reduced by apamin (Keef *et al.*, 1993).

There is clear morphological evidence for nitrergic enteric nerves (see Table 1). In the guinea-pig colon, cell bodies of nitrergic neurons are present in myenteric and submucous ganglia, and axons project to both the neuronal plexuses, to circularly and longitudinally arranged smooth muscle, and to the muscularis mucosa (Young *et al.*, 1992; McConalogue and Furness, 1993). All the axons projecting to the circular muscle are directed anally (McConalogue and Furness, 1993). This is in accord with the finding that nitrergic transmission plays an essential role in descending inhibition in the colon (see Section 12.4.5).

12.4.5 Nitrergic involvement in enteric reflexes

Field stimulation of segments of the gastrointestinal tract will activate all the intrinsic neurons resulting in the release of a multitude of transmitters with diverse actions, and even though part or even all of the response is shown to be nitrergically mediated, the physiological function of nitrergic transmission can only be a matter of conjecture. However, reflexes activate selectively specific neural pathways, so studies on the participation of nitrergic transmission in reflexly induced responses are of special importance. The following paragraph summarizes the evidence for it in reflexes through the entire length of the gastrointestinal tract.

In swallowing, which is controlled by both intrinsic enteric and extrinsic vagal parasympathetic nerves, oesophageal peristalsis (Tøttrup, 1993) and relaxation of the lower oesophageal sphincter induced reflexly by distension of the tract were reduced by NOS inhibitors (Yamato *et al.*, 1992a,b; Paterson *et al.*, 1992; Tøttrup, 1993). In the guinea-pig isolated stomach, an increase in intragastric pressure resulted in an adaptive relaxation involving a local reflex activation of nitrergic nerves (Desai *et al.*, 1991a). In the small intestine, blocking nitrergic nerve activity increased the ascending cholinergic contraction produced by stimulation of the anal end of an isolated segment of rat ileum (Allescher *et al.*, 1992b), motility of rat jejunum was increased (Calignano *et al.*, 1992), and anally directed distension was blocked (Kanada *et al.*, 1992). Descending inhibition in the rat proximal colon elicited by distension of an intraluminal balloon (Hata *et al.*, 1990) or stretch (Grider, 1993) was blocked by L-NMA. The rectoanal inhibitory reflex relaxation *in situ* was inhibited by L-NNA (Rattan *et al.*, 1992).

12.5 Mechanisms of nitrergic transmission

The simplest scheme for nitrergic transmission envisages: (1) activation of intraneuronal NOS by a rise in the cytosolic Ca^{2+} concentration; (2) conversion of L-arginine to NO; (3) passive diffusion of NO into the extracellular fluid and thence into the effector cells; (4) binding to the haem group of soluble guanylyl cyclase and activation of the enzyme; and finally (5) increased production of cGMP which mediates the relaxation (Sanders and Ward, 1992). Despite our fondness for Occam's razor, we bear in mind the late Jimmy Graham's precept that God does not shave with Occam's razor (unpublished discussion at a meeting of the British Pharmacological Society), and consider that this is a premature oversimplification.

It has been suggested by some investigators (e.g. Ignarro *et al.*, 1990) that an unknown substance may be the primary transmitter in what we have termed nitrergic transmission, and this unknown substance in turn releases NO from, for example, the effector smooth muscle cells. Support for this contention comes from the work of Grider and his colleagues who suggested that VIP is a primary, NO-releasing transmitter (see Section 12.5.6.1). The nitrergic transmitter can be released from neurons by a number of other transmitters (see Section 12.5.7), but these findings still leave the nitrergic transmitter as the final mediator.

There have been suggestions that the interstitial cells of Cajal may be involved in nitrergic transmission in the gastrointestinal tract. These cells are often associated with enteric neuron terminals and are electrically coupled to smooth muscle cells (Publicover *et al.*, 1993). Huizinga *et al.* (1992) suggested that VIP released NO from the interstitial cells of the dog colon. Publicover *et al.* (1993), using isolated enzymatically dispersed interstitial cells from the dog colon, showed that NO increased the intracellular Ca^{2+} concentration, apparently by mobilizing Ca^{2+} from an intracellular store. Interstitial cells have NADPH diaphorase activity (Xue *et al.*, 1993), and presumably also NOS activity. When the intracellular Ca^{2+} concentration in interstitial cells was raised by a microinjection technique, that in neighbouring smooth muscle cells was decreased and this effect was blocked by NOS inhibitors and oxyhaemoglobin, indicating a diffusible NO-like mediator was involved. If this mediator also acted back on the interstitial cells, it would form a positive feedback loop. It was suggested that this mechanism provided a means of amplifying nitrergically mediated relaxations.

12.5.1 Nature of the nitrergic transmitter

Although the nitrergic transmitter is initially generated by NOS, there is clear evidence that free NO does not meet the criteria for acceptance as the nitrergic transmitter since the action of NO but not that of the transmitter was blocked by various agents. Thus relaxant responses to NO but not those to nitrergic nerve stimulation were blocked by hydroquinone in the bovine retractor penis muscle (Gillespie and Sheng, 1990) and mouse anococcygeus muscle (Hobbs *et al.*, 1991), by hydroxocobalamin in rat anococcygeus muscle (Li and Rand, 1993a; Rand and Li, 1993b), and by LY-83583 in the rat gastric fundus (Barbier and Lefebvre, 1992b). Furthermore, the response to NO, but not that to nitrergic nerve stimulation was potentiated by cysteine (Rand and Li, 1992c).

The possibility that the transmitter is a nitrosothiol, possibly nitrosocysteine, is most favoured (Thornbury *et al.*, 1991; Kitamura *et al.*, 1993; Li and Rand, 1993a; Rand and Li, 1993b). However, it is not nitrosocysteine in the mouse anococcygeus muscle since hydroquinone greatly reduced responses to it but did affect nitrergic transmission (Gibson *et al.*, 1992). What can be said about the nitrergic transmitter is that it is an NO donor having the ability to activate soluble guanylyl cyclase (see Section 12.2.7) and that it can be denitrosylated by oxyhaemoglobin but it cannot cross the erythrocyte membrane (see Section 12.2.6.2). It is entirely possible that the nitrergic transmitter is not identical in all tissues, and its nature may depend on local availability of compounds that can form NO adducts.

12.5.2 Is there exocytosis in nitrergic transmission?

The available morphological evidence suggests that the ultrastructure of NOS axon varicosities is similar to that of other autonomic nerve terminals in containing vesicles (see Martin and Gillespie, 1991). However, it is not evident so far from ultrastructural studies on NOS localization in peripheral nerves whether or not NOS is in subcellular organelles that might correspond to transmitter storage vesicles. NOS activity in homogenates of bovine retractor penis and anococcygeus muscles is present in both soluble and particulate fractions (Mitchell *et al.*, 1991; Sheng *et al.*, 1992), but the nature of the particulate fraction has not been determined. The frequently quoted statement that neuronal NOS is cytosolic is not meaningful unless the methodology for concentrating it has been specifically designed to avoid disruption of subcellular organelles.

If exocytosis of NOS occurs as an essential step in nitrergic transmission, the L-arginine substrate must also be released with it since, in isolated tissues, the response does not depend on the previous presence of L-arginine in the bathing medium. An indication that L-arginine is released comes from the observations that NOS inhibitors increased the level of extracellular L-arginine released in rat cerebellum *in situ* by the glutamate receptor agonist NMDA (Sim and Sim, 1993a,b).

To accommodate the possibility of exocytosis, the scheme for nitrergic transmission shown in Figure 1 would need revision to the extent that the enzymatic reactions shown in the axon terminals would be shifted into the neuroeffector junction.

12.5.2.1 Effects of L-arginine

L-Arginine has been reported to decrease tone slightly in the rat anococcygeus muscle (Gillespie *et al.*, 1989) and gastric fundus (Li and Rand 1990a; Boeckxstaens *et al.*, 1991b), increase nitrergic relaxations of anococcygeus muscles from the rat (Li and Rand, 1989b) and mouse (Gibson *et al.*, 1990), increase the release of NO, measured by bioassay, from the dog ileocolonic junction (Boeckxstaens, 1991c), enhance the stimulation-induced relaxation of the pig trigone (Persson and Andersson, 1992) and rabbit urethra (Andersson *et al.*, 1991, 1992), and reduce the bronchoconstriction elicited in guinea-pigs by vagus nerve stimulation indicating increased nitrergic relaxation (Lei *et al.*, 1993). However, it was reported that L-arginine did not enhance stimulation-induced relaxations of rat anococcygeus or bovine retractor penis muscles (Gillespie *et al.*, 1989; Ramagopal and Leighton, 1989; Liu *et al.*, 1991; Martin *et al.*, 1993), dog duodenal longitudinal muscle (Toda *et al.*, 1990a, 1991), jejunum (Stark *et al.*, 1993) or urethra (Hashimoto *et al.*, 1993) or human ileal smooth muscle (Maggi *et al.*, 1991).

The K_m of brain NOS for L-arginine has been reported to be 1.5–2.9 μM (Bredt and Snyder, 1990; Knowles *et al.*, 1990; Schmidt and Murad, 1991), and presumably would be similar for NOS of peripheral nitrergic neurons. The concentration of L-arginine in the biophase of NOS is not known; however, in those cases in which responses to nitrergic stimulation were not increased by L-arginine, it was apparently at or above the K_m, whereas when L-arginine increased responses to nitrergic stimulation the biophase concentration may have been below the K_m. The disparity in the findings could possibly be due to differences in levels of endogenous NOS inhibitors (see Section 12.2.2.1), or to the previous history of nitrergic nerve stimulation in the various preparations which may have partly depleted L-arginine stores or may have affected the amount of NOS released by exocytosis.

12.5.2.2 Dependence of nitrergic transmission on Ca^{2+}

The previously accepted mechanism of excitation–secretion coupling in transmitter release is that the action potential triggers the influx of Ca^{2+} and the rise in the cytosolic Ca^{2+} concentration is essential for the docking of transmitter storage vesicles with the plasma membrane and exocytotic release of their contents (Kandel, 1991).

There is good evidence that nitrergic transmission depends on Ca^{2+} influx. Exocytotic transmitter release is attenuated in a low Ca^{2+}, high Mg^{2+} medium, and in such a medium ([Ca^{2+}] lowered from 2.5 to 0.5 mM and [Mg^{2+}] raised from 1.2 to 15 mM) the IJP amplitude in response to nitrergic stimulation in isolated circular of dog jejunum was significantly reduced but hyperpolarization produced by NO was not affected (Stark *et al.*, 1991). Furthermore, the inhibition of influx of Ca^{2+} through N-type channels by ω-conotoxin was as effective in reducing nitrergic as it was reducing autonomic cholinergic and noradrenergic transmission (Li and Rand, 1990c; De Luca *et al.*, 1990; Altiere *et al.*, 1992).

The enhancement of responses induced by stimulation of cholinergic and noradrenergic nerves by tetraethylammonium chloride (TEA) is attributed to prolongation of the action potential in nerve terminals that allowed a consequent greater influx of Ca^{2+}, hence greater release of transmitter by exocytosis. In rat anococcygeus muscles, TEA potentiated not only stimulation-induced noradrenergic contractions but also relaxations that are now known to be nitrergically mediated

(Gillespie and Tilmisany, 1976). In the dog ileocolonic junction, TEA enhanced relaxations elicited by nitrergic nerve stimulation but not those elicited by NO (De Man *et al.*, 1993).

These findings could be construed as evidence for exocytotic release in nitrergic transmission; however, since NOS is calcium and calmodulin dependent, the consequences of increased or decreased Ca^{2+} influx may be different degrees of activation of NOS intraneuronally.

12.5.3 Spontaneous release of the nitrergic transmitter

NOS inhibitors have been reported to increase the basal tone of a number of nitrergically innervated smooth muscle preparations, including anococcygeus muscle of the rat (Li and Rand, 1989b; Gillespie *et al.*, 1989; Liu *et al.*, 1991) and rabbit (Graham and Sneddon, 1993), rat gastric fundus (Li and Rand, 1990a; Boeckxstaens *et al.*, 1991b) and distal colon (Middleton *et al.*, 1993), dog gastric antrum (Ozaki *et al.*, 1992) and pyloric sphincter *in situ* (Allescher *et al.*, 1992a), dog isolated ileocolonic junction (Boeckxstaens, 1990a), and urethra (Hashimoto *et al.*, 1993), bovine retractor penis (Liu *et al.*, 1991), human ileum (Maggi *et al.*, 1991) and taenia coli (Tam and Hillier, 1992) and guinea-pig colon (Briejer *et al.*, 1992). There is also electrophysiological evidence in dog ileocolonic sphincter for tonic release (Ward *et al.*, 1992a). However, it has been reported that NOS inhibitors did not affect tone in rat anococcygeus muscles (Ramagopal and Leighton, 1989), duodenum (Irie *et al.*, 1991) and colon longitudinal smooth muscle (Suthamnatpong *et al.*, 1993) or dog ileum (Boeckxstaens *et al.*, 1990a), lower oesophageal sphincter (De Man *et al.*, 1991) and jejunum (Stark *et al.*, 1991) or opossum lower oesophageal sphincter *in situ* (Tøttrup *et al.*, 1991a,b).

Other agents that impair nitrergic transmission have also been reported to increase basal tone: these include oxyhaemoglobin in dog ileocolonic junction (Boeckxstaens, 1990a), guinea-pig colon (Briejer *et al.*, 1992) and rat anococcygeus (Li and Rand, 1993a), methylene blue in guinea-pig colon (Briejer *et al.*, 1992) and tetrodotoxin in rat distal colon but not after tone had already been raised by L-NMMA (Middleton *et al.*, 1993).

In those cases in which attenuation or blockade of the nitrergic transmission results in an increase in tone or motility, it is reasonable to assume that there is a spontaneous release. However, Sanders and Ward (1992) discussed the possibility that EDNO from blood vessels in the walls of the gut might provide a tonic inhibitory drive, not only to the vascular smooth muscle, but also to the adjacent gastrointestinal smooth muscle cells.

12.5.4 Modulation of nitrergic transmission

Prejunctional modulation has been demonstrated for many modes of transmission (for review, see Westfall and Martin, 1991). Nitrergic transmission is no exception.

12.5.4.1 Automodulation

In most types of neurons there are mechanisms for feedback control of transmitter release whereby the transmitter acts on specific receptors associated with the sites of release and usually mediates inhibition of further release: this process has been termed automodulation of transmission. It is generally thought that the receptors involved in automodulation are coupled to second messenger systems that probably affect the exocytotic mechanism in transmission release (Rand *et al.*, 1988). If there is inhibitory automodulation of nitrergic transmission, it would probably be through a mechanism based on the finding that the activity of NOS was enhanced by oxyhaemoglobin and decreased by NO and NO donors (Rogers and Ignarro, 1992; Assreuy *et al.*, 1993; Rengasamy and Johns, 1993). It was postulated that the inhibitory effect of NO was due to its combination with a haem group associated with the enzyme. However, to our knowledge no functional studies to explore this possibility have been carried out.

12.5.4.2 Inhibitory modulation of nitrergic transmission

For all of the established modes of transmission, the terminals are endowed with a range of different types of receptors on which agonists can act to inhibit transmitter release. There is evidence for prejunctional inhibitory muscarinic cholinoceptors, α_2-adrenoceptors and opioid receptors at nitrergic nerve terminals.

Muscarinic cholinoceptor agonists inhibited NANC (since identified as nitrergic) relaxations in the rat anococcygeus (Li and Rand, 1989a) and this effect was also suggested for the nitrergic innervation of dog mesenteric arteries (Toda *et al.*, 1990b).

The α_2-adrenoceptor agonist UK-14304 inhibited NANC relaxations in the rat gastric fundus (MacDonald *et al.*, 1990), both the nitrergic and VIP components being reduced (Lefebvre and Smits, 1992). However, clonidine was not effective (Lefebvre and Bogaert, 1986). In substance P-contracted dog ileocolonic junction both UK-14304 and clonidine reduced nitrergic relaxations, but in noradrenaline-contracted preparations clonidine augmented relaxations (Boeckxstaens *et al.*, 1993b). In the noradrenaline-contracted tissue prejunctional α_2-adrenoceptors would already be occupied, resulting in a degree of inhibition of transmitter release, and in fact a stronger stimulation was required to get equal relaxation to those in SP-contracted tissues: clonidine, being a partial agonist, was probably manifesting itself as an α_2-adrenoceptor antagonist, blocking the inhibitory effect of noradrenaline (Medgett and Rand, 1981).

Opioid receptor agonists inhibited nitrergic relaxations in the dog lower oesophageal sphincter (Barnette *et al.*, 1990) and nitrergically mediated component of IJPs in strips of dog pyloric sphincter (Bayguinov and Sanders, 1993b).

12.5.5 Co-transmitters, independent transmitter or tandem transmitters

Many studies have shown that there are at least two components in electrophysiological and mechanical responses to nitrergic nerve stimulation. For example, in the presence of NOS inhibitors or oxyhaemoglobin there were residual IJPs in the dog jejunum (Stark *et al.*, 1991) and residual relaxations in dog lower oesophageal sphincter (De Man *et al.*, 1991), rat gastric fundus (Li and Rand, 1990a; Boeckxstaens *et al.*, 1992), dog ileocolonic junction and ileum (Boeckxstaens *et al.*, 1990a, 1991a; Ward *et al.*, 1992a) and colon (Dalziel *et al.*, 1991), and rabbit anococcygeus muscle (Graham and Sneddon, 1993). The residual responses are due to at least one other transmitter. The question is whether the components are mediated by transmitters from different neurons, or whether there is co-transmission.

12.5.5.1 VIP

There is functional evidence for the participation of both the nitrergic transmitter and VIP in stimulation-induced relaxations of the rat gastric fundus (see Section 12.4.4.2), guinea-pig trachea (see Section 12.3.3), opossum internal anal sphincter (Rattan and Chakder, 1992) and guinea-pig uterine artery (Morris, 1993). In general, the VIP-mediated component of stimulation-induced relaxations requires a higher frequency and a longer train of stimulation to elicit it than does the nitrergic component: this might indicate the existence of co-transmission since such a differential release has been observed with peptide and classical co-transmitters (Bartfai *et al.*, 1988). In addition, with both exogenous and endogenous substances, the response to VIP develops more slowly and is more persistent than the response to NO. A possible explanation for these findings is that the nitrergic transmitter released VIP. However, it more probably reflects differences in the transduction and second messenger systems employed by the two transmitters since, in the rat gastric fundus, stimulation of NANC inhibitory nerves led to a rapid nitrergically mediated rise in the cGMP content, peaking after 20 s, at which time the VIP-mediated rise in the cAMP content began and progressed slowly (Ito *et al.*, 1990).

NOS inhibitors slightly reduced the relaxant responses to VIP in rat gastric fundus (Li and Rand, 1990a) and opossum internal anal sphincter (Rattan and Chakder, 1992) and VIP-induced vasodilatation in the cat submandibular gland (Edwards and Garrett, 1993), but not the relaxant responses in the lower oesophageal sphincter of dog (De Man *et al.*, 1991) or opossum (Tøttrup *et al.*, 1991a,b) or in various airways preparations (see Barnes and Belvisi, 1993). In those cases in which NOS inhibition reduced responses to VIP, the effect is not necessarily due to NO-mediated release of VIP (Huizinga *et al.*, 1992): an alternative explanation is synergism between the two mediators, as suggested below.

Morphological evidence for nitrergic/VIPergic co-transmission is that all neurons with NOS immunoreactivity in the myenteric plexus of the guinea-pig small intestine and taenia caeci also have VIP immunoreactivity (Costa *et al.*, 1992;

Furness *et al.*, 1992a). NOS and VIP are co-localized in some perivascular nerves in guinea-pigs (Kummer *et al.*, 1992).

In addition to simultaneous nitrergic and VIPergic transmission or co-transmission, a more complex relationship has been proposed based on experiments with guinea-pig and rabbit stomach (Grider *et al.*, 1992; Jin *et al.*, 1993; Murphy *et al.*, 1993) and rat colon preparations (Grider, 1993), including isolated smooth muscle cells. In this scheme, neuronally released NO facilitates the neuronal release of VIP, and the VIP not only activates adenylyl cyclase in the smooth muscle cells but also stimulates the release of NO from them, and this in turn further facilitates neuronal release of VIP. The NO of both neuronal and smooth muscle origin activates guanylyl cyclase, and the resultant relaxation is mediated by both cAMP and cGMP, which act synergistically. Activation of NOS was abolished by L-NNA but VIP release and relaxation were only partly inhibited, which accommodates the findings of VIP release and mediation of part of stimulation-induced relaxations (see Grider, 1993). If NO (or an NO-like factor) is generated in smooth muscle cells, the NOS involved is immunologically different from neuronal NOS.

12.5.5.2 ATP

It is possible that there may be nitrergic/purinergic co-transmission. In human colonic circular muscle, part of the stimulation-induced response is nitrergically mediated (see Section 12.4.4.8), and ATP may mediate non-nitrergic, apamin-sensitive, fast hyperpolarization and relaxation (Keef *et al.*, 1993; Boeckxstaens *et al.*, 1993a). There is also evidence that both mechanisms are involved in neurogenic inhibitory transmission in the rabbit portal vein (Brizzolara *et al.*, 1993). In dog terminal ileum and ileocolonic junction ATP produced relaxations that, like those to nitrergic stimulation, were inhibited by tetrodotoxin, L-NMMA, L-NNA and oxyhaemoglobin (Boeckxstaens *et al.*, 1990c, 1991a); hence, the responses to ATP appear to be due to activation of nitrergic nerves. It remains to be determined whether ATP of neuronal origin activates nitrergic nerves, and if so whether this is another example of feedback modulation by a co-transmitter.

12.5.6 Excitation of nitrergic transmission by agonists

12.5.6.1 Nicotinic agonists

Nicotine and other agonists act on nicotinic receptors in peripheral tissue associated with terminals of autonomic and sensory nerves resulting in release of the respective transmitters (Rand, 1989): the same occurs with nitrergic nerves. NOS inhibitors blocked relaxations elicited by nicotine and other nicotinic agonists (including acetylcholine) in rat anococcygeus muscles (Rand and Li, 1992a, 1993c), duodenum (Irie *et al.*, 1991) and gastric fundus strips (McLaren *et al.*, 1993), guinea-pig gastric fundus (Kojima *et al.*, 1993), dog duodenum (Toda *et al.*, 1992) and ileocolonic junction (Pelckmans *et al.*, 1989), opossum isolated internal and sphincter

(Chakder and Rattan, 1993a) and cerebral arteries (see Chapter 11). The release of an NO-like substance from the dog ileocolonic junction by the nicotinic agonist DMPP was detected by bioassay (Boeckxstaens *et al.*, 1991c). DMPP also released VIP and activated NOS in isolated ganglia from the myenteric plexus (Grider and Jin, 1993). The threshold pressure for adaptive relaxation of the guinea-pig stomach was decreased by DMPP (Desai *et al.*, 1991a) and nicotine (A. McLaren, C.G. Li and M.J. Rand, unpublished observations), and these effects were abolished by NOS inhibitors.

12.5.6.2 GABA

GABA also activated nitrergic nerves of the dog ileocolonic junction by acting on bicuculline-sensitive $GABA_A$ receptors (Boeckxstaens *et al.*, 1990b,c).

12.5.6.3 Serotonin (5-hydroxytryptamine)

Serotonin acts on $5\text{-}HT_1$ receptors of enteric nitrergic neurons in dog terminal ileum and ileocolonic junction (Bogers *et al.*, 1991) and guinea-pig colon (Briejer *et al.*, 1992; Knudsen and Tøttrup, 1992a) and stomach (Meulemans *et al.*, 1993). Evidence, for example, is that the relaxant action of serotonin on the guinea-pig colon was inhibited by tetrodotoxin, L-NNA, oxyhaemoglobin and methylene blue (Briejer *et al.*, 1992).

12.6 Interactions between NO and other autonomic transmitters

12.6.1 Noradrenaline

In tissues with both a noradrenergic and a nitrergic innervation, and under conditions in which reactions to both can be manifested, noradrenergic contractions were enhanced by NOS inhibitors but responses to noradrenaline were not affected: this has been observed in anococcygeus muscles (Li and Rand, 1989b; Liu *et al.*, 1991; Gibson *et al.*, 1990; Vila *et al.*, 1992; Brave *et al.*, 1993a), retractor penis muscle (Liu *et al.*, 1991), and various blood vessels (Toda and Okamura, 1990, 1992a; Zhang *et al.*, 1993). These findings indicate that field stimulation activated both noradrenergic and nitrergic transmission, and the nitrergic transmitter had an inhibitory modulatory effect. This is due to a counteracting postjunctional effect and not to prejunctional inhibition of noradrenergic transmission since NOS inhibitors had no effect on stimulation-induced release of noradrenaline (Toda and Okamura, 1992a; Brave *et al.*, 1993a; Rand & Li, 1993a). Other agents that impair nitrergic transmission at various steps in the process also enhanced noradrenergic contractions, as has been demonstrated in the rat anococcygeus with oxyhaemoglobin (Li and Rand, 1993a) and DPI (Rand and Li, 1993a), and with oxyhaemoglobin and hypoxia in the bovine retractor penis (Bowman and McGrath, 1985).

In endothelium-intact rat tail arteries, in which there is no evidence for a nitrergic innervation, vasoconstrictor responses to perivascular nerve stimulation and noradrenaline were enhanced by NOS inhibitors which blocked the counteracting effect of the endothelium-derived NO-like relaxing factor, and again noradrenaline release was not affected (Vo *et al.*, 1992; Bucher *et al.*, 1992). However, an endothelial factor inhibited noradrenaline release from rabbit carotid (Cohen and Weisbrod, 1985) and dog mesenteric and pulmonary arteries (Greenberg *et al.*, 1991). On the other hand, it has been suggested that NO may facilitate sympathetic response to hypogastric nerve stimulation in the opossum internal anal sphincter (Rattan and Thatikunta, 1993).

12.6.2 Acetylcholine

Cholinergically mediated neurogenic contractions were increased by NOS inhibitors in the rat gastric fundus (Lefebvre *et al.*, 1992c). In preparations from guinea-pig trachea (Brave *et al.*, 1991; Belvisi *et al.*,, 1991, 1993) and human trachea and bronchi (Ward *et al.*, 1993) nerve stimulation-induced contractions were enhanced by NOS inhibitors but responses to exogenously applied acetylcholine and the efflux of radioactivity after incubation with [^{3}H]choline were not affected (Brave *et al.*, 1991). The consensus was that the nitrergic transmitter, which would also be released by the field stimulation used in these studies, functionally counteracted the contractile effect of the cholinergic transmitter at the level of the smooth muscle. However, prejunctional inhibition of acetylcholine release remains a possibility as suggested by studies with guinea-pig and rat trachea (Belvisi *et al.*, 1991; Sekizawa *et al.*, 1993), guinea-pig taenia caeci (Knudsen and Tøttrup, 1992b), rat gastric fundus (Lefebvre *et al.*, 1992c), dog duodenum (Toda *et al.*, 1991) and detrusor of the pig bladder (Andersson and Persson, 1993). Complex interactions between acetylcholine and nitrergic nerves in the rat anococcygeus muscle were reported by Rand and Li (1993c).

NO inhibits the activity of acyl transferase because of formation of NOCoA (Tu *et al.*, 1984). Since coenzyme A is a cofactor for choline acetyltransferase, its loss may result in a slowly developing impairment of cholinergic transmission.

12.7 Concluding remarks

There has been a virtual explosion of knowledge about nitrergic transmission in the autonomic and enteric nervous systems in the three years since the end of 1989. However, there are several disparities in observations that remain to be resolved. More importantly, additional evidence is required for the physiological and pathophysiological roles of nitrergic neuroeffector transmission, the extent to which it is integrated with other neuroeffector control mechanisms and, of course, the implications for therapeutics. The resolution of some of these matters amounts to housekeeping, but some call for more basic studies. In our opinion, the most pressing fundamental questions to be answered are the exact natures of the nitrergic transmitter and the transmission process. We are eagerly awaiting new developments in the saga, and hope to contribute to them.

Acknowledgements

The writing of this chapter and work from the authors' laboratory reported in it were supported by programme grants from the National Health & Medical Research Council and the Australian Tobacco Research Foundation. We are grateful to our colleagues Dr George Harris, Dr Julianne Reid, Dr Margot Story and Professor David Story for their critical appraisal of drafts.

References

Ahlner, J., Ljusegren, M.E., Grundström, N. & Axelsson, K.L. (1991) *Circ. Res.* **68**, 756–762.
Aimi, Y., Kimura, H., Konoshita, T., Minami, Y., Fujimura, M. & Vincent, S.R. (1993) *Neuroscience* **53**, 553–560.
Allescher, H.-D., Tougas, G., Vergara, P., Lu, S. & Daniel, E.E. (1992a) *Am. J. Physiol.* **262**, G695–G702.
Allescher, H.-D., Sattler, D., Piller, C., Schusdziarra, V. & Classen, M. (1992b) *Eur. J. Pharmacol.* **217**, 153–162.
Allescher, H.-D., Lu, S., Daniel, E.E. & Classen, M. (1993) *Can. J. Physiol. Pharmacol.* **71**, 525–530.
Alm, P., Larsson, B., Ekblad, E., Sundler, F. & Andersson, K.-E. (1993) *Acta Physiol. Scand.* **148**, 421–429.
Altiere, R.J., Diamond, L. & Thompson, D.C. (1992) *J. Pharmacol. Exp. Ther.* **260**, 98–103.
Ambache, N., Killick, S.W. & Zar, M.A. (1975) *Br. J. Pharmacol.* **54**, 409–410.
Andersson, K.-E., and Persson, K. (1993) *Genet. Pharmacol.* **24**, 833–839.
Andersson, K.-E., Garcia-Pascual, A., Forman, A. & Tøttrup, A. (1991) *Acta Physiol. Scand.* **141**, 133–134.
Andersson, K.-E., Garcia-Pascual, A., Persson, K., Forman, A. & Tøttrup, A. (1992) *J. Urol.* **147**, 253–259.
Andriantsitohaina, R. & Surprenant, A. (1992) *J. Physiol.* **453**, 493–502.
Aoki, E., Takeuchi, I.K., Shoji, R. & Sembra, R. (1993) *Brain Res.* **620**, 142–145.
Archer, S. (1993) *FASEB J.* **7**, 349–360.
Arvola, P., Pörsti, I., Vuorinen, P., Huhtala, H., Metsä-Ketalä, T. & Vaapatalo, H. (1992) *Eur. J. Pharmacol.* **214**, 289–292.
Assreuy, J., Cunha, F.Q., Liew, F.W. & Moncada, S. (1993) *Br. J. Pharmacol.* **108**, 833–837.
Bai, T.R. & Bramley, A.M. (1993) *Am. J. Physiol.* **264**, L425–L430.
Baker, R.A., Saccone, G.T.P. & Toouli, J. (1991) *Proc. Austral. Physiol. Pharmacol. Soc.* **22**, 102P.

Barbier, A.J. & Lefebvre, R.A. (1992a) *Eur. J. Pharmacol.* **210**, 315–323.
Barbier, A.J. & Lefebvre, R.A. (1992b) *Eur. J. Pharmacol.* **219**, 331–334.
Barbier, A.J. & Lefebvre, R.A. (1993) *J. Pharmacol. Exp. Ther.* **266**, 172–178.
Barbiers, M., Timmermans, J.P., Scheuermann, D.W., Adriaensen, D., Mayer, B. & Degroodtlasseel, M.H.A. (1993) *Histochemistry* **100**, 27–34.
Barnes, P.J. & Belvisi, M.G. (1993) *Thorax* **48**, 1034–1043.
Barnette, M.S., Torphy, T.J., Grous, M., Fine, C. & Ormsbee, H.S. (1989) *J. Pharmacol. Exp. Ther.* **249**, 524–526.
Barnette, M.S., Grous, M., Manning, C.D., Callahan, J.F. & Barone, F.C. (1990) *Eur. J. Pharmacol.* **182**, 363–368.
Barnette, M.S., Barone, F.C., Fowler, P.J., Grous, M., Price, W.J. & Ormsbee, H.S. (1991) *Gut* **32**, 4–9.
Bartfai, T., Iverfeldt, K., Fisone, G. & Serfözö, P. (1988) *Annu. Rev. Pharmacol. Toxicol.* **28**, 285–310.
Bartho, L., Koczan, G., Petho, G. & Maggi, C.A. (1992) *Neurosci. Lett.* **145**, 43–46.
Bates, J.N., Harrison, D.G., Myers, P.R. & Minor, R.L. (1991) *Basic Res. Cardiol.* **86** (Suppl. 2), 17–26.
Bayguinov, O. & Sanders, K.M. (1993a) *Am. J. Physiol.* **264**, G975–G983.
Bayguinov, O. & Sanders, K.M. (1993b) *Br. J. Pharmacol.* **108**, 1024–1030.
Bayguinov, O., Vogalis, F., Morris, B. & Sanders, K.M. (1992) *Am. J. Physiol.* **263**, G887–G894.
Beart, P.M., Westergaard, N. & Schousboe, A. (1993) *J. Neurochem.* **61** (Suppl.), S4.
Belai, A., Schmidt, H.H.H.W., Hoyle, C.H., Hassall, C.J., Saffrey, M.J., Moss, J., Förstermann, U., Murad, F. & Burnstock, G. (1992) *Neurosci. Lett.* **143**, 60–64.
Belvisi, M.G., Stretton, D. & Barnes, P.J. (1991) *Eur. J. Pharmacol.* **198**, 219–221.
Belvisi, M.G., Stretton, C.D., Miura, M., Verleden, G.M., Tadjkarimi, S., Yacoub, M.H. & Barnes, P.J. (1992a) *J. Appl. Physiol.* **73**, 2505–2510.
Belvisi, M.G., Stretton, C.D., Yacoub, M. & Barnes, P.J. (1992b) *Eur. J. Pharmacol.* **210**, 221–222.
Belvisi, M.G., Miura, M., Stretton, D. & Barnes, P.J. (1993) *Eur. J. Pharmacol.* **231**, 9–102.
Boeckxstaens, G.E., Pelckmans, P.A., Bult, H., De Man, J.G., Herman, A.G. & Van Maercke, Y.M. (1990a) *Eur. J. Pharmacol.* **190**, 239–246.
Boeckxstaens, G.E., Pelckmans, P.A., Rampart, M., Ruytjens, I.F., Verbeuren, T.J., Herman, A.G. & Van Maercke, Y.M. (1990b) *Br. J. Pharmacol.* **101**, 460–464.
Boeckxstaens, G.E., Pelckmans, P.A., Bult, H., De Man, J.G., Herman, A.G. & Van Maercke, Y.M. (1990c) *Br. J. Pharmacol.* **102**, 434–438.
Boeckxstaens, G.E., Pelckmans, P.A., Rampart, M., Verbeuren, T.J., Herman, A.G. & Van Maercke, Y.M. (1991a) *Arch. Int. Pharmacodyn. Ther.* **303**, 270–281.
Boeckxstaens, G.E., Pelckmans, P.A., Bogers, J.J., Bult, H., De Man, J.G., Oosterbosch, L., Herman, A.G. & Van Maercke, Y.M. (1991b) *J. Pharmacol. Exp. Ther.* **256**, 442–447.
Boeckxstaens, G.E., Pelckmans, P.A., Ruytjens, I.F., Bult, H., De Man, J.G., Herman, A.G. & Van Maercke, Y.M. (1991c) *Br. J. Pharmacol.* **103**, 1085–1091.
Boeckxstaens, G.E., Bult, H. & Pelckmans, P.A. (1991d) *J. Cardiovasc. Pharmacol.* **17** (Suppl. 3), S238–S242.
Boeckxstaens, G.E., Pelckmans, P.A., De Man, J.G., Bult, H., Herman, A.G. & Van Maercke, Y.M. (1992) *Arch. Int. Pharmacodyn. Ther.* **318**, 107–115.
Boeckxstaens, G.E., Pelckmans, P.A., Herman, A.G. & Van Maercke, Y.M. (1993a) *Gastroenterology* **104**, 690–697.
Boeckxstaens, G.E., De Man, J.G., Pelckmans, P.A., Hermann, A.G. & Van Maercke, Y.M. (1993b) *Br. J. Pharmacol.* **109**, 1079–1084.
Bogers, J.J., Pelckmans, P.A., Boeckxstaens, G.E., De Man, J.G., Herman, A.G. & Van Maercke, Y.M. (1991) *Naunyn-Schmiedebergs Arch. Pharmacol.* **344**, 716–719.
Bogle, R.G., Moncada, S., Pearson, J.D. & Mann, G.E. (1992) *Br. J. Pharmacol.* **105**, 768–770.

Bowman, A. & Drummond, A.H. (1984) *Br. J. Pharmacol.* **81**, 665–674.
Bowman, A. & Gillespie, J.S. (1982) *J. Physiol.* **328**, 11–26.
Bowman, A. & McGrath, J.C. (1985) *Br. J. Pharmacol.* **85**, 869–875.
Bowman, A., Gillespie, J.J. & Pollock, D. (1982) *Eur. J. Pharmacol.* **85**, 221–224.
Bowman, W.C. & Rand, M.J. (1980) *Textbook of Pharmacology*, pp. 9.9–9.18. Oxford, Blackwell Scientific Publications.
Brave, S.R., Hobbs, A.J., Gibson, A. & Tucker, J.F. (1991) *Biochem. Biophys. Res. Commun.* **179**, 1017–1022.
Brave, S.R., Bhat, S., Hobbs, A.J., Tucker, J.F. & Gibson, A. (1993a) *J. Auton. Pharmacol.* **13**, 219–225.
Brave, S.R., Tucker, J.F., Gibson, A., Bishop, A.E., Riveros-Moreno, V., Moncada, S. & Polak, J.M. (1993b) *Neurosci. Lett.* **161**, 93–96.
Bredt, D.S. & Snyder, S.H. (1990) *Proc. Natl. Acad. Sci. USA* **87**, 682–685.
Bredt, D.S., Hwang, P.M. & Snyder, S.H. (1990) *Nature* **347**, 768–770.
Bredt, D.S., Huang, P.M., Glatt, C.E., Lowenstein, C., Reed, R.R. & Snyder, S.H. (1991) *Nature* **351**, 714–718.
Briejer, M.R., Akkermans, L.M.A., Meulemans, A.L., Lefebvre, R.A. & Schuurkes, J.A. (1992) *Br. J. Pharmacol.* **107**, 756–761.
Brizzolara, A.L., Crowe, R. & Burnstock, G. (1993) *Br. J. Pharmacol.* **109**, 606–608.
Brown, J.F., Tepperman, B.L., Hanson, P.J., Whittle, B. & Moncada, S. (1992) *Biochem. Biophys. Res. Commun.* **184**, 680–685.
Brüne, B. & Lapetina, E.G. (1990) *Arch. Biochem. Biophys.* **279**, 286–290.
Brunton, T.L. (1867) *Lancet* **ii**, 97–98.
Bucher, B., Ouedraogo, S., Tschöpl, M., Paya, D. & Stoclet, J.-C. (1992) *Br. J. Pharmacol.* **107**, 976–982.
Bult, H., Boeckxstaens, G.E., Pelckmans, P.A., Jordaens, F.H., Van Maercke, Y.M. & Herman, A.G. (1990) *Nature* **345**, 346–347.
Burleigh, D.E. (1992) *Gastroenterology* **102**, 679–683.
Burnstock, G., Campbell, G. & Rand, M.J. (1966) *J. Physiol.* **182**, 504–526.
Byrne, N.G. & Muir, T.C. (1984) *J. Auton. Pharmacol.* **4**, 261–271.
Bywater, R.E.R. & Taylor, G.S. (1991) In *Novel Peripheral Neurotransmitters* (ed. Bell, C.), pp. 247–291. New York, Pergamon Press.
Calignano, A., Whittle, B.J.R., Di Rosa, M. & Moncada, S. (1992) *Eur. J. Pharmacol.* **229**, 273–276.
Campbell, G. (1991) In *Novel Peripheral Neurotransmitters* (ed. Bell, C.), pp. 1–7. New York, Pergamon Press.
Chakder, S. & Rattan, S. (1992) *J. Pharmacol. Exp. Ther.* **260**, 1113–1118.
Chakder, S. & Rattan, S. (1993a) *Am. J. Physiol.* **264**, G7–12.
Chakder, S. & Rattan, S. (1993b) *Am. J. Physiol.* **264**, G702–G707.
Christinck, F., Jury, J., Cayabyab, F. & Daniel, E.E. (1991) *Can. J. Physiol. Pharmacol.* **69**, 1448–1458.
Cohen, M.L. & Weisbrod, R.M. (1985) *Am. J. Physiol.* **254**, H871–H878.
Conklin, J.L. & Du, C. (1992) *Am. J. Physiol.* **263**, 687–690
Conklin, J.L., Du, C., Murray, J.A. & Bates, J.N. (1993) *Gastroenterology* **104**, 1439–1444.
Costa, M., Furness, J.B., Pompolo, S., Brookes, S.J., Bornstein, J.C., Bredt, D.S. & Snyder, S.H. (1992) *Neurosci. Lett.* **148**, 121–125.
Dail, W.G., Galloway, B. & Bordegaray, J. (1993) *Neurosci. Lett.* **160**, 17–20.
Dale, H.H. (1933) *J. Physiol.* **80**, 10P–11P.
Dalziel, H.H., Thornbury, K.D., Ward, S.M. & Sanders, K.M. (1991) *Am. J. Physiol.* **260**, G789–G792.
D'Amato, M., Curro, D. & Montuschi, P. (1992) *J. Auton. Nerv. Syst.* **37**, 175–186.
Dawson, T.M., Bredt, D.S., Fotuhi, M., Hwang, P.M. & Snyder, S.H. (1991) *Proc. Natl. Acad. Sci. USA* **88**, 7797–7801.

de Belder, A.F., Radomski, M.W., Why, H.J., Richardson, P.J., Bucknall, C.A., Salas, E., Martin, J.F. & Moncada, S. (1993) *Lancet* **341**, 84–85.
Delaforge, M., Servent, D., Wirsta, P., Ducrocq, C., Mansuy, D. & Lefant, M. (1993) *Chem. Biol. Interact.* **86**, 103–117.
De Luca, A., Li, C.G., Rand, M.J., Reid, J.J., Thaina, P. & Wong-Dusting, H.K. (1990) *Br. J. Pharmacol.* **101**, 437–447.
De Man, J.G., Pelckmans, P.A., Boeckxstaens, G.E., Bult, H., Oosterbosch, L., Herman, A.G. & Van Maercke, Y.M. (1991) *Br. J. Pharmacol.* **103**, 1092–1096.
Desai, K.M., Sessa, W.C. & Vance, J.R. (1991a) *Nature* **351**, 477–479.
Desai, K.M., Zembowicz, A., Sessa, W.C. & Vane, J.R. (1991b) *Proc. Natl. Acad. Sci. USA* **88**, 11490–11494.
Dey, H.U., Mayer, B. & Said, S.I. (1993) *Neuroscience* **54**, 839–843.
Diaz de Rada, O., Villaro, A.C., Montuenga, L.M., Martinez, A., Springall, D.R. & Polack, J.M. (1993) *Neurosci. Lett.* **162**, 121–124.
Di Iulio, J.L., Gude, N.M., Li, C.G., Rand, M.J. & King, R.G. (1993) *Clin. Exp. Pharmacol. Physiol. Suppl.* **1**, 20.
Dimmeler, S. & Brüne, B. (1991) *Biochem. Biophys. Res. Commun.* **178**, 848–855.
Dokita, S., Morgan, W.R., Wheeler, M.A., Yoshida, M., Latifpour, J. & Weiss, R.M. (1991) *Life Sci.* **48**, 2429–2436.
Du, C., Murray, J., Bates, J.N. & Conklin, J.N. (1991) *Am. J. Physiol.* **261**, G1012–G1016.
Dupuy, P.M., Shore, S.Z., Drazen, J.M., Frostell, C., Hill, W.A. & Zapol, W.M. (1992) *J. Clin. Invest.* **90**, 421–428.
Edwards, A.V. & Garrett, J.R. (1993) *J. Physiol.* **464**, 379–392.
Ellis, J.L. & Undem, B.J. (1992) *Am. Rev. Respir. Dis.* **146**, 1543–1547.
Faussone-Pellegrini, M.S., Bacci, S., Pantalone, D. & Cortesini, C. (1993) *Neurosci. Lett.* **157**, 135–139.
Feelisch, M. (1991) *J. Cardiovasc. Pharmacol.* **17** (Suppl.), S25–S33.
Fischer, A., Mundel, P., Mayer, B., Preissler, U., Philippin, B. & Kummer, W. (1993) *Neurosci. Lett.* **149**, 157–160.
Fisher, J.T., Anderson, J.W. & Waldron, M.A. (1993) *J. Appl. Physiol.* **74**, 31–39.
Forster, E.R. & Southam, E. (1993) *NeuroReport* **4**, 275–278.
Förstermann, U., Pollock, J.S., Schmidt, H.H.H.W., Heller, M. & Murad, F. (1991) *Proc. Natl. Acad. Sci. USA* **88**, 1788–1792.
Fukuto, J.M., Wallace, G.C., Hszieh, R. & Chaudhuri, G. (1992) *Biochem. Pharmacol.* **43**, 607–613.
Furchgott, R.F. (1988) In *Vasodilatation: Vascular Smooth Muscle, Peptides, Autonomic Nerves, and Endothelium* (ed. P.M. Vanhoutte), pp. 401–414. Raven Press, New York.
Furchgott, R.F. & Jothianandan, D. (1991) *Blood Vess.* **28**, 52–61.
Furness, J.B., Pompolo, S., Shuttleworth, C.W. & Burleigh, D.E. (1992a) *Cell Tiss. Res.* **270**, 125–137.
Furness, J.B., Bornstein, J.C., Murphy, R. & Pompolo, S. (1992b) *Trends Neurosci.* **15**, 66–71.
Gao, Y. & Vanhoutte, P.M. (1993) *Br. J. Pharmacol.* **109**, 887–891.
Garcia-Pascual, A., Costa, G., Garcia-Sacristan, A. & Andersson, K.-E. (1991) *Acta Physiol. Scand.* **141**, 531–539.
Gibson, A. & Mirzazadeh, S. (1989) *Br. J. Pharmacol.* **98**, 617P.
Gibson, A. & Tucker, J.F. (1982) *Br. J. Pharmacol.* **77**, 97–103.
Gibson, A., Mirzazadeh, S., al-Swayeh, O.A., Chong, N.W.S. & Moore, P.K. (1989) *Br. J. Pharmacol.* **98**, 904P.
Gibson, A., Mirzazadeh, S., Hobbs, A.J. & Moore, P.K. (1990) *Br. J. Pharmacol.* **99**, 602–606.
Gibson, A., Babbedge, R., Brave, S.R., Hart, S.L., Hobbs, A.J., Tucker, J.F., Wallace, P. & Moore, P.K. (1992) *Br. J. Pharmacol.* **107**, 715–721.
Gibson, Q.H. & Roughton, F.J.W. (1957) *J. Physiol.* **136**, 507–526.

Gillespie, J.S. (1987). In *Pharmacology* (eds. M.J. Rand and C. Raper), pp. 161–170. Exerpta Medica, Amsterdam.
Gillespie, J.S. & Sheng, H. (1988) *Br. J. Pharmacol.* **95**, 1151–1165.
Gillespie, J.S. & Sheng, H. (1989) *Br. J. Pharmacol.* **98**, 445–450.
Gillespie, J.S. & Sheng, H. (1990) *Br. J. Pharmacol.* **99**, 194–196.
Gillespie, J.S. & Tilmisany, A.K. (1976) *Br. J. Pharmacol.* **58**, 47–55.
Gillespie, J.S., Hunter, J.C. & McKnight, A.T. (1982) *Br. J. Pharmacol.* **75**, 189–198.
Gillespie, J.S., Liu, X.R. & Martin, W. (1989) *Br. J. Pharmacol.* **98**, 1080–1082.
Gillespie, J.S., Liu, X. & Martin, W. (1990) In *Nitric Oxide from L-Arginine. A Bioregulatory System* (ed. S. Moncada), pp. 147–164. Elsevier, Amsterdam.
Girard, P. & Potier, P. (1993) *Trends Neurosci.* **14**, 293–299.
Goy, M.F. (1991) *Trends Neurosci.* **14**, 293–299.
Graham, A.M. & Sneddon, P. (1993) *Eur. J. Pharmacol.* **237**, 93–99.
Greenberg, S.S., Peevy, K. & Tanaka, T.P. (1991) *Am. J. Hyperten.* **4**, 464–467.
Grider, J.R. (1993) *Am. J. Physiol.* **264**, G334–G340.
Grider, J.R. & Jin, J.-G. (1993) *Neuroscience* **54**, 521–526.
Grider, J.R., Murthy, K.S., Jin, J.-G. & Makhlouf, G.M. (1992) *Am. J. Physiol.* **262**, G774–G778.
Grous, M., Joslyn, A.F., Tompson, W. & Barnette, M.S. (1991) *J. Gastrointest. Motility* **3**, 46–52.
Grozdanovic, Z., Baumgarten, H.G. & Bruning, G. (1992) *Neuroscience* **48**, 225–235.
Grundy, D., Gharibnaseri, M.K. & Hutson, D. (1993) *J. Auton. Nerv. Syst.* **43**, 241–246.
Gryglewski, R.J., Palmer, R.M.J. & Moncada, S. (1986) *Nature* **320**, 454–456.
Gryglewski, R.J., Zembowicz, A., Salvemini, D., Taylor, G.W. & Vane, J.R. (1992) *Br. J. Pharmacol.* **106**, 838–845.
Gustafsson, B.I. & Delbro, D.S. (1993) *J. Auton. Nerv. Syst.* **44**, 179–187.
Gustafsson, L.E., Wiklund, C.U., Wiklund, N.P., Persson, M.G. & Moncada, S. (1990) *Biochem. Biophys. Res. Commun.* **173**, 106–110.
Hashimoto, S., Kigoshi, S. & Muramatsu, L. (1993) *Eur. J. Pharmacol.* **231**, 209–214.
Hassall, C.J., Saffrey, M.J., Belai, A., Hoyle, C.H., Moules, E.W., Moss, J., Schmidt, H.H.H.W., Murad, F., Förstermann, U. & Burnstock, G. (1992) *Neurosci. Lett.* **143**, 65–68.
Hassall, C.J., Saffrey, M.J. & Burnstock, G. (1993) *NeuroReport* **4**, 49–52.
Hata, F., Ishii, T., Kanada, A., Yamano, N., Kataoka, T., Takeuchi, T. & Yagasaki, O. (1990) *Biochem. Biophy. Res. Commun.* **172**, 1400–1406.
He, X.D. & Goyal, R.K. (1993) *J. Physiol.* **461**, 485–499.
Henry, Y., Lepoivre, M., Drapier, J-C., Ducrocq, C., Boucher, J-L. & Guissani, A. (1993) *FASEB J.* **7**, 1124–1134.
Hobbs, A.J. & Gibson, A. (1990) *Br. J. Pharmacol.* **100**, 749–752.
Hobbs, A.J., Tucker, J.F. & Gibson, A. (1991) *Br. J. Pharmacol.* **104**, 645–650.
Högman, M., Frostell, C., Arnberg, H. & Hedenstierna, G. (1993) *Eur. Respir. J.* **16**, 177–180.
Hope, G.T. & Vincent, S.R. (1989) *J. Histochem. Cytochem.* **37**, 653–661.
Hope, G.T., Michael, G.J., Knigge, K.M. & Vincent, S.R. (1991) *Proc. Natl. Acad. Sci. USA* **88**, 2811–2814.
Hoyle, C.H.V. & Burnstock, G. (1989) In *Handbook of Physiology, The Gastrointestinal System I, Motility and Circulation* (ed. J.D. Wood), pp. 435–464. American Physiological Society, Bethesda, MD.
Huizinga, J.D., Tomlinson, J. & Pintin-Quezada, J. (1992) *J. Pharmacol. Exp. Ther.* **260**, 803–808.
Ignarro, L.J. (1989) *Pharmacol. Res.* **6**, 651–659.
Ignarro, L.J. (1990) *Hypertension* **16**, 477–483.
Ignarro, L.J., Lippton, F., Edwards, J.C., Baricos, W.H., Hyman, A.L., Kadowitz, P.Z. & Gruetter, C.A. (1981) *J. Pharmacol. Exp. Ther.* **218**, 739–749.
Ignarro, L.J., Bush, P.A., Buga, G.M. & Rajfer, J. (1990) *Nature* **347**, 131–132.

Irie, K., Muraki, T., Furukawa, K. & Nomoto, T. (1991) *Eur. J. Pharmacol.* **202**, 285–288.
Ito, S., Kurokawa, A., Ohga, A., Ohta, T. & Sawabe, K. (1990) *J. Physiol.* **430**, 337–353.
James, M.J., Birmingham, A.T. & Hill, S.J. (1993) Br. J. Clin. Pharmacol. **35**, 366–372.
Jansen, A., Drazen, J., Osborne, J.A., Brown, R., Loscalzo, J.P. & Stamler, J.S. (1992) *J. Pharmacol. Exp. Ther.* **261**, 154–160.
Jin, J.G., Murthy, K.S., Grider, J.R. & Makhlouf, G.M. (1993) *Am. J. Physiol.* **264**, G470–G477.
Jury, J., Admedzadeh, N. & Daniel, E.E. (1992) *Can. J. Physiol. Pharmacol.* **70**, 1182–1189.
Kanada, A., Hata, F., Suthamnatpong, N., Maehara, T., Ishii, T., Takeuchi, T. & Yagasaki, O. (1992) *Eur. J. Pharmacol.* **216**, 287–292.
Kandel, E.R. (1991) In *Principles of Neural Science* (eds. E.R. Kandel, J.H. Schwartz & T.M. Jessell), pp. 194–212. Elsevier, New York.
Kannan, M.S. & Johnson, D.E. (1992a) *J. Pharmacol. Exp. Ther.* **260**, 1180–1184.
Kannan, M.S. & Johnson, D.E. (1992b) *Am. J. Physiol.* **262**, L511–L514.
Katsuki, S., Arnold, W. & Murad, F. (1977) *J. Cycl. Nucleotide Res.* **3**, 239–247.
Kaufman, H.S., Shermak, M.A., May, C.A., Pitt, H.A. & Lillemoe, K.D. (1993) *Am. J. Surg.* **165**, 74–80.
Keef, K.D., Du, C., Ward, S.M., McGregor, B. & Sander, K.M. (1993) *Gastroenterology* **105**, 1009–1016.
Kelm, M. & Schrader, J. (1990) *Circ. Res.* **66**, 1561–1575.
Kemp, M.C., Kuonen, D.R., Sutton, A. & Roberts, P.J. (1988) *Biochem. Pharmacol.* **37**, 3063–3070.
Kerr, S.W., Buchanan, L.V., Bunting, S. & Mathews, W.R. (1992) *J. Pharmacol. Exp. Ther.* **263**, 285–292.
Kimoto, M., Tsuji, H., Ogawa, T. & Sasaoka, K. (1993) *Arch. Biochem. Biophys.* **300**, 657–662.
Kitamura, K., Lian, Q., Carl, A. & Kuriyama, H. (1993) *Br. J. Pharmacol.* **109**, 415–423.
Klimaschewski, L., Kummer, W., Mayer, B., Couraud, J.Y., Preissler, U., Philippin, B. & Heym, C. (1992) *Circ. Res.* **71**, 1533–1537.
Knowles, R.G., Palacios, M., Palmer, R.M. & Moncada, S. (1990) *Biochem. J.* **269**, 207–210.
Knudsen, M.A. & Tøttrup, A. (1991a) *Br. J. Pharmacol.* **107**, 756–761.
Knudsen, M.A. & Tøttrup, A. (1992b) *Br. J. Pharmacol.* **107**, 837–941.
Knudsen, M.A., Svane, D. & Tøttrup, A. (1991) *Digest. Dis.* **9**, 365–370.
Knudsen, M.A., Svane, D. & Tøttrup, A. (1992) *Am. J. Physiol.* **262**, G840–G846.
Kobzik, L., Bredt, D.S., Lowenstein, C.J., Drazen, J., Gaston, B., Sugarbaker, D. & Stamler, J.S. (1993) *Am. J. Respir. Cell Mol. Biol.* **9**, 371–377.
Kojima, S., Ishizaki, R. & Shimo, Y. (1993) *Eur. J. Pharmacol.* **241**, 171–175.
Kostka, P., Jang, E., Watson, E.G., Stewart, J.L. & Daniel, E.E. (1993) *J. Pharmacol. Exp. Ther.* **264**, 234–239.
Kowaluk, E.A. & Fung, H.L. (1990) *J. Pharmacol. Exp. Ther.* **255**, 1256–1264.
Kowaluk, E.A. & Fung, H.L. (1991) *J. Pharmacol. Exp. Ther.* **259**, 519–525.
Kowaluk, E.A., Seth, P. & Fung, H.-L. (1992) *J. Pharmacol. Exp. Ther.* **262**, 916–922.
Krammer, H.J., Karahan, S.I., Mayer, B., Zhang, M. & Kuhnel, W. (1993) *Ann. Anat.* **175**, 225–230.
Kummer, W. & Mayer, B. (1993) *Histochemistry* **99**, 175–179.
Kummer, W., Fischer, A., Mundel, P., Mayer, B., Hoba, B., Philippin, B. & Preissler, U. (1992) *NeuroReport* **3**, 653–655.
Lammers, J.W., Barnes, P.J. & Chung, K.F. (1992) *Eur. Respir. J.* **5**, 239–246.
Larsson, B., Alm, P., Persson, K. & Andersson, K.-E. (1992) *Neurourol. Urodyn.* **11**, 444–445.
Leckström, A., Ahlner, J., Grundström, N. & Axelsson, K.I. (1993) *Pharmacol. Toxicol.* **72**, 194–198.
Lefebvre, R.A. (1993) *Gen. Pharmacol.* **24**, 257–266.
Lefebvre, R.A. & Bogaert, M.G. (1986) *J. Pharm. Pharmacol.* **38**, 621–622.

Lefebvre, R.A. & Smits, G.J.M. (1992) *Br. J. Pharmacol.* **107**, 256–261.

Lefebvre, R.A., Hasrat, J. & Gobert, A. (1992a) *Br. J. Pharmacol.* **105**, 315–320.

Lefebvre, R.A., Baert, E. & Barbier, A.J. (1992b) *Br. J. Pharmacol.* **106**, 173–179.

Lefebvre, R.A., De Vriese, A. & Smits, G.J. (1992c) *Eur. J. Pharmacol.* **221**, 235–242.

Lei, S.Z., Pan, Z.-H., Aggarwal, S.K., Chen, H.-S.V., Hartman, J., Sucher, N.J. & Lipton, S.A. (1992) *Neuron* **8**, 1087–1099.

Lei, Y.-Y., Barnes, P.J. & Rogers, D.F. (1993) *Br. J. Pharmacol.* **108**, 228–235.

Li, C.G. & Rand, M.J. (1989a) *Eur. J. Pharmacol.* **168**, 107–110.

Li, C.G. & Rand, M.J. (1989b) *Clin. Exp. Pharmacol. Physiol.* **16**, 933–938.

Li, C.G. & Rand, M.J. (1990a) *Eur. J. Pharmacol.* **191**, 303–309.

Li, C.G. & Rand, M.J. (1990b) *Proc. Aust. Neurosci. Soc.* **1**, 100.

Li, C.G. & Rand, M.J. (1990c) *Clin. Exp. Pharmacol. Physiol.* Suppl. **16**, 86.

Li, C.G. & Rand, M.J. (1991) *Br. J. Pharmacol.* **102**, 91–94.

Li, C.G. & Rand, M.J. (1993a) *Clin. Exp. Pharmacol. Physiol.* **20**, 633–640.

Li, C.G. & Rand, M.J. (1993b) *Clin. Exp. Pharmacol. Physiol.* Suppl. **1**, 43.

Li, C.G., Brosch, S.F. & Rand, M.J. (1994) *Clin. Exp. Pharmacol. Physiol.* **21**, 293–299.

Lincoln, T.M. & Cornwell, T.L. (1993) *FASEB J.* **7**, 328–338.

Liu, X., Gillespie, J.S. & Martin, W. (1994) *Br. J. Pharmacol.* **111**, 1287–1295.

Liu, S.F., Crawley, D.E., Rohde, J.A., Evans, T.W. & Barnes, P.J. (1992) *Br. J. Pharmacol.* **107**, 861–866.

Liu, X., Gillespie, J.S., Gibson, I.F. & Martin, W. (1991) *Br. J. Pharmacol.* **104**, 53–58.

Llewellyn-Smith, I.J., Song, Z.M., Costa, M., Bredt, D.S. & Snyder, S.H. (1992) *Brain Res.* **577**, 337–342.

Lüscher, T. & Vanhoutte, P.M. (1990) *The Endothelium: Modulator of Cardiovascular Function.* Boca Raton, FL, CRC Press.

Lyster, D.J., Bywater, R.A., Taylor, G.S. & Watson, M.J. (1992) *J. Autonom. Nerv. Syst.* **41**, 187–196.

MacDonald, A., Kelly, J. & Dettmar, P.W. (1990) *J. Pharm. Pharmacol.* **42**, 752–757.

Maggi, C.A. & Giuliani, S. (1993) *Naunyn-Schmiedebergs Arch. Pharmacol.* **347**, 830–834.

Maggi, C.A., Barbanti, G., Turini, D. & Giuliani, S. (1991) *Br. J. Pharmacol.* **103**, 1970–1972.

Marczin, N., Ryan, U.S. & Catravas, J.D. (1992) *J. Pharmacol. Exp. Ther.* **263**, 170–179.

Marks, G.S., McLaughlin, B.E., Brown, L.B., Beaton, D.E., Booth, B.P., Nakatsu, K. & Brien, J.F. (1991) *Can. J. Physiol. Pharmacol.* **69**, 889–892.

Martin, W. & Gillespie, J.S. (1991) In *Novel Peripheral Neurotransmitters* (ed. C. Bell), pp. 65–79. Pergamon Press, New York.

Martin, W., Smith, J.S., Lewis, M.J. & Henderson, A.H. (1988) *Br. J. Pharmacol.* **93**, 579–586.

Martin, W., Gillespie, J.S. & Gibson, L.F. (1993) *Br. J. Pharmacol.* **108**, 242–247.

Matsumoto, T., Okamura, T., Kinoshita, M. & Toda, N. (1993) *Eur. J. Pharmacol.* **237**, 31–37.

Mayer, B., Brunner, F. & Schmidt, K. (1993) *Biochem. Pharmacol.* **45**, 367–374.

McConalogue, K. & Furness, J.B. (1993) *Cell Tiss. Res.* **271**, 545–553.

McKirdy, H.C. (1991) In *Novel Peripheral Neurotransmitters* (ed. C. Bell), pp. 221–246. Pergamon Press, New York.

McKirdy, H.C., McKirdy, M.L., Lewis, M.J. & Marshall, R.W. (1992) *Exp. Physiol.* **77**, 509–511.

McLaren, A., Li, C.G. & Rand, M.J. (1993) *Clin. Exp. Pharmacol. Physiol.* **20**, 451–457.

McNeill, D.L., Traugh, N.E., Vaidya, A.M., Hua, H.T. & Papka, R.E. (1992) *Neurosci. Lett.* **147**, 33–36.

Medgett, I.C. & Rand, M.J. (1981) *Clin. Exp. Pharmacol. Physiol* **8**, 503–507.

Meulemans, A.L., Helsen, L.F. & Schuurkes, J.A. (1993) *Naunyn-Schmiedebergs Arch. Pharmacol.* **347**, 225–230.

Middleton, S.J., Cuthberg, A.W., Shorthouse, M. & Hunter, J.O. (1993) *Br. J. Pharmacol.* **108**, 974–979.

Mirzazadeh, S., Hobbs, A.J., Tucker, J.F. & Gibson, A. (1991) *J. Pharm. Pharmacol.* **43**, 247–257.
Mitchell, J.A., Sheng, H., Förstermann, U. & Murad, F. (1991) *Br. J. Pharmacol.* **104**, 289–291.
Molina y Vedia, L., McDonald, B., Reep, B., Brüne, B., Di Silvio, M., Billiar, T.R. & Lapetina, E.G. (1992) *J. Biol. Chem.* **267**, 24929–24932.
Moncada, S., Palmer, M.J. & Higgs, E.A. (1991) *Pharmacol. Rev.* **43**, 109–142.
Morley, D. & Keefer, L.K. (1993) *J. Cardiovasc. Pharmacol.* **22** (Suppl. 7), S3–S9.
Morris, J.L. (1993) *J. Autonom. Nerv. Syst.* **42**, 11–21.
Mourelle, M., Cuarner, F., Molero, X., Moncada, S. & Malagalada, J.R. (1993) *Gut* **34**, 911–915.
Mülsch, A., Busse, R., Liebau, S. & Förstermann, U. (1988) *J. Pharmacol. Exp. Ther.* **247**, 283–288.
Murad, F., Arnold, W.P., Mittal, C.K. & Braughler, J.M. (1979) *Adv. Cycl. Nucleotide Res.* **11**, 175–204.
Murphy, K.S., Zhang, K.-M., Jin, J.-G., Grider, J.R. & Makhlouf, G.M. (1993) *Am. J. Physiol.* **265**, G660–G671.
Murray, J., Du, C., Ledlow, A., Bates, J.N. & Conklin, J.L. (1991) *Am. J. Physiol.* **261**, G401–G406.
Murray, L.A., Du, C., Ledlow, A., Mantemach, P.L. & Conklin, J.L. (1992) *Am. J. Physiol.* **263**, G97–G101.
Myers, P.R., Minor, Jr., R.L., Guerra, R., Bates, J.N. & Harrison, D.G. (1990) *Nature* **345**, 161–163.
Nathan, C. (1992) *FASEB J.* **6**, 3051–3064.
Needleman, P., Jakschik, B. & Johnson, Jr., E.M. (1973) *J. Pharmacol. Exp. Ther.* **187**, 324–331.
Nichols, K., Krantis, A. & Staines, W. (1992) *Neuroscience* **51**, 791–799.
Nijkamo, F.P., Van Der Linde, H.J. & Folkerts, G. (1993) *Am. Rev. Respir. Dis.* **148**, 727–734.
Niklasson, L.G., Fasth, S., Hulten, L. & Delbro, D.S. (1992) *Acta Physiol. Scand.* **144**, 489–490.
Okelly, T., Brading, A & Mortenson, N. (1993) *Gut* **34**, 689–693.
Osawa, Y. & Davila, J.C. (1993) *Biochem. Biophys. Res. Commun.* **194**, 1435–1439.
Osthaus, L.E. & Galligan, J.J. (1992) *J. Pharmacol. Exp. Ther.* **260**, 140–145.
Ozaki, H., Blondfield, D.P., Hori, M., Publicover, N.G., Kato, I. & Sanders, K.M. (1992) *J. Physiol.* **445**, 231–247.
Papasova, M. (1989) In *Handbook of Physiology, The Gastrointestinal System I, Motility and Circulation* (ed. J.D. Wood), pp. 987–1023. American Physiological Society, Bethesda, MD.
Papka, R.E. & McNeill, D.L. (1992) *Neurosci. Lett.* **147**, 224–228.
Parlani, M., Conte, B. & Manzini, S. (1993) *J. Pharmacol. Exp. Ther.* **265**, 713–719.
Paterson, W.G., Anderson, M.A. & Anand, N. (1992) *Can. J. Physiol. Pharmacol.* **70**, 1011–1015.
Pauletzki, J.G., Sharkey, K.A., Davison, J.S., Bomzon, A. & Shaffer, E.A. (1993) *Eur. J. Pharmacol.* **232**, 263–270.
Pelckmans, P.A., Boeckxstaens, G.E., Van Maercke, Y.M., Herman, A.G. & Verbeuren, T.J. (1989) *Eur. J. Pharmacol.* **170**, 235–242.
Persson, K. & Andersson, K.E. (1992) *Br. J. Pharmacol.* **106**, 416–422.
Persson, K., Igawa, Y., Mattiasson, A. & Andersson, K.-E. (1992) *Br. J. Pharmacol.* **107**, 178–184.
Persson, A., Alm, P., Johansson, K., Larsson, B. & Andersson, K.-E. (1993) *Br. J. Pharmacol.* **110**, 521–530.
Publicover, N.G., Hammond, E.M. & Sanders, K.M. (1993) *Proc. Natl. Acad. Sci. USA* **90**, 2087–2091.
Rajanayagam, M.A., Li, C.G. & Rand, M.J. (1993) *Br. J. Pharmacol.* **108**, 3–5.

Ramagopal, M.V. & Leighton, H.J. (1989) *Eur. J. Pharmacol.* **174**, 297–299.
Rand, M.J. (1989) In *Nicotinic Receptors in the CNS, Vol. 79, Progress in Brain Research* (eds. A. Nordberg, K. Fuxe, B. Holmstedt and A. Sundwall), pp. 3–11. Elsevier, Amsterdam.
Rand, M.J. (1992a) *Clin. Exp. Pharmacol. Physiol.* **19**, 147–169.
Rand, M.J. (1992b) *Proc. Austral. Physiol. Pharmacol. Soc.* **23**, 1–16.
Rand, M.J. & Li, C.G. (1990) *Eur. J. Pharmacol.* **183**, 1144.
Rand, M.J. & Li, C.G. (1992a) *Clin. Exp. Pharmacol. Physiol.* **19**, 103–111.
Rand, M.J. & Li, C.G. (1992b) *Clin. Exp. Pharmacol. Physiol.* **19**, 331–334.
Rand, M.J. & Li, C.G. (1992c) *Proc. Austral. Physiol. Pharmacol. Soc.* **23**, 175P.
Rand, M.J. & Li, C.G. (1993a) *Clin. Exp. Pharmacol. Physiol.* **20**, 141–148.
Rand, M.J. & Li, C.G. (1993b) *Eur. J. Pharmacol.* **241**, 249–254.
Rand, M.J. & Li, C.G. (1993c) *Br. J. Pharmacol.* **110**, 1479–1482.
Rand, M.J. & Li, C.G. (1993d) *Clin. Exp. Pharmacol. Physiol.* Suppl. **1**, 59.
Rand, M.J. & Li, C.G. (1994) *Br. J. Pharmacol.* **111**, 1089–1094.
Rand, M.J., Ishac, E.J.N., De Luca, A., Majewski, H., Musgrave, I.F. & Story, D.F. (1988) *Proc. Austral. Soc. Exp. Biol., Canberra, February*, pp. 145–148.
Rapoport, R.M. & Murad, F. (1983) *Gen. Pharmacol.* **19**, 61–65.
Rattan, S. & Chakder, S. (1992) *Am. J. Physiol.* **262**, G107–G112.
Rattan, S. & Thatikunta, P. (1993) *Gastroenterology* **105**, 827–836.
Rattan, S., Sarkar, A. & Chakder, S. (1992) *Gastroenterology* **103**, 43–50.
Renaud, J.P., Boucher, J.L., Vadon, S., Delaforge, M. & Mansuy, D. (1993) *Biochem. Biophys. Res. Commun.* **192**, 53–60.
Rengasamy, A. & Johns, R.A. (1991) *J. Pharmacol. Exp. Ther.* **259**, 310–316.
Rengasamy, A. & Johns, R.A. (1993) *Mol. Pharmacol.* **44**, 124–128.
Rogers, N.E. & Ignarro, L.J. (1992) *Biochem. Biophys. Res. Commun.* **189**, 242–249.
Rubanyi, G.M. & Vanhoutte, P.M. (1990) *Endothelium-Derived Relaxing Factors.* Karger, Basle.
Rubanyi, G.M., Johns, A., Wilcox, D., Bates, F.N. & Harrison, D.G. (1991) *J. Cardiovasc. Pharmacol.* **17** (Suppl. 3), S41–S45.
Saffrey, M.J., Hassall, C.J.S., Hoyle, C.H.V., Belai, A., Moss, J., Schmidt, H.H.H.W., Förstermann, U., Murad, F. & Burnstock, G. (1992) *NeuroReport* **3**, 333–336.
Sanders, K.M. & Ward, S.M. (1992) *Am. J. Physiol.* **262**, G379–G392.
Santer, R.M. & Symons, D. (1993) *Cell Tiss. Res.* **271**, 115–121.
Schachter, M., Matthews, B. & Bhoola, K.D. (1992) *Agents and Actions* **38** (Suppl.), 366–370.
Schmidt, H.H.H.W. & Murad, F. (1991) *Biochem. Biophys. Res. Commun.* **181**, 1372–1377.
Schmidt, H.H.H.W., Smith, R.M., Nakane, M. & Murad, F. (1992a) *Biochemistry* **31**, 3243–3249.
Schmidt, H.H.H.W., Gagne, G.D., Nakane, M., Pollock, J.S., Miller, M.F. & Murad, F. (1992b) *J. Histochem. Cytochem.* **40**, 1439–1456.
Schmidt, H.H.H.W., Lohmann, S.M. & Walter, U. (1993) *Biochim. Biophys. Acta* **1178**, 153–175.
Sekizawa, K., Fukushima, T., Ikarashi, Y., Maruyama, Y. & Sasaki, H. (1993) *Br. J. Pharmacol.* **110**, 816–820.
Servent, D., Delaforge, M., Ducrocq, C., Mansuy, D. & Lenfant, M. (1989) *Biochem. Biophys. Res. Commun.* **163**, 1210–1216.
Sheng, H. (1991) *Br. J. Pharmacol.* **104**, 137P.
Sheng, H., Schmidt, H.H.H.W., Nakane, M., Mitchell, J.A., Pollock, J.S., Förstermann, U. & Murad, F. (1992) *Br. J. Pharmacol.* **106**, 768–773.
Sheng, H., Gagne, G.D., Matsumoto, T., Miller, M.F., Förstermann, U. & Murad, F. (1993) *J. Neurochem.* **61**, 1120–1126.
Shew, R.L., Papka, R.E., McNeill, D.L. & Yee, J.A. (1993) *Peptides* **14**, 637–641.
Shimamura, K., Fujisawa, A., Toda, N. & Sunano, S. (1993) *Eur. J. Pharmacol.* **231**, 103–109.
Shuttleworth, C.W., Murphy, R. & Furness, J.B. (1991) *Neurosci. Lett.* **130**, 77–80.

Shuttleworth, C.W., Sanders, K.M. & Keef, K.D. (1993) *Br. J. Pharmacol.* **109**, 739–747.
Sim, M.K. & Sim, E.M. (1993a) *Asia Pacific J. Pharmacol.* **8**, S16.
Sim, M.K. & Sim, E.M. (1993b) *Asia Pacific J. Pharmacol.* **8** (Suppl. 2), S23–S24.
Sneddon, P. & Graham, A. (1992) *J. Autonom. Pharmacol.* **23**, 445–456.
Song, Z.-M., Brookes, S.J.H. & Costa, M. (1993) *Neurosci. Lett.* **158**, 221–224.
Springall, D.R., Riveros-Moreno, V., Buttery, L., Suburo, A., Bishop, A.E., Merrett, M., Moncada, S. & Polak, J.M. (1992) *Histochemistry* **98**, 259–266.
Stamler, J.S., Singel, D.J. & Loscalzo, J. (1992) *Science* **258**, 1898–1902.
Stark, M.E. & Szurszewski, J.H. (1992) *Gastroenterology* **103**, 1928–1949.
Stark, M.E., Bauer, A.J. & Szurszewski, J.H. (1991) *J. Physiol.* **444**, 743–761.
Stark, M.E., Bauer, A.J., Sarr, M.G. & Szurszewski, J.H. (1993) *Gastroenterology* **104**, 398–409.
Stretton, D. (1991) *Clin. Exp. Pharmacol. Physiol.* **18**, 675–684.
Stuehr, D.J. & Griffith, O.W. (1992) *Adv. Enzymol. Rel. Areas Mol. Biol.* **65**, 287–346.
Stuehr, D.J., Fasehun, O.A., Kwon, N.S., Gross, S.S., Gonzalez, J.A., Levi, R. & Natiian, G.F. (1991a) *FASEB J.* **5**, 98–103.
Stuehr, D.J., Kown, N.S., Nathan, C.F., Griffith, O.W., Feldman, P.L. & Wiseman, J. (1991b) *J. Biol. Chem.* **266**, 6259–6263.
Suthamnatpong, N., Hata, F., Kanada, A., Takeuchi, T. & Yagasaki, O. (1993) *Br. J. Pharmacol.* **108**, 348–355.
Takeuchi, K., Ohuchi, T., Miyake, H., Niki, S. & Okabe, S. (1993) *Eur. J. Pharmacol.* **231**, 135–138.
Talmage, E.K. & Mawe, G.M. (1993) *J. Autonom. Nerv. Syst.* **43**, 83–90.
Tam, S-F. & Hillier, K. (1992) *Life Sci.* **51**, 1277–1284.
Tanaka, K., Hassall, G.J.S. & Burnstock, G. (1993) *Cell Tiss. Res.* **273**, 293–300.
Taylor, T.W.J. & Baker, W. (1945) *Sidgwick's Organic Chemistry of Nitrogen.* Oxford University Press, Oxford.
Thornbury, K.D., Ward, S.M., Dalziel, H.H., Carl, A., Westfall, D.P. & Sanders, K.M. (1991) *Am. J. Physiol.* **261**, G553–G557.
Thornbury, K.D., Hollywood, M.A. & McHale, N.G. (1992) *J. Physiol.* **451**, 133–144.
Toda, N. & Okamura, T. (1990) *Jpn. J. Pharmacol.* **52**, 170–173.
Toda, N. & Okamura, T. (1992a) *Hypertension* **19**, 161–166.
Toda, N. & Okamura, T. (1992b) *News Physiol. Sci.* **7**, 148–152.
Toda, N., Baba, H. & Okamura, T. (1990a) *Jpn. J. Pharmacol.* **53**, 281–284.
Toda, N., Inoue, S., Okunishi, H. & Okamura, T. (1990b) *Naunyn-Schmiedebergs Arch. Pharmacol.* **341**, 30–36.
Toda, N., Tanobe, Y. & Baba, H. (1991) *Jpn. J. Pharmacol.* **57**, 527–534.
Toda, N., Baba, H., Tanobe, Y. & Okamura, T. (1992) *J. Pharmacol. Exp. Ther.* **260**, 697–701.
Torphy, T.J., Fine, C.F., Burman, M., Barnette, M.S. & Ormsbee, H.H. III (1986) *Am. J. Physiol.* **251**, G786–G793.
Tøttrup, A. (1993) *Dis. Esophagus* **6**, 2–10.
Tøttrup, A., Knudsen, M.A. & Gregersen, H. (1991a) *Br. J. Pharmacol.* **104**, 113–116.
Tøttrup, A., Svane, D. & Forman, A. (1991b) *Am. J. Physiol.* **260**, G385–G389.
Tøttrup, A., Glavind, E.B. & Svane, D. (1992) *Gastroenterology* **102**, 409–415.
Triguero, D., Prieto, D. & Garcia-Pascual, A. (1993) *Neurosci. Lett.* **163**, 93–96.
Tu, S.-I., Byler, D.M. & Cavanaugh, J.R. (1984) *J. Agri. Food Chem.* **32**, 1057–1060.
Tucker, J.K., Brave, S.R., Charalambous, L., Hobbs, A.J. & Gibson, A. (1990) *Br. J. Pharmacol.* **100**, 663–664.
Vallance, P., Leone, A., Calver, A., Collier, J. & Moncada, S. (1992a) *Lancet* **339**, 572–575.
Vallance, P., Leone, A., Calver, A., Collier, J. & Moncada, S. (1992b) *J. Cardiovasc. Pharmacol.* **20** (Suppl. 12), S60–S62.
Vedernikov, Y.P., Mordvintcev, P.I., Malenkova, I.V. & Vanin, A.F. (1992) *Eur. J. Pharmacol.* **211**, 313–317.
Vila, E., Tabernero, A., Fernandes, F. & Salaices, M. (1992) *Br. J. Pharmacol.* **107**, 66–72.

Vizzard, M.A., Erdman, S.L. & Degroat, W.C. (1993) *J. Auton. Nerv. Syst.* **44**, 85–90.
Vo, P.A., Reid, J.J. & Rand, M.J. (1992) *Br. J. Pharmacol.* **107**, 1121–1128.
Waldman, S.A. & Murad, F. (1987) *Pharmacol. Rev.* **39**, 163–196.
Wang, Y.-X., Poon, C.I., Poon, K.S. & Pang, C.C.Y. (1993) *Br. J. Pharmacol.* **110**, 1232–1238.
Ward, J.K., Belvisi, M.G., Fox, A.J., Miura, M., Tadjkarimi, S., Yacoub, M.H. & Barnes, P.J. (1993) *J. Clin. Invest.* **92**, 736–742.
Ward, S.M., McKeen, E.S. & Sanders, K.M. (1992a) *Br. J. Pharmacol.* **105**, 776–782.
Ward, S.M., Xue, C., Shuttleworth, C.W., Bredt, D.S., Snyder, S.H. & Sanders, K.M. (1992b) *Am. J. Physiol.* **263**, G277–G284.
Ward, S.M., Dalziel, H.H., Bradley, M.E., Buxton, I.L., Keef, K., Westfall, D.P. & Sanders, K.M. (1992c) *Br. J. Pharmacol.* **107**, 1075–1082.
Ward, S.M., Dalziel, H.H., Thornbury, K.D., Westfall, D.P. & Sanders, K.M. (1992d) *Am. J. Physiol.* **262**, G237–G243.
Warren, J.B., Loi, R., Rendel, N.B. & Taylor, G.W. (1990) *Biochem. J.* **266**, 921–923.
Westfall, T.C. & Martin, J.R. (1991) In *Presynaptic Regulation of Neurotransmitter Release: a Handbook, Vol. 1* (eds. J. Feigenbaum and M. Hanani), pp. 311–370. Freund Publishing House, London.
White, K.A. & Marletta, M.A. (1992) *Biochemistry* **31**, 6627–6631.
White, T.D. (1988) *Pharmacol. Ther.* **38**, 129–168.
Wiklund, C.U., Wiklund, N.P. & Gustafsson, L.E. (1993) *Eur. J. Pharmacol.* **240**, 235–242.
Will, S., Bieger, D. & Triggle, C.R. (1990) *Eur. J. Pharmacol.* **183**, 2419–2420.
Williams, D.L.H. (1985) *Chem. Soc. Rev.* **14**, 171–196.
Wolin, M.S., Cherry, P.D., Rodenburg, J.M., Messina, E.J. & Kaley, G. (1990) *J. Pharmacol. Exp. Ther.* **254**, 872–876.
Woolsey, T.A. & Van der Loos, H. (1970) *Brain Res.* **17**, 205–242.
Xue, C., Ward, S.M., Shuttleworth, C.W. & Sander, K.M. (1993) *Histochemistry* **99**, 373–384.
Yamato, S., Saha, J.K. & Goyal, R.K. (1992a) *Life Sci.* **50**, 1263–1272.
Yamato, S., Spechler, S.J. & Goyal, R.K. (1992b) *Gastroenterology* **103**, 197–204.
Yoshida, K., Okamura, T., Kimura, H., Bredt, D.S., Snyder, S.H. & Toda, N. (1993) *Brain Res.* **629**, 67–72.
Young, H.M., Furness, J.B., Shuttleworth, C.W., Bredt, D.S. & Snyder, S.H. (1992) *Histochemistry* **97**, 375–378.
Young, H.M., McConalogue, K., Furness, J.B. & De Vente, J. (1993) *Neuroscience* **55**, 583–596.
Zembowicz, A., Hecker, M., Macarthur, H., Sessa, W.C. & Vane, J.R. (1991) *Proc. Natl. Acad. Sci. USA* **88**, 11172–11176.
Zembowicz, A., Swierkosz, T.A., Southan, G.J., Hecker, M. & Vane, J.R. (1992) *Br. J. Pharmacol.* **107**, 1001–1007.
Zhang, J. & Snyder, S.H. (1992) *Proc. Natl. Acad. Sci. USA* **89**, 9382–9385.
Zhang, J.-X., Okamura, T. & Toda, N. (1993) *Hypertension Res.* **16**, 29–32.

CHAPTER 13

NITRIC OXIDE AND THE NEURAL REGULATION OF THE PENIS

Nils O. Sjöstrand* and Erik Klinge†

**Departments of Physiology and Pharmacology, Division of Physiology I, Karolinska Institutet, S-171 77 Stockholm, Sweden, and †Department of Pharmacy, Division of Pharmacology and Toxicology, University of Helsinki, P.O. Box 15, Kirkkokatu 20, SF-00014 Helsinki, Finland*

Table of Contents

13.1 Introduction 282
13.2 Basic mechanisms of penile erection, smooth muscle effectors and haemodynamics 282
13.3 Efferent nerves 284
13.4 The debate on the neurotransmission of penile erection and the search for a transmitter 286
13.4.1 Historical remarks 286
13.4.2 The concept of non-adrenergic, non-cholinergic (NANC) innervation of smooth muscle effectors of penile erection 286
13.4.3 Vasoactive intestinal peptide; a launched and rejected candidate 288
13.4.4 Towards nitric oxide as a plausible transmitter candidate 290
13.5 Nitric oxide as the principal transmitter candidate 291
13.5.1 Pharmacological and biochemical evidence 291
13.5.2 Presence of nitric oxide synthase in penile nerves 293
13.6 Nitric oxide as likely mediator of cholinergic effects 294
13.6.1 Muscarinic relaxation and the endothelium 294
13.6.2 Nicotinic relaxation 298
13.7 Final considerations 299
13.7.1 The solution of the problem 299
13.7.2 Nitric oxide is possibly not the only transmitter in the erectile fibres 300
13.7.3 Clinical implications 301
13.7.4 Concluding remarks 302

Acknowledgements 302
References 303

Nitric Oxide in the Nervous System
ISBN 0-12-721985-4

13.1 Introduction

The physiology of penile erection, especially the haemodynamics and its underlying basic neuromuscular events, has been a topic of scientific discussion for a very long time (see Klinge and Sjöstrand, 1974). During the last decades there has been a markedly increased interest in the matter. One reason for this is the growing interest in the therapy of impotence and the realization that this condition often is caused by vascular and/or peripheral neural disturbances and does not merely reflect mental disturbances.

The fundamental studies of penile erection were made in the second half of the 19th century. As it is still nowadays the dog was the favourite experimental animal, but a few studies were also made on the rabbit and cat. von Kölliker (1852), the discoverer of smooth muscle cells, was the first to realize that relaxation of the smooth muscle in the penile arteries, the penile cavernous tissue and the retractor penis must be a crucial event in penile erection. Eckhard (1863) demonstrated that stimulation of the pelvic nerves, which he called the nervi erigentes, produced penile erection and increased blood flow through the penis. Lovén (1866) showed that stimulation of the pudendal nerve produced penile vasoconstriction, which was important in maintaining the relaxed state of the organ. Eckhard (1876) further found that stimulation of the hypogastric nerves could cause erection in the rabbit but not in the dog. Later work by François-Franck (1895) and Langley and Anderson (1895, 1896) further consolidated the view although Langley and Anderson refuted the concept of erectile fibres in the hypogastric nerve.

13.2 Basic mechanisms of penile erection, smooth muscle effectors and haemodynamics

The main smooth muscle effectors of penile erection (and penile shrinkage and retraction) are the retractor penis muscle, the muscle of the arteries supplying the cavernous spaces and the muscle in the walls and trabeculae of these spaces. The retractor penis muscle is present in most mammalian orders but lacking in primates and lagomorphs (rabbit). The muscle takes its origin on the coccygeal vertebrae (e.g. ox) or the tissues around the anal orifice (e.g. dog) and has its insertion either directly on the distal part of the penis (e.g. ox) or on preputial connective tissue (e.g. dog). When the paired muscle is contracted it retains the penis under the skin. When it relaxes the penis protrudes (Figure 1).

All mammals have smooth muscle cells in the arteries supplying the erectile tissue but the detailed anatomy of the vascular system differs. For penile erection the corpora cavernosa penis are the most important, while the less rigid compartments of the corpus spongiosum and the venous glans penis are less important. In a 'vascular' type of penis (e.g. human, dog, horse and rabbit) the main blood supply filling the sinusoids of the corpus cavernosum penis comes from the paired deep penile

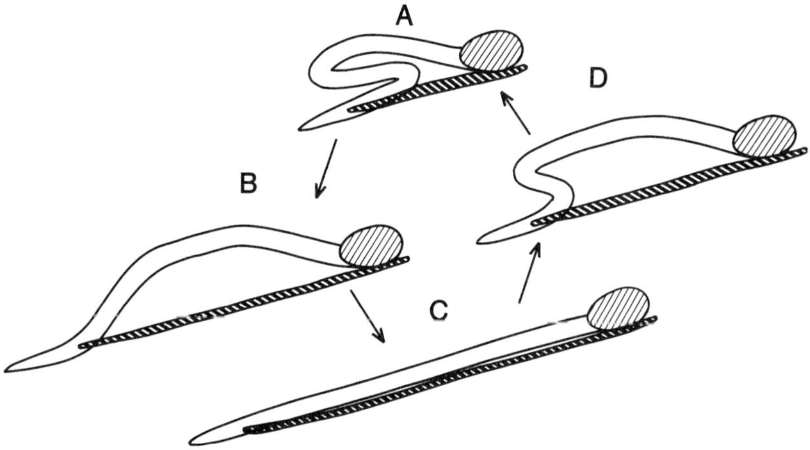

Figure 1 Schematic drawings of the function of the retractor penis muscle (darkly striated). The ischiocavernosus muscle covering the basal parts of the penis is indicated by light striation. (A) The contracted retractor penis keeps the relaxed penis withdrawn in a characteristic sigmoid flexure. The localization and size of the flexure varies. In the bull where the corpora cavernosa penis confluence distally, the flexure is situated in the medio-distal part of the penis, whereas in a species like dog with long ossification (os penis) of the distal part of the median septum, the flexure lies closer to the root of the penis. (B) When erection starts the retractor penis relaxes and the penis protrudes. (C) In full erection the stiff penis is straightened and the retractor penis is fully relaxed and may in several species more than double its length. (D) When erection ceases the retractor penis recontracts and the sigmoid curve is gradually re-established.

artery, which runs centrally in the cavernous body. The artery has intimal ridges and thick muscular walls where the muscle cells run in both circular and longitudinal bundles. From the central artery branches, the helicine arteries leave to the cavernous spaces. The final precavernous segments of the helicine arteries that open in the spaces have thick media, often composed of one layer of smooth muscle cells that may be epitheloid in shape (Fujimoto and Takeshige, 1975, rabbit). The cavernous tissue is lined with endothelium and smooth muscle cells are a main constituent of the trabeculae separating the lacunae. Underneath the thick tunica albuginea which surrounds the corpus cavernosum penis there is a plexus of small post-cavernous veins that drain the cavernous spaces. The venulae confluence in deep veins, which via circumflex veins communicate with the large, in humans usually unpaired, dorsal, subfascial vein. This vein and the deep penile veins leave the penis at its base. The collecting venous plexus may be compressed when the cavernous tissue is engorged.

In an 'avascular' or 'fibroblastic' type of penis (e.g. ox, sheep, goat and swine) the cavernous bodies are essentially devoid of smooth muscle and the corpus cavernosum penis has a very thick fibrous tunica. Fibrous septa and cords are present in the cavernous tissue, which is crossed by a system of longitudinal canals. The deep penile artery divides after its entrance in the crus penis into short thick-walled branches that open directly into the lacunae and canal system of the cavernous body. The venous outlet is at the base of the penis (Beckett, 1983).

In a 'vascular' penis the main haemodynamic events in penile erection seem to be as follows. (1) Dilatation of arteries and increased blood flow through the organ possibly mainly via corpus spongiosum and glans. (2) Then blood is 'shunted' to the sinusoidal spaces of the corpora cavernosa, which can be filled due to relaxation of their smooth muscle and the dilatation of the precavernous arterioli. During filling venous outflow is lower than arterial inflow but when the cavernous bodies are filled the venous outflow equals arterial inflow. (3) The outflow resistance of the corpora cavernosa may increase in full erection particularly due to compression of post-cavernous veins. The importance of venous occlusion mechanisms seems to depend on the magnitude of the arterial inflow resistance (Hanyu *et al.*, 1992; Vardi and Siroky, 1993). (4) During the height of the act compression of the crura and the bulb, caused by contraction of the ischio- and bulbo-cavernosus muscles (striated), produces vaso-compression and a further increase in pressure within the cavernous bodies (Beckett, 1983). This mechanism seems to be much more important in a fibroblastic penis where (1) and (2) seem to coincide.

When penile erection ceases the tone of the arteries increases; hence inflow resistance exceeds outflow resistance which leads to emptying of the erectile tissue. In a vascular type of penis contraction of the muscles in the cavernous tissue strongly reduces its capacity to store blood which is thereby expelled. Schematic drawings of filling and emptying of a cavernous space model are presented in Figure 2.

The smooth muscles of the penile arteries, the cavernous tissue and the retractor penis have many properties in common. They are spontaneously active. They have an excitatory adrenergic innervation and contract vigorously to noradrenaline. Furthermore, they have the same type of inhibitory innervation. From a physiological point of view the contracted state of the smooth muscle effectors of erection corresponds to the flaccid and retracted state of the penis while the relaxed state corresponds to the erect and protruded state of the organ. For further references on this section see Klinge and Sjöstrand (1974), Sjöstrand and Klinge (1979), Beckett (1983), De Groat and Steers (1988), Andersson (1993) and De Groat and Booth (1993).

13.3 Efferent nerves

The classical pathway for erectile fibres is the sacral parasympathetic outflow via the pelvic nerves. In humans most fibres leave at S2–S4 but there are individual (and species) variations. The fibres have their synaptic (cholinergic) relay in the pelvic plexus, at the base of the penis or maybe even within penile smooth muscle. Another supply of erectile fibres derives from the thoraco-lumbar sympathetic outflow and reaches the penis via the hypogastric nerves. In humans the spinal level

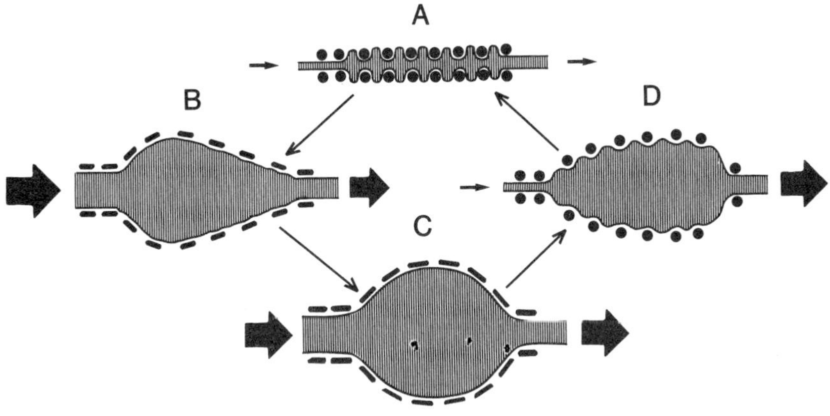

Figure 2 Drawings of a model of a cavernous space with a precavernous arteriole. Contracted smooth muscle cells are indicated by black dots and relaxed cells by black stripes. (A) At the flaccid state of the penis, the smooth muscle cells are contracted. The space has low capacity for storing blood. Inflow resistance is higher than outflow resistance. Flow is essentially determined by inflow resistance. (B) At the start of erection the smooth muscle relaxes. Inflow resistance becomes lower than outflow resistance. Capacity for storing blood increases and, as long the space is being filled with blood, the inflow exceeds the outflow. (C) When the space is filled, inflow equals outflow and is determined by outflow resistance. (Due to venous occlusion centripetally to illustrated section, flow may even become smaller than during the flaccid condition. This is, however, not illustrated in the figure, which only illustrates the function of the smooth muscle in the precavernous arterioli and the walls of the cavernous spaces). (D) When erection ceases, the smooth muscle cells contract. Inflow resistance increases and exceeds outflow resistance. As long as the capacity of the space for storing blood decreases outflow exceeds inflow. When capacity is back to the initial level the stage in A is regained.

seems to be Thll–L2. Also these fibres have a peripheral (cholinergic) synaptic relay. Presumably at least some postganglionic neurons receive excitatory presynaptic input from the pelvic as well as from the hypogastric nerves and therefore by definition belong to both the parasympathetic and the sympathetic system. From the pelvic plexus nerves proceed to the penis. The cavernous nerves convey most of the erectile fibres to the cavernous bodies. Figure 3A presents a scheme of the paths for erectile fibres to the penis.

The excitatory, vasoconstrictor nerves derive from the thoraco-lumbar sympathetic outflow. Most, if not all, fibres have their synaptic relay in the sympathetic chains. The classical path from the chain is via the pudendal nerve but some fibres may follow the hypogastric or the pelvic nerves (Figure 3B). For references see papers listed above (Section 13.2) and Dail (1993).

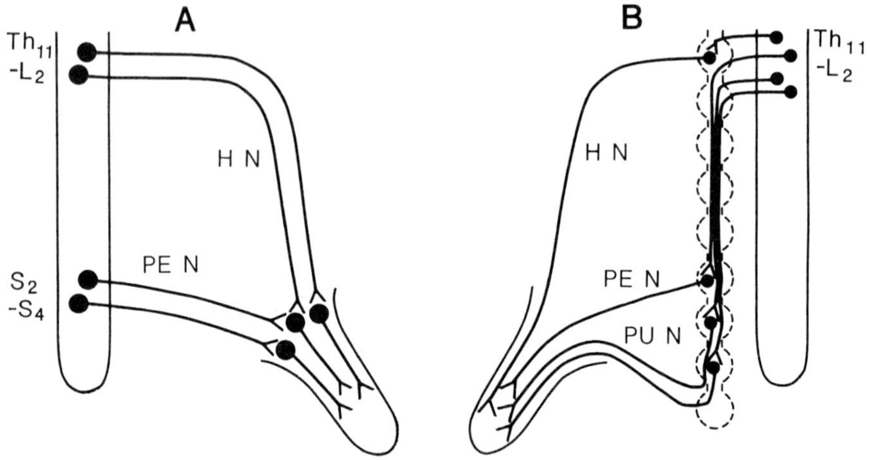

Figure 3 Schematic drawings of the paths for erectile fibres (A) and their opponents (B). The figures indicate site of spinal origin and synaptic relay of presynaptic neuron (situation in humans). Sympathetic chain is indicated by dotted lines. The prevertebral inferior mesenteric ganglion, which is vestigial in humans, is not illustrated. HN, Hypogastric nerve; PEN, pelvic nerve; PUN, pudendal nerve.

13.4 The debate on the neurotransmission of penile erection and the search for a transmitter

13.4.1 Historical remarks

The debate on the neurotransmission of erection is much older than the concept of cholinergic transmission. It goes back to an early dispute on the effect of atropine on the erectile response to pelvic nerve stimulation in the dog. Nikolsky (1879) in Kazan stated that atropine 'paralysed' the pelvic nerve as it 'paralysed' the vagus nerve. This concept was, however, refuted by von Anrep and Cybulski (1884) in St Petersburg and Piotrowski (1894) in Lemberg (Lvov). Thereafter the effects of anticholinergic compounds, direct cholinergic drugs and acetylcholine esterase inhibitors were investigated in several *in vivo* and *in vitro* studies with conflicting results [for review, scrutiny and summary of earlier studies see Klinge and Sjöstrand (1974) Sjöstrand and Klinge (1979)]. By 1970 the situation was not much clearer than at the turn of the century.

13.4.2 The concept of non-adrenergic, non-cholinergic (NANC) innervation of smooth muscle effectors of penile erection

A reinvestigation of the innervation of, and the responses to autonomic drugs, of penile smooth muscle started. *In vitro* studies of the retractor penis muscle and the penile artery of the bull revealed that the muscles, besides an essentially adrenergic excitatory innervation, also had an inhibitory innervation. The inhibitory response to field stimulation was not blocked by antimuscarinic or ganglion blocking drugs. Furthermore, it was not blocked by hemicholinium or botulinum toxin, and it was not enhanced by acetylcholine esterase inhibitors. Consequently, its nature seemed to be non-cholinergic. The inhibitory response to field stimulation was not blocked by adrenergic neuron blocking or adrenoceptor blocking agents. Therefore it seemed to be non-adrenergic, but it was blocked by tetrodotoxin or local anaesthetic drugs (Klinge and Sjöstrand, 1974). Comparative studies of the effect of various contenders as inhibitory transmitters ruled out histamine, 5-hydroxytryptamine, various amino acids, prostaglandins, ATP and peptides as the mediator of the non-adrenergic, non-cholinergic (NANC) relaxation in penile smooth muscle of different species (Klinge and Sjöstrand, 1974, 1977a; Sjöstrand *et al.*, 1981).

In the retractor penis muscle the inhibitory response was characterized by a very steep frequency–response relationship and a rapidly achieved maximum (Figure 4A). Furthermore, in the frequency range 0.5–10 Hz the response was mainly dependent on the number of pulses, not on their frequency provided that stimulation voltage was maximal. These findings suggested (1) an intimate relationship

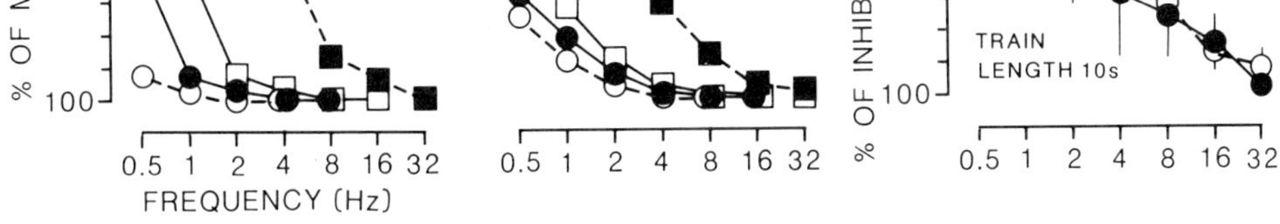

Figure 4 Frequency–response relationship of the inhibitory response to field stimulation of NANC nerves. Supramaximal stimulation voltage (Klinge and Sjöstrand, 1977b). (A) and (B) illustrate retractor penis and corpus spongiosum from the same cat. Preparations consist of whole organs. Guanethidine 13 μM present. Diamond shows relaxation to single pulse. Family of curves to stimulation with different train length. In the retractor penis the curves are essentially determined by number of pulses, i.e. 5 or 10 pulses give the same relaxation regardless of frequency (the 20, 10 and 5 s trains show parallel shift). In the corpus spongiosum the curves are more frequently dependent. (C) Spiral strips of bovine deep penile artery and longitudinal segments of human 'central' arteries (material from transsexual operations). Preparations in tone after guanethidine (6.5 or 13 μM). Mean and SD of five preparations from different specimens. Note similarity of the curves and their difference from curves in (A) and (B).

between inhibitory nerves and effector cells, e.g. a rather dense innervation (see Figure 5A), and (2) a limited facilitation of the secretion of the transmitter from the inhibitory nerves when the interval between the pulses was shortened. The frequency–response curve to NANC nerve stimulation was similar although somewhat less steep in the corpus spongiosum (Figures 4B and 6) and still less steep in the bovine penile artery (Figure 4C) suggesting a less intimate relation between the inhibitory nerves and the effector cells (Figure 5B). The frequency–response curves show almost no species variation (Figures 4C and 6).

The penile smooth muscle joined the group of smooth muscle furnished by NANC inhibitory nerves that was gradually defined in the course of the 1960s and 1970s (see reviews by Sanders and Ward, 1992; Sneddon and Graham, 1992 and Chapter 12). Of these muscles the anococcygeus muscle studied by Gillespie and his group (Gillespie, 1972; Martin and Gillespie, 1991) deserves special attention in our context. In many species the anococcygeus muscle has a similar origin as the retractor penis muscle (Klinge and Sjöstrand, 1974; Martin and Gillespie, 1991).

13.4.3 Vasoactive intestinal peptide; a launched and rejected candidate

Around 1980 vasoactive intestinal peptide (VIP) was launched as the candidate for the transmitter of erection (Willis *et al.*, 1981, 1983, Virag *et al.*, 1982, Ottesen *et al.*, 1984; for reviews see Ottesen *et al.*, 1988; Wagner and Sjöstrand, 1988; Andersson and Holmquist, 1990; Andersson, 1993). Significant amounts of VIP immunoreactivity are present in the penis of humans and experimental animals. The cavernous tissue has VIP immunofluorescent nerve fibres in its smooth muscle,

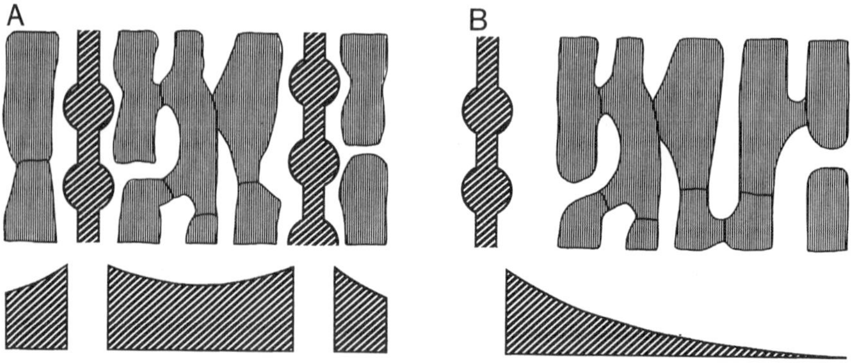

Figure 5 Schematic illustration of supposed relation between muscle cells and inhibitory nerves in the retractor penis (A) and the penile artery (B). Below tentative concentration curves of the inhibitory transmitter following moderate stimulation are indicated on an arbitrary scale. In the densely innervated retractor penis a high degree of receptor occupancy is likely to occur even with moderate stimulation frequencies but this is not likely to occur in the penile artery, which therefore requires high frequency stimulation for relaxation close to maximum (cf. Figure 4).

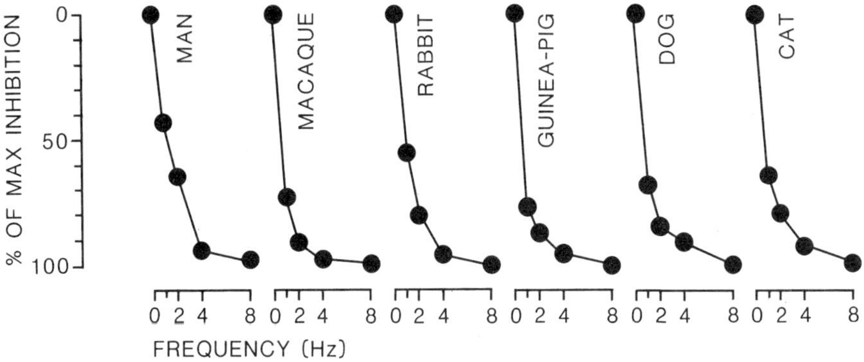

Figure 6 Frequency–response relationship of NANC relaxation in the corpus spongiosum of different mammals (10 s trains, supramaximal voltage). Guanethidine 6.5–26 μM present. In order to secure tone adrenaline (1.5 μM) was added to the human strips and noradrenaline (1–3 μM) to the rabbit strips. Each point represents the mean of three to five preparations from different specimens. Macaque=cynomolgus monkey (*M. irus s. fascicularis*). Note great similarity of the curves.

and such fibres are also present in penile arteries and veins. The VIP immunoactivity is mainly confined to nerve fibres of the cholinergic type where it is localized in the large ('peptidergic') vesicles. That is, as in other tissues, VIP seems to be a co-transmitter in cholinergic neurons. Many studies showed that VIP relaxed penile smooth muscle *in vitro* and produced penile tumescence *in vivo*. Furthermore, following pelvic nerve stimulation VIP immunoreactivity increases in penile venous effluent. Consequently much evidence spoke in favour of VIP as the mediator of the neurogenic relaxation of penile smooth muscle. There were, however, also early critical comments and doubts. Sjöstrand *et al.* (1981) found that, although VIP relaxed the retractor penis muscle of ox, dog and cat, this effect displayed pronounced tachyphylaxis, a phenomenon also seen in human penile smooth muscle (Figure 7). The relaxant effect of VIP on the corpus spongiosum of dog, cat, rabbit and guinea-pig was inconsistent and lacking in the bovine penile artery. Also Bowman and Gillespie (1983), working with perfused penile arteries of dog and ox, doubted that VIP was the mediator of neurogenic vasodilatation.

The accumulated pharmacological evidence is against VIP as the major mediator of NANC relaxation in penile smooth muscle. Thus for example in the human corpus cavernosum penis VIP antiserum (Adaikan *et al.*, 1986) and chymotrypsin (Pickard *et al.*, 1993) do block the VIP-induced relaxation but not the neurogenic one. On the other hand, in the bovine retractor penis relaxations due to field stimulation of intramural nerves are blocked by ethanol, oxyhaemoglobin, hypoxia and *N*-methylhydroxylamine, while relaxations caused by VIP are unaffected (Bowman *et al.*, 1982; Bowman and Drummond, 1985; Bowman and McGrath, 1985). VIP seems to use cAMP as second messenger while cGMP is second messenger of the NANC relaxation (see below). Today after a decade of

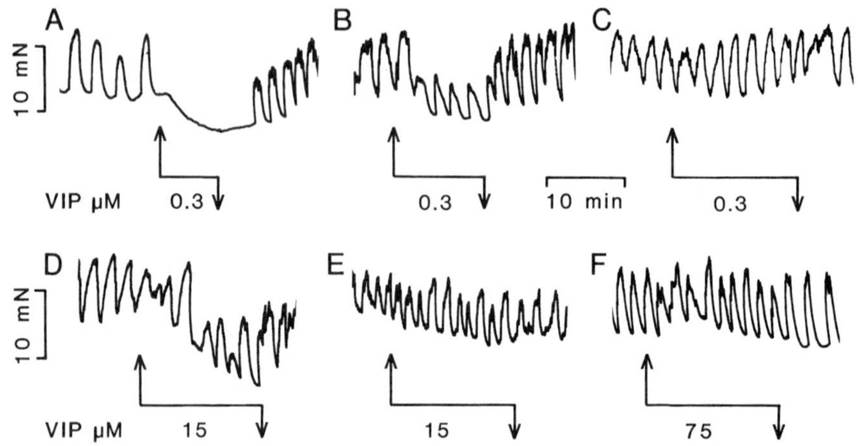

Figure 7 Segment of central (deep penile) artery of human. Spontaneous rhythmicity, longitudinal contractions. VIP is given six times at 60 min intervals (A to F). Note the marked tachyphylaxis of the relaxations seen in the onset and magnitude as well as recovery following washout of the compound.

intense research most investigators seem to agree with the concept put forward in 1981 by Sjöstrand *et al.* according to which it is unlikely that VIP serves as a major mediator of neurogenic relaxation of penile smooth muscle.

13.4.4 Towards nitric oxide as a plausible transmitter candidate

For the further search for the mediator responsible for neurogenic relaxation of penile smooth muscle certain developments in related areas were crucial. These were: (1) the increased knowledge and understanding of the relaxant action of nitrovasodilators such as glyceryl trinitrate, nitroprusside, and nitric oxide (NO) and (2) the discovery of endothelium-dependent vasodilatation and the endothelium-derived relaxing factor (EDRF; Furchgott and Zawadski, 1980; see Furchgott, 1984, 1990). In the final progress of the work studies of NO production in the CNS and by macrophages were also of importance. Nitrovasodilators were found to act via stimulation of guanylyl cyclase and increased levels of cGMP (Arnold *et al.*, 1977; Gruetter *et al.*, 1979; Murad *et al.*, 1979) and NO seemed to be the active principle of all nitrovasodilators (see Rapoport and Murad, 1983; Murad, 1986; Waldman and Murad, 1987). EDRF was also found to operate via guanylyl cyclase activation and formation of the second messenger cGMP. The NANC responses of the bovine retractor penis muscle were found to be related to increased levels of cGMP (Bowman and Drummond, 1984, 1985) and blocked by methylene blue, *N*-methylhydroxylamine and oxyhaemoglobin (Bowman *et al.*, 1982), which inhibit the action of nitrovasodilators (Murad, 1986), and by hypoxia (Bowman and McGrath, 1985) which inhibits EDRF release from endothelium (see Furchgott, 1990).

Ambache *et al.* (1975) prepared a crude extract from the bovine retractor penis muscle. The extract contained an active principle that was thermolabile, ether insoluble and activated by acid, and relaxed the retractor penis muscle. Further investigations of the extract and its inhibitory substance, called the inhibitory factor (IF), were performed by Gillespie and his coworkers in Glasgow (Gillespie and Martin, 1980; Bowman *et al.*, 1979, 1981, 1982; Bowman and Gillespie, 1981, 1982; Gillespie *et al.*, 1981; Bowman and Drummond, 1984; Bowman and McGrath, 1985; Byrne and Muir, 1985; see Gillespie *et al.*, 1990; Martin and Gillespie, 1991).

The IF was not specific for the retractor penis muscle but could be extracted from many tissues. It relaxed the retractor penis muscle and its homologue the anococcygeus muscle as well as various arterial preparations. The inhibitory material existed in a stable but inactive form, that was activated by brief exposure to acid. Its action was susceptible to haemoglobin, methylene blue and *N*-methylhydroxylamine. It activated guanylyl cyclase and increased cGMP levels in the relaxing smooth muscle. Hence IF used the same second messenger as the NANC nerves and EDRF and nitrovasodilators. The similarity of IF and EDRF was apparent and it became gradually established that the major active principle of the EDRF (Palmer *et al.*, 1987; Ignarro *et al.*, 1987; Furchgott, 1990) and the IF (Furchgott, 1988; Martin *et al.*, 1988; Yui *et al.*, 1989) was in fact NO or a compound delivering NO.

A further step forward was the elucidation of the biochemical path for NO formation and the introduction of nitric oxide synthase (NOS) inhibitors. NO was found to be formed from the terminal guanidino atom(s) of L-arginine with oxygen by the enzyme NOS. L-Citrulline is the other reaction product. The first NOS inhibitor presented was N^{G}-monomethyl-L-arginine (L-NMMA; Rees *et al.*, 1989). It inhibited NANC relaxation in the rat anococcygeus muscle (Gillespie *et al.*, 1989). But even in high concentration it had no distinct effect on NANC relaxation in the bovine retractor penis muscle (Gillespie and Xiaorong, 1989) or the corpus cavernosum penis of humans (Sjöstrand *et al.*, 1990) and rabbits (Kim *et al.*, 1990). These findings were puzzling in view of the homology of the retractor penis and anococcygeus muscles and the identity of their innervation and that of erectile tissue and penile arteries. Hence there were some transient doubts on the possibility that NO formed from L-arginine was the NANC transmitter in smooth muscle effectors of penile erection. However, new and more potent NOS inhibitors rapidly became available. When these were introduced the opinion changed.

13.5 Nitric oxide as the principal transmitter candidate

13.5.1 Pharmacological and biochemical evidence

The potent new NOS inhibitors were N^{G}-nitro-L-arginine (L-NOARG, L-NNA, L-NA), its methyl ester (N^{G}-nitro-L-arginine methyl ester, L-NAME) and N^{G}-

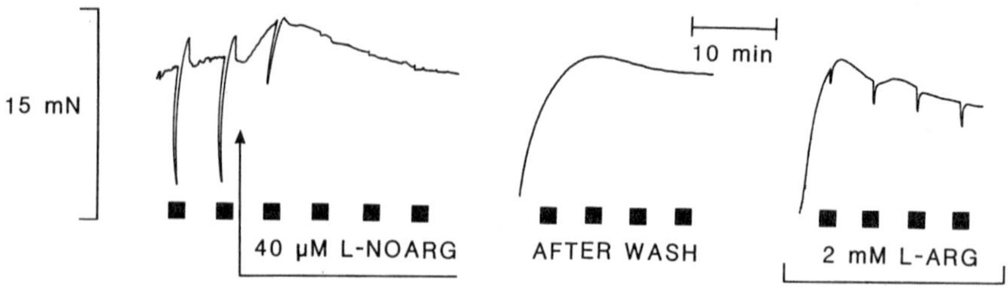

Figure 8 Strip of human corpus cavernosum penis. Guanethidine 13 μM present. Preparation contracted by adrenaline 1.5 μM. At squares field stimulation (8 Hz, 0.2 ms, 12 V, 10 s). Left panel: L-N^G-nitroarginine produces transient increase in tone and abolishes the NANC relaxation. Middle panel: after four washings and 20 min no inhibitory response is seen when preparation is contracted by adrenaline. Right panel: after 20 min incubation with high concentration of L-arginine weak relaxations reappear.

amino-L-arginine. Since the problem was defined, and the experimental methods were developed and the analytical tools were available and efficient, there was a rapid progress in a number of reports showing that these NOS inhibitors blocked the response to NANC nerve stimulation in isolated penile smooth muscle. NANC relaxation of strips from the corpus cavernosum of rabbit and humans is thus effectively blocked (Ignarro *et al.*, 1990; Holmquist *et al.*, 1991a,b, 1992; Pickard *et al.*, 1991; Adaikan *et al.*, 1991; Knispel *et al.*, 1992a,b; Bush *et al.*, 1992b; Rajfer *et al.*, 1992; Kim *et al.*, 1993; Kirkeby *et al.*, 1993; Figure 8). Also the neurogenic relaxations of the bovine retractor penis and penile artery are blocked (Liu *et al.*, 1991; Martin *et al.*, 1993). In many of these studies the appropriate control experiments proving the specific enzymatic path and oxygen dependency were also undertaken. Thus the blocking action of the NOS inhibitors could be counteracted by high concentration of the proper substrate L-arginine but not D-arginine. Further, D-isomers of the NOS inhibitors were without effect on the NANC relaxations. Martin *et al.* (1993) further found that L-NMMA counteracted L-NOARG in the bovine retractor penis muscle. NOS inhibitors also block erectile responses *in vivo* to cavernous nerve stimulation (Holmqvist *et al.*, 1991b, rabbit; Burnett *et al.*, 1992, rat), sacral cord stimulation (Finberg *et al.*, 1993, rat), pelvic nerve stimulation (Trigo-Rocha *et al.*, 1993b, dog; Sjöstrand, Beckett and Klinge, unpublished, goat) and hypogastric nerve stimulation (Sjöstrand, Beckett and Klinge, unpublished, goat).

Following field stimulation of the isolated corpus cavernosum penis there are increased tissue levels of nitrite and citrulline in the rabbit (Ignarro *et al.*, 1990; Bush *et al.*, 1992a) and in humans an increased overflow of nitrite and nitrate can be detected (Leone *et al.*, 1994). After inhibition of NOS or blockade of axonal conduction, the field-stimulated increase in the NO metabolites is blocked. NO and NO donors relax isolated erectile tissue of humans and rabbit (Bush *et al.*, 1992b; Rajfer *et al.*, 1992; Holmqvist *et al.*, 1993) and produce tumescence in the anaesthetized dog (Trigo-Rocha *et al.*, 1993a) and cat (Wang *et al.*, 1994). Thus, the accumulated evidence agrees well with the concept that NO is a main mediator of neurogenic relaxation of penile smooth muscle.

13.5.2 Presence of nitric oxide synthase in penile nerves

NOS was isolated from cerebellum of rat (Bredt and Snyder, 1990) and swine (Mayer *et al.*, 1990). The enzyme is dependent on calmodulin, Ca^{2+}, NADPH and tetrahydrobiopterin. The purification of the enzyme made it possible to produce antisera to it and to localize it with immunohistology (Bredt *et al.*, 1990). A further important contribution was the demonstration that neuronal NADPH diaphorase is an NOS (Hope *et al.*, 1991; Dawson *et al.*, 1991). Consequently, also the NADPH diaphorase histochemical technique (Vincent, 1986) can be used for visualization of NOS in organs.

NOS has been localized in nerves surrounding the deep penile artery and its helicine branches as well as in nerves running to the smooth muscle of the walls of the cavernous spaces in rat by Burnett *et al.* (1992) using the technique of Bredt *et al.* (1990) and by Alm *et al.* (1993) using antibodies directed towards synthetic C- and N-terminals of a cloned enzyme from cerebellum. Similar findings have been made in penis from dog (Burnett *et al.*, 1992) and humans (Burnett *et al.*, 1993; Leone *et al.*, 1994; Figure 9). NOS activity seems to be lacking in penile veins (Alm *et al.*, 1993) which agrees with the finding that human penile circumflex veins are devoid of NANC innervation (Kirkeby *et al.*, 1993). NADPH diaphorase staining gives similar results in rat (Keast, 1992; Dail *et al.*, 1992; Vizzard *et al.*, 1993a,b) and humans (Burnett *et al.*, 1993; Leone *et al.*, 1994; Figure 9). On the basis of the staining of neurons in the rat major pelvic ganglion, Keast (1992) found it likely that many, if not all, penile neurons contain NOS. This implies that acetylcholine

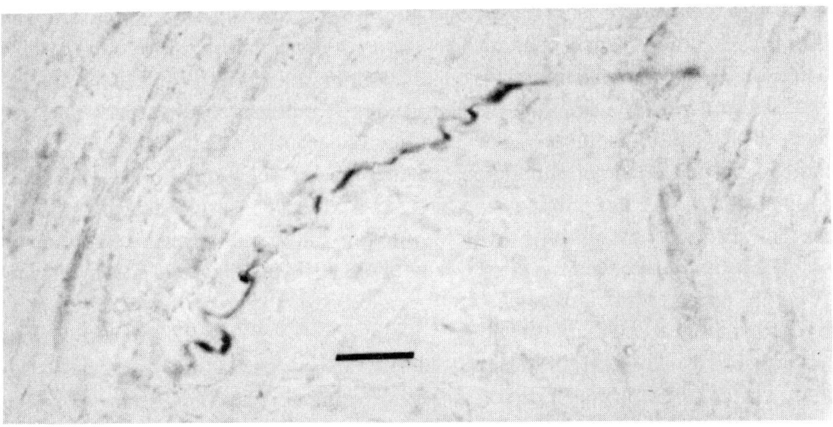

Figure 9 NADPH diaphorase staining of nerve in human corpus cavernosum penis, bar 50 μM. Courtesy of Dr N. Peter Wiklund.

esterase-positive neurons and neurons displaying VIP immunoreactivity also may be NOS positive. Another interesting finding is that afferent nerves – at least from the bladder – also may contain NOS (Vizzard *et al.*, 1993b).

In the rat retractor penis muscle NOS-positive nerve fibres have been detected with NADPH diaphorase histochemistry by Dail *et al.* (1993) while Sheng *et al.* (1992) demonstrated such fibres in the bovine retractor penis muscle with immunohistology. Sheng *et al.* (1992) further characterized a partly purified enzyme from the bovine retractor penis muscle and concluded that it was similar to cerebellar NOS. A similar conclusion was drawn by Bush *et al.* (1992a) concerning NOS in the rabbit corpus cavernosum. Thus, with the aid of different specific techniques the presence of NOS in penile nerves supplying the smooth muscle effectors of penile erection is well established.

13.6 Nitric oxide as likely mediator of cholinergic effects

13.6.1 Muscarinic relaxation and the endothelium

The early literature gives a diverse and conflicting picture of the effects of actylcholine and other cholinomimetics on penile smooth muscle (Klinge and Sjöstrand, 1974). Klinge and Sjöstrand (1974, 1977a,b) concluded that the direct action of muscarinic agonists on penile smooth muscle was contractile not relaxant. Relaxant effects of acetylcholine seemed to be due to nicotinic stimulation of inhibitory nerves (see below). However, later muscarinic vasodilatation was demonstrated in the perfused rabbit penis (Sjöstrand and Klinge, 1979). In the 1980s studies of isolated strips of human penile smooth muscle, especially from the corpus cavernosum penis, started in many laboratories. This revived interest in muscarinic relaxation of penile smooth muscle because such relaxations were a common finding in human material (e.g. Benson *et al.*, 1980; Adaikan *et al.*, 1983; Andersson *et al.*, 1983; Hedlund *et al.*, 1984; Hedlund and Andersson, 1985) although contraction was also reported (Benson *et al.*, 1980; Adaikan *et al.*, 1983). The fundamental discovery of the endothelium dependency of cholinergic vasodilatation (see Section 13.4.4) hinted at an explanation of the dual muscarinic effect. Thus, it was reported that de-endothelialization produced by rubbing or scraping prevented muscarinic relaxation in strips of the human corpus cavernosum penis (Saenz de Tejada *et al.*, 1988) and deep penile artery (Sjöstrand *et al.*, 1988; Figure 10). Intracavernous injection of acetylcholine produces erectile responses in the dog (Stief *et al.*, 1989a) which can be abolished by chemical destruction of the endothelium (Trigo-Rocha *et al.*, 1993b). The muscarinic receptors of the endothelium in the human corpus cavernosum penis have been characterized (Traish *et al.*, 1990). The muscarinic relaxation is blocked by NOS inhibitors and methylene blue (Kim *et al.*, 1991; Azadzoi *et al.*, 1992; Holmquist *et al.*, 1991a, 1992; Knispel *et al.*, 1991; Liu *et al.*, 1991;

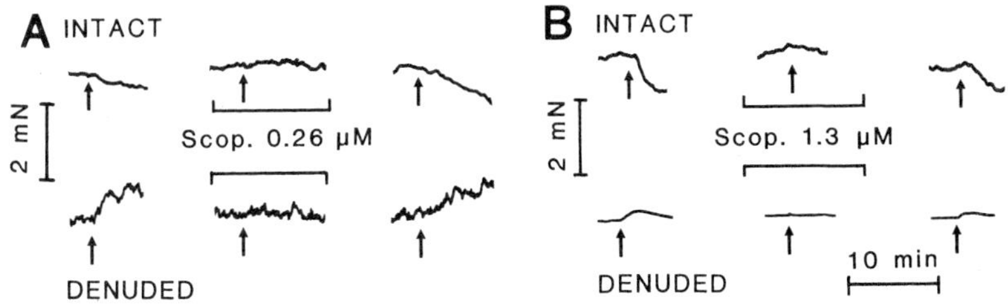

Figure 10 (A) Two consecutive longitudinal segments of a human deep penile artery. The segments were opened by a longitudinal section with fine scissors. In the upper segment no further preparation was performed, while in the lower, the endothelium was removed by scraping with the blunt side of a fine scalpel and thereafter pressing the intimal side on a dry paper for 10 s. At arrow 27.5 µM of carbachol is added. This relaxes the intact preparation but contracts the denuded one. Both effects are abolished by a low concentration of scopolamine but reappear 20 min after wash. Mecamylamine (5µM) and lidocaine (74 µM) were present in order to eliminate nicotinic effects. (B) Strips from human glans penis; the upper is intact, the lower is rubbed according to Saenz de Tejada *et al.* (1988). At arrow acetylcholine (80 µM) is added. The intact preparation is relaxed while the denuded one is contracted. Both effects are abolished by scopolamine but reappear after washing. Mecamylamine and lidocaine present as in (A).

Martin *et al.*, 1993; Trigo-Rocha *et al.*, 1993a,b). Also L-NMMA is an effective blocker although the newer NOS inhibitors are more potent (Figure 11). Muscarinic relaxation has not been demonstrated in the retractor penis muscle, only in vascular penile smooth muscle, which agrees with the concept that it is endothelium dependent. According to our own experience muscarinic relaxation is most consistent in strips from the human glans penis (Figures 10B and 11).

With respect to the topic of this chapter the crucial questions are: (1) can acetylcholine released from penile efferent nerves induce NO secretion from the endothelium that produces an erectile response? and (2) if this is so, what is the relative importance of NO secreted from the endothelium versus that secreted from nitrergic nerves? Many studies have demonstrated acetylcholine and nerves of the cholinergic type in penile smooth muscle (Klinge and Sjöstrand, 1974, 1977b; Sjöstrand and Klinge, 1979; Sjöstrand *et al.*, 1993). Neurogenic release of labelled acetylcholine has been demonstrated in the human corpus cavernosum penis (Saenz de Tejada *et al.*, 1988). One action of endogenous acetylcholine in penile smooth muscle, i.e. muscarinic suppression of the excitatory adrenergic neurotransmission, is well documented with various techniques (e.g. Klinge and Sjöstrand, 1977b; Sjöstrand and Klinge, 1979; Samuelson *et al.*, 1983; Byrne and Muir, 1984; Kinekawa *et al.*, 1984; Hedlund and Andersson, 1985; Saenz de Tejada *et al.*, 1988; Sjöstrand *et al.*, 1993). In this case morphological studies have shown intimate juxtapositions of adrenergic and cholinergic nerves (Eränkö *et al.*, 1976; Klinge *et al.*, 1978). In the case of the penile endothelium no electron microscopic study has hitherto reported an intimate relationship between nerve terminals and the endothelial cells.

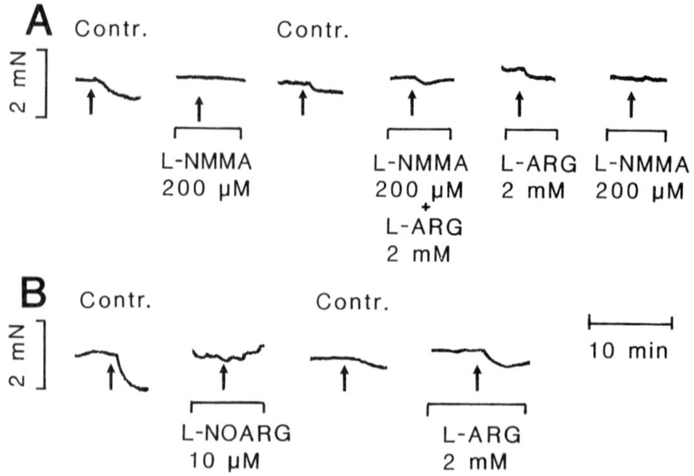

Figure 11 (A),(B) Strips from human glans penis. At arrows 11 μM acetylcholine is given at 40 min intervals. Mecamylamine and lidocaine present as in Figure 10. (A) L-NMMA blocks the muscarinic relaxation, which reappears after wash. L-Arginine counteracts the effect of L-NMMA. (B) L-N^G-nitroarginine is a more potent blocker of the muscarinic relaxation. L-Arginine restores relaxation.

Most vascular nerves are located at the border between the media and the adventitia (Burnstock, 1980). Thus, the muscular media and the internal elastica separates the intimal endothelium from the nerves (Figure 12). Acetylcholine released from the nerves will have to diffuse through the separating barriers in order to produce NO release from the endothelium. NO secreted from the endothelial cells will have to pass the same barriers in the opposite direction to reach the external layers of smooth muscle in the media, which must be relaxed in order to secure a substantial vasodilatation. It is apparent that, as exemplified in Figure 12, the conditions vary in different sections of the vascular bed depending on the thickness of the barriers, especially the media. In addition, the minimal separation between the nerves at the adventitial side of the media and the closest muscle cells decreases with decreasing diameter of the vessel (Burnstock, 1980). Consequently it can be deduced that the chances for a neurogenic muscarinic endothelium-dependent vasodilatation to occur would be greater, for example, in the precavernous arterioli and the cavernous sinusoids than, for example, in the central artery. In this context it is of particular interest that Andersson *et al.* (1984) found that, in the dog, muscarinic blockade caused no significant decrease in the penile blood flow response induced by pelvic nerve stimulation but curtailed the filling of the cavernous bodies, i.e. the erectile volume response proper. That this response can be impaired by muscarinic blockade has also been shown in other recent studies (Sjöstrand and Klinge, 1979, rabbit, hypogastric nerve stimulation; Andersson *et al.*, 1987, cat, pelvic and hypogastric nerve stimulation; Stief *et al.*, 1989a, macaque,

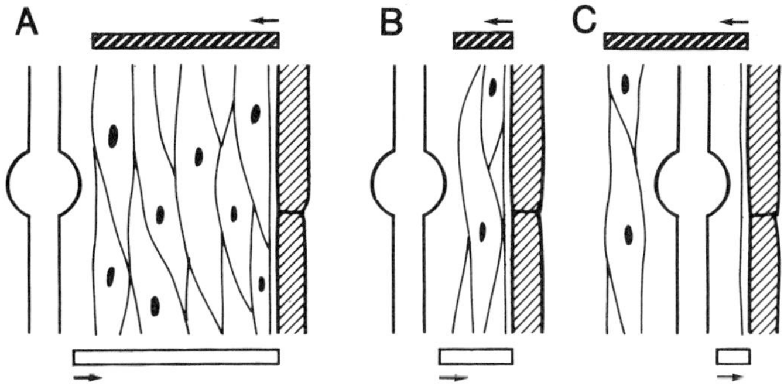

Figure 12 Schematic illustrations of the possible relation between cholinergic nerves and endothelium in penile vessels. (A) Situation in larger artery, for example the deep penile artery. Nerves are at the adventential side of the media. Several layers of smooth muscle cells and the (white) elastica interna separate the terminal from the endothelial cells. Acetylcholine will have to diffuse through the distance indicated by white bar in order to produce NO secretion from the endothelium. NO will have to diffuse in the opposite direction (striated bar) to relax the smooth muscle. In this case the chances for the occurrence of an endothelium-dependent neurogenic vasodilatation must be regarded as small. (B) Situation in arteriole or sinusoid. The distance is much smaller and there are fewer cell barriers. Consequently the prerequisites are much better for an endothelium-dependent, neurogenic vasodilatation. (C) Illustrates the hypothetical situation where the nerve terminal is located between the muscle cells and the endothelium.

cavernous nerve stimulation and cavernous pressure recording; Stief *et al.*, 1989b dog, pelvic nerve stimulation, cavernous pressure recording). Trigo-Rocha *et al.* (1993b) further found that in dog chemical destruction of the sinusoidal endothelium reduced the cavernous pressure response to pelvic nerve stimulation by 25%. Of the many recent studies of isolated cavernous bodies only one presents evidence of a muscarinic component in the relaxation due to activation of intramural nerves where an action on adrenergic nerves was excluded: Saenz de Tejada *et al.* (1988) found that in bretylium-treated strips from humans physostigmine increased the duration of neurogenic relaxations while atropine counteracted physostigmine and reduced the duration to below control values.

In conclusion, compiled evidence from the literature suggests that part of the neurogenic erectile response may be due to muscarinic stimulation of endothelial cells. Via NO release from these cells, cholinergic nerves may act as indirect inhibitors of penile smooth muscle. NO release from the endothelium seems to be supplementary to NO release from nitrergic nerves, which seems to be obligatory for a full erectile response. A better understanding of the role of endothelial NO in penile erection will be obtained when a compound blocking NO release from nerves, but leaving neuronal release of acetylcholine and endothelial release of NO unimpaired, is developed.

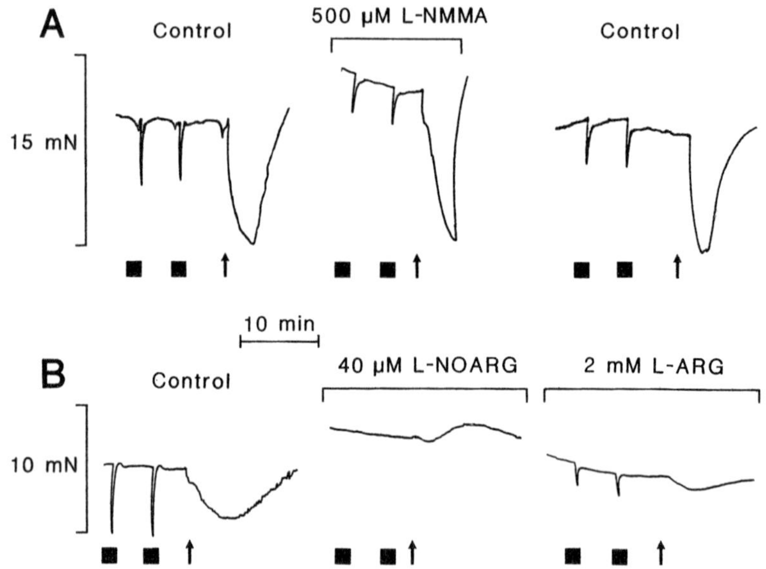

Figure 13 Effect of NOS inhibitors on relaxation of strips of human corpus cavernosum penis induced by field stimulation (4 Hz, 0.2 ms, 12 V, 10 s) at squares or nicotine (10 μM) at arrows. (A) Even a very high concentration of L-NMMA has little blocking action on the response to field stimulation or nicotine. (B) In another strip L-N^G-nitroarginine abolishes both types of relaxation. After wash and incubation with L-arginine weak relaxations reappear to both kinds of stimuli. Ninety minutes between each panel. Precontraction with adrenaline (1.5 μM).

13.6.2 Nicotinic relaxation

As mentioned above, acetylcholine can relax penile smooth muscle via activation of nicotinic receptors (Klinge and Sjöstrand, 1974, 1977b; Adaikan *et al.*, 1983; Sjöstrand *et al.*, 1988). The nicotinic relaxation has been analysed in more detail in the bovine retractor penis muscle where it is very consistent and reproducible (Klinge *et al.*, 1988) and can be used for assessment of ganglionic blocking activity of drugs (Alaranta *et al.*, 1990; Klinge *et al.*, 1993). The relaxation is very similar to that produced by stimulation of NANC nerves. Thus it is blocked by hypoxia, and guanylyl cyclase inhibitors (Klinge *et al.*, 1988). Furthermore, it is blocked by L-NOARG while L-NMMA is ineffective (Kostianen *et al.*, 1991). Similar results are obtained in the human corpus cavernosum penis (Figure 13). The nicotinic relaxation is very sensitive to lidocaine but is partially resistant to tetrodotoxin (Klinge *et al.*, 1988; Figure 14B). This is puzzling because the NANC relaxation is very sensitive to tetrodotoxin but less sensitive to lidocaine. It is, however, most likely that the relaxation is due to excitation of nicotinic receptors on nitrergic nerves and release of NO from their terminals, if appropriate concentrations of nicotinic

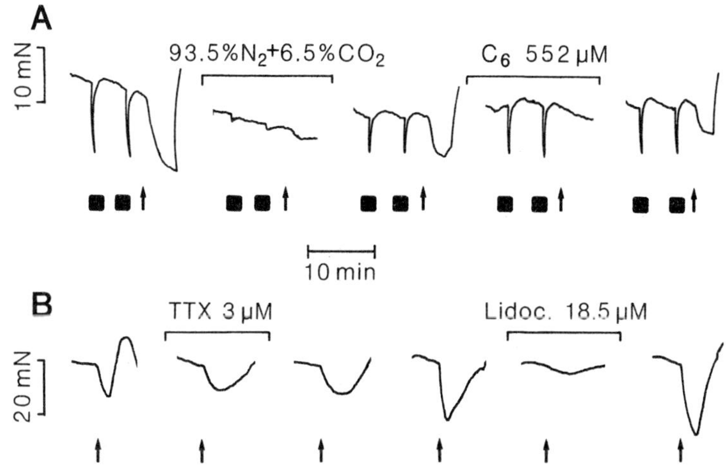

Figure 14 Nicotinic relaxations of strips of the human corpus cavernosum penis precontracted by adrenaline (2.5 μM). (A) At squares, electric field stimulation (8 Hz, 0.2 ms, 15 V, 10 s) at arrow, nicotine (20 μM). First panel: control. Second panel: hypoxia reduces field stimulation and effect of nicotine. Third panel: control. Fourth panel: a high concentration of hexamethonium has no effect on the relaxation to field stimulation but eliminates the response to nicotine. Fifth panel: after washing the nicotinic relaxation reappears. Interval between nicotinic relaxations 60 min. (B) In another preparation nicotine (20 μM) is added at 60 min intervals. First control the rapid relaxation is followed by rebound contraction (cf. Figure 8). Tetrodotoxin in high concentration has little effect on the nicotinic relaxation while it is essentially eliminated by a low concentration of lidocaine.

agonists are used. Since small ganglia are present in penile smooth muscle (see Alaranta *et al.*, 1989; Sjöstrand *et al.*, 1993) it is likely that exogenous nicotinic agonists act on the nerve cell bodies in a similar way as acetylcholine released from preganglionic fibres. However, some additional actions on axons and terminals cannot be excluded. Figure 15 gives schematic illustrations of possible muscarinic activation (B) leading to NO release and nicotinic activation (C) as well as direct release from nitrergic nerves (A).

13.7 Final considerations

13.7.1 The solution of the problem

This chapter has illustrated the stepwise elucidation of an old problem. Many details still remain obscure, for example the mechanism behind and the significance of the inhibitory junction potentials in the retractor penis muscle (Samuelson *et al.*, 1983;

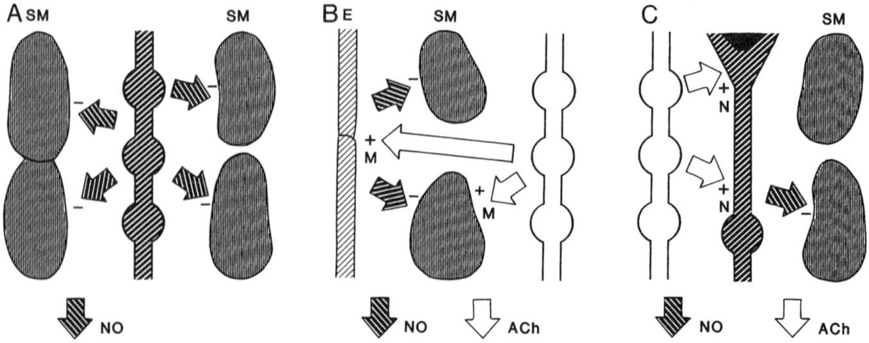

Figure 15 Schematic illustration of various types of nerve mediated relaxation. (A) Direct nitrergic relaxation. (B) Possible cholinergic muscarinic indirect endothelium-dependent relaxation. (C) Indirect nicotinic relaxation. SM, smooth muscle; E, endothelium; M, muscarinic effect; N, nicotinic effect. Cholinergic fibre white. Nitrergic fibre striated.

Byrne and Muir, 1984), which seem not to be related to guanylyl cyclase activation (Bowman and Drummond, 1985) but nevertheless may be due to action of NO (Sneddon and Graham, 1992). Yet there is no doubt about the fact that the discovery of the NOS pathway brought the problem to a new state of clarity. In many ways the solution was classical. The first step was the definition of the problem and the NANC neurotransmission. Essential prerequisites for the development of this concept were the new autonomic drugs as the adrenergic neuron blocking agents and the new techniques for electric field stimulation of intramural nerves in isolated smooth muscles that were introduced around 1960. In the second stage, developments in related fields were crucial. The problems of these fields were also old, e.g. the mechanism of action of nitrovasodilators and the explanation of muscarinic vasodilatation versus muscarinic vasoconstriction. Finally, the introduction of specific inhibitors based on the classical analogue principle rapidly convinced scientists who were engaged in the problem.

13.7.2 Nitric oxide is possibly not the only transmitter in the erectile fibres

As mentioned in Section 13.5.2 it is possible that at least some NOS-positive fibres also may be acetylcholine esterase positive and/or VIP immunoreactive (Keast, 1992). Possibly, NO may be a co-transmitter in cholinergic nerves. Previously, evidence has been accumulated indicating that ATP (Burnstock, 1986) and VIP (Lundberg and Hökfelt, 1986) could be co-transmitters in such nerves. Therefore, the possibility exists that some erectile fibres may have two or more, possibly four, transmitters: acetylcholine and ATP preferentially stored in small vesicles, VIP stored in large vesicles and NO that is not stored in vesicles but formed when the impulse reaches the terminal. According to presented evidence NO is the principal transmitter of erection; acetylcholine may participate as an indirect transmitter via stimulation of endothelial cells and suppression of adrenergic fibres; VIP may participate as a direct (and indirect via endothelial release of NO) transmitter but is hardly of major importance (Section 13.4.3); ATP may have similar actions as VIP but it is unlikely that ATP is of major importance for penile erection (Klinge and Sjöstrand, 1974, 1977a).

The relative importance of each individual transmitter may depend on the spatial relationship between the nerve terminals secreting the transmitter and the receptors for the transmitter on the target cells (Hammarström and Sjöstrand, 1984a,b). A large and rather unstable molecule such as ATP most likely requires a very narrow gap between the site of release and the site of action, while, for example, NO could be active even if the gap is wide. Also a temporal factor such as the impulse frequency is likely to be important not only by the local transmitter concentration gradients it creates but also by changes in the relative proportion of different transmitters. Thus 'classical' transmitters such as acetylcholine usually have a secretory maximum (amount per impulse) around 15 Hz while peptides such as VIP are supposed to have maxima at higher frequencies. According to data presented in Section 13.4.2, NO secretion (NANC secretion) – at least in the retractor penis muscle – seems not to be very frequency dependent. Thus, the relative proportion of the transmitters released per impulse may vary with the firing frequency in a multisecretory nerve. The combination of spatial and temporal factors may explain why Sjöstrand and Klinge (1979) found a muscarinic component in the erectile response to hypogastric nerve stimulation but not in the corresponding response to pelvic nerve stimulation in the rabbit. The hypogastric response had its maximum at 16 Hz, the pelvic at 2 Hz. If the hypogastric nerve innervated fewer postganglionic neurons than the pelvic nerve, then one would have to compensate for the paucity of fibres by adopting a higher stimulation frequency. However, by doing this one might also reach a frequency where the ratio of acetylcholine/NO secreted per pulse would be greater. In this context it should, however, be kept in mind that we do not know the physiological firing pattern in the erectile nerves and, in experimental physiology, a similar result is likely to occur when a nerve with presynaptic fibres to mixed efferent neurons (e.g. nitrergic and cholinergic) is stimulated.

13.7.3 Clinical implications

At present a few clinical trials with NO donors used as substitute for papaverine–phentolamine or prostaglandin E_1 in the treatment of impotence have been published. Stief *et al.* (1992) found that linsidomine injected intracavernosally could be an attractive alternative to papaverine–phentolamine injection due to decreased risk for prolonged erection and low therapy cost. However, Porst (1993) considered linsidomine not to be a genuine alternative to prostaglandin E_1. Linsidomine seems to have few side effects in contrast to sodium nitroprusside with which severe hypotension and only mild tumescence have been reported (Brock *et al.*, 1993).

Nitroglycerin given topically may be of some benefit to impotent men (Heaton *et al.*, 1990; Meyhoff *et al.*, 1992) but the reported effect is not impressive. Thus, for the time being, our new knowledge of the neurotransmission of erection has not resulted in a real therapeutic breakthrough. However, it should be recalled that neurotransmitters such as acetylcholine and noradrenaline have limited use in current therapy, accordingly, it may be naive to expect that the transmitter NO itself would be of great clinical use. But in analogy with noradrenaline and acetylcholine, it is likely that the increased knowledge of nitrergic transmitter mechanisms will lead to the future development of valuable drugs.

13.7.4 Concluding remarks

The simple molecule NO and the fact that it is not stored in synaptic vesicles as well as other differences from the 'classical' neurotransmitters seem to have created certain obstacles for the acceptance of NO as a neurotransmitter. However, in the historical perspective of the development of the concept of chemical neurotransmission, there is nothing surprising about the fact that such a molecule turned out to be a neurotransmitter. When, in 1877, the great 19th century physiologist Emil Du Bois-Reymond discussed the possibility of chemical transmission between nerve and muscle his first example of a possible constituent of the 'reizende Secretion' was ammonia, which is not very far away from NO. Finally, there is something attractive in the idea that a basic physiological act, that is crucial for the survival of the species, is based on a very simple biological molecule. The complexity in the neural regulation of penile erection seems not to lie in its main mediator NO, or its main target, the guanylyl cyclase or the second messenger cGMP, or its action on the contractile process, but on other matters. For example, there are two sites for NO production: nerves and endothelium, and two sets of erectile fibres: parasympathetic and sympathetic, i.e. there are alternative pathways for activating the basic mechanism.

Acknowledgements

The authors gratefully acknowledge the financial support of the Swedish Medical Research Council project 14X-07918 and the Finnish Cultural Foundation. We wish to thank Ms Annika Rosén for producing the figures and Ms Ulla Lindren for preparing the manuscript. Dr J. Eldh at the Department of Plastic Surgery, Karolinska Hospital, kindly supplied us with human material (transsexual operations).

References

Adaikan, P.G., Karim, S.M.M., Kottegoda, S.R. & Ratnam, S.S. (1983) *J. Autonom. Pharmacol.* **3**, 107–111.
Adaikan, P.G., Karim, S.M.M. & Ratnam, S.S. (1986) *J. Urol.* **135**, 636–640.
Adaikan, P.G., Lau, L.C., Ng, S.C. & Ratnam, S.S. (1991) *Asian Pacific J. Pharmacol.* **6**, 213–217.
Alaranta, S., Uusitalo, H., Klinge, E., Palkama, A. & Sjöstrand, N.O. (1989) *Neuroscience* **32**, 823–827.
Alaranta, S., Klinge, E., Pätsi, T. & Sjöstrand, N.O. (1990) *Br. J. Pharmacol.* **101**, 472–476.
Alm, P., Larsson, B., Edblad, E., Sundler, F. & Andersson, K.-E. (1993) *Acta Physiol. Scand.* **148**, 421–429.
Ambache, N., Killick, S.W. & Aboo Zar, M. (1975) *Br. J. Pharmacol.* **54**, 409–410.
Andersson, K.-E. (1993) *Pharmacol. Rev.* **45**, 253–308.
Andersson, K.-E. & Holmquist, F. (1990) *Int. J. Impotence Res.* **2**, 209–225.
Andersson, K.-E., Hedlund, H., Mattiasson, A., Sjögren, C. & Sundler, F. (1983) *World J. Urol.* **1**, 203–208.
Andersson, P.-O., Bloom, S.R. & Mellander, S. (1984) *J. Physiol. (Lond.)* **350**, 209–224.
Andersson, P.-O., Björnberg, J., Bloom, S.R. & Mellander, S. (1987) *J. Urol.* **138**, 419–422.
Arnold, W.P., Mittal, C.K., Katsuki, S. & Murad, R. (1977) *Proc. Natl. Acad. Sci. USA* **74**, 3203–3207.
Azadzoi, K.M., Kim, N., Goldstein, J., Cohen, R.A. & Saenz de Tejada, I. (1992) *J. Urol.* **147**, 220–225.
Beckett, S.D. (1983) In *Handbook of Physiology, Vol. III, Section II. The Cardiovascular System, Part 1* (eds Shepherd, J.T. & Abboud, F.M.) American Physiological Society, Bethesda, pp. 271–283.
Benson, G.S., McConnell, J., Lipshulz, L.J. & Corriere, J.N. (1980) *Clin. Invest.* **65**, 506–513.
Bowman, A. & Drummond, A.H. (1984) *Br. J. Pharmacol.* **81**, 665–674.
Bowman, A. & Drummond, A.H. (1985). In *Trends in Autonomic Pharmacology* (ed. Kalsner, S.) vol. 3. Taylor & Francis, Philadelphia, pp. 319–330.
Bowman, A. & Gillespie, J.S. (1982) *J. Physiol. (Lond.)* **328**, 11–25.
Bowman, A. & Gillespie, J.S. (1983) *J. Physiol. (Lond.)* **341**, 603–616.
Bowman, A. & McGrath, J.C. (1985) *Br. J. Pharmacol.* **85**, 869–875.
Bowman, A., Gillespie, J.S. & Martin, W. (1979) *Br. J. Pharmacol.* **67**, 327–328.
Bowman, A., Gillespie, J.S. & Martin, W. (1981) *Br. J. Pharmacol.* **72**, 365–372.
Bowman, A., Gillespie, J.S. & Pollock, D. (1982) *Eur. J. Pharmacol.* **85**, 221–224.
Bredt, D.S. & Snyder, S.H. (1990) *Proc. Natl. Acad. Sci. USA* **87**, 682–685.
Bredt, D.S., Hwang, P.M. & Snyder, S.H. (1990) *Nature* **347**, 768–770.
Brock, G., Breza, J. & Lue, T.F. (1993) *J. Urol.* **150**, 864–867.
Burnett, A.L., Lowenstein, C.J., Bredt, D.S., Chang, T.S.K. & Snyder, S.H. (1992) *Science* **257**, 401–403.
Burnett, A.L., Tillman, S.L., Chang, T.S.K., Epstein, J.I., Lowenstein, C.J., Bredt, D.S., Snyder, S.H. & Walsh, P.C. (1993) *J. Urol.* **150**, 73–76.
Burnstock, G. (1980) In *Handbook of Physiology vol. II, Section II. The Cardiovascular System* (eds Bohr, D.F., Somloy, A.P., Sparks, Jr. H.V. & Geiger, S.R.) American Physiological Society, Bethesda, pp. 567–612.
Burnstock, G. (1986) In *Progress in Brain Research* (eds Hökfelt, T., Fuxe, K. & Pernow, B.) vol. 68, Elsevier, Amsterdam, pp. 193–203.
Bush, P.A., Gonzales, N.E. & Ignarro, L.J. (1992a) *Biochem. Biophys. Res. Commun.* **186**, 308–314.
Bush, P.A., Aronson, W.J., Buga, G.M., Rajfer, J. & Ignarro, L.J. (1992b) *J. Urol.* **147**, 1650–1655.

Byrne, N.G. & Muir, T.C. (1984) *J. Auton. Pharmacol.* **4**, 261–271.
Byrne, N.G. & Muir, T.C. (1985) *Br. J. Pharmacol.* **85**, 149–161.
Dail, W.G. (1993) In *The Autonomic Nervous System* (ed. Maggi, C.A.) vol. 3. Harwood Academic Publishers, London, pp. 69–102.
Dail, W.G., Galloway, B., Bordegarary, J. & Walton, G. (1992) *Soc. Neurosci. Abstract.* **18**, 128.
Dail, W.G., Galloway, B. & Bordegaray, J. (1993) *Neurosci. Lett.* **160**, 17–20.
Dawson, T.M., Bredt, D.S., Fotuhi, M., Hwang, P.M. & Snyder, S.H. (1991) *Proc. Natl. Acad. Sci. USA* **88**, 7797–7801.
De Groat, W.C. & Booth, A.M. (1993) In *The Autonomic Nervous System* (ed. Maggi C.A.) vol. 3. Harwood Academic Publishers, London, pp. 467–523.
De Groat, W.C. and Steers, W.D. (1988) In *Contemporary Management of Impotence and Infertility* (eds Tanagho, E.A., Lue, T.F. & McClure, R.D.) Williams & Wilkins, Baltimore, pp. 3–27.
Du Bois-Reymond, E. (1877) *Ges. Abhandl.z.Muskel-u. Nervenphysik. II. Bd. Veit,* Leipzig p. 700.
Eckhard, C. (1863) *Beitr.z.Anat.u.Physiol. von C. Eckhard III Bd.* Giessen, pp. 123–162.
Eckhard, C. (1876) *Beitr.z.Anat.u.Physiol. von C. Eckhard, VII Bd.* Giessen, pp. 68–80.
Eränkö, O., Klinge, E. & Sjöstrand, N.O. (1976) *Experiencia* **32**, 1135–1137.
Finberg, J.P.M., Levy, S. & Vardi, Y. (1993) *Br. J. Pharmacol.* **108**, 1038–1042.
François-Franck, M.(C.E.) (1895) *Arch. Physiol. norm. Path. (Paris)* **7**, 122–153.
Fujimoto, S. & Takeshige, Y. (1975) *Anat. Rec.* **181**, 641–658.
Furchgott, R.F. (1984) *Ann. Rev. Pharmacol. Toxicol.* **24:** 175–197.
Furchgott, R.F. (1988) In *Vascular Smooth Muscle, Peptides, Autonomic Nerves and Endothelium* (eds Vanhoutte, P.M. & Lausen, J.) Raven Press, New York, pp. 401–414 (Vasodilatation vol. 4).
Furchgott, R.F. (1990) *Acta Physiol. Scand.* **139**, 257–270.
Furchgott, R.F. & Zawadski, J.V. (1980) *Nature* **288**, 373–376.
Gillespie, J.S. (1972) *Br. J. Pharmacol.* **45**, 404–416.
Gillespie, J.S. & Martin, W. (1980) *J. Physiol. (Lond.)* **309**, 55–64.
Gillespie, J.S. & Xiaorong, L. (1989) *Br. J. Pharmacol.* **97**, 453P.
Gillespie, J.S., Hunter, J.C. & Martin, W. (1981) *J. Physiol. (Lond.)* **315**, 111–125.
Gillespie, J.S., Liu, X. & Martin, W. (1989) *Br. J. Pharmacol.* **98**, 1080–1082.
Gillespie, J.S., Liu, X. & Martin, W. (1990) In *Nitric Oxide from L-Arginine: A Bioregulatory System.* (eds Moncada, S. & Higgs, E.A.) Excerpta Medica, Amsterdam, pp. 147–164.
Gruetter, C.A., Barry, B.K., McNamara, D.B., Gruetter, D.Y., Kadowitz, P.J. & Ignarro, L.J. (1979) *J. Cyclic. Nucleotide. Res.* **5**, 211–224.
Gu, J., Polak, J.M., Probert, L., Islam, K.N., Marangos, P.J., Mina, S., Adrian, T.E., McGregor, G.P., O'Shaughnessy, D.J. & Bloom, S.R. (1983) *J. Urol.* **130**, 386–391.
Hammarström, M. & Sjöstrand, N.O. (1984a) *Acta Physiol. Scand.* **122**, 465–474.
Hammarström, M. & Sjöstrand, N.O. (1984b) *Acta Physiol. Scand.* **122**, 475–481.
Hanyu, S., Iwanaga, T., Tamaki, M., Kano, K. & Sato, S. (1992) *Urol. Int.* **48**, 58–63.
Heaton, J.P.W., Morales, A., Owen, J., Saunders, F.W. & Fenemore, J. (1990) *J. Urol.* **143**, 729–731.
Hedlund, H. & Andersson, K.-E. (1985) *J. Auton. Pharmacol.* **5**, 81–88.
Hedlund, H., Andersson, K.-E. & Mattiasson, A. (1984) *J. Auton. Pharmacol.* **4**, 241–249.
Holmquist, F., Hedlund, H. & Andersson, K.-E. (1991a) *Acta Physiol. Scand.* **141**, 441–442.
Holmquist, F., Stief, C.G., Jonas, U. & Andersson, K.-E. (1991b) *Acta Physiol. Scand.* **143**, 299–304.
Holmquist, F., Hedlund, H. & Andersson, K.-E. (1992) *J. Physiol. (Lond.)* **449**, 295–311.
Holmquist, F., Fridstrand, M., Hedlund, H. & Andersson, K.-E. (1993) *J. Urol.* **150**, 1310–1315.
Hope, B.T., Michael, G.J., Knigge, K.M. & Vincent, S.R. (1991) *Proc. Natl. Acad. Sci. USA* **88**, 2811–2814.
Ignarro, L.J., Buga, G.M., Wood, K.S., Byrns, R.E. & Chaudhuri, G. (1987) *Proc. Natl. Acad. Sci. USA* **84**, 9265–9269.

Ignarro, L.J., Bush, P.A., Buga, G.M., Wood, K.S., Fukoto, J.M. & Rajfer, J. (1990) *Biochem. Biophys. Res. Commun.* **170**, 843–850.
Keast, J.R. (1992) *Neurosci. Lett.* **143**, 69–73.
Kerr, S.W., Buchanan, L.V., Bunting, S. & Mathews, W.R. (1992) *J. Pharmacol. Exp. Ther.* **263**, 285–292.
Kim, N.N., Goldstein, I., Krane, R.J. & Saenz de Tejada, I. (1990) *J. Urol.* **143**, 244A.
Kim, N., Azadzoi, K.M., Goldstein, I. & Saenz de Tejada, I. (1991) *J. Clin. Invest.* **88**, 112–118.
Kim, N., Vardi, Y., Padma-Nathan, H., Daley, J., Goldstein, I. & Saenz de Tejada, I. (1993) *J. Clin. Invest.* **91**, 437–442.
Kinekawa, F., Komori, S. & Ohashi, H. (1984) *Jpn. J. Pharmacol.* **34**, 343–352.
Kirkeby, H.J., Svane, D., Poulsen, J., Tøttrup, A., Forman, A. & Andersson, K.-E. (1993) *Acta Physiol. Scand.* **149**, 385–392.
Klinge, E. & Sjöstrand, N.O. (1974) *Acta Physiol. Scand.* (Suppl 420) **93**, 1–88.
Klinge, E. and Sjöstrand, N.O. (1977a) *Acta Physiol. Scand.* **100**, 354–367.
Klinge, E. & Sjöstrand, N.O. (1977b) *Acta Physiol. Scand.* **100**, 368–376.
Klinge, E., Eränkö, O. & Sjöstrand, N.O. (1978) *Experientia* **34**, 1624–1626.
Klinge, E., Alaranta, S. & Sjöstrand, N.O. (1988) *J. Pharmacol. Exp. Ther.* **245**, 280–286.
Klinge, E., Alaranta, S., Parkkisenniemi, U.M., Kostiainen, E. & Sjöstrand, N.O. (1993) *J. Pharmacol. Toxicol. Meth.* **30**, 197–202.
Knispel, H.H., Goessl, C. & Beckmann, R. (1991) *J. Urol.* **146**, 1429–1433.
Knispel, H.H., Goessl, C. & Beckmann, R. (1992a) *Invest. Urol.* **40**, 471–476.
Knispel, H.H., Goessl, C. & Beckmann, R. (1992b) *Urol. Res.* **20**, 253–257.
Kostianen, E., Klinge, E., Alaranta, S. & Sjöstrand, N.O. (1991) *Br. J. Pharmacol.* **104**, 396P.
Langley, J.N. & Anderson, H.K. (1895) *J. Physiol. (Lond.)* **19**, 85–139.
Langley, J.N. & Anderson, H.K. (1896) *J. Physiol. (Lond.)* **20**, 372–406.
Larsen, J.-J., Ottesen, B., Fahrenkrug, J. & Fahrenkrug, L. (1981) *Invest. Urol.* **19**, 211–213.
Leone, A.M., Wiklund, N.P., Hokfekt, T., Brundin, L. & Moncada, S. (1994) *Neuroreport* **6**, 733–736.
Liu, X., Gillespie, J.S., Gibson, I.F. & Martin, W. (1991) *Br. J. Pharmacol.* **104**, 53–58.
Lovèn, Ch. (1866) *Berichte ü.d. Verhandl. d. Kgl. Sächs. Gesellsch. d. Wissenschaften z. Lpz, mathem.-physische Classe.* **18**, 85–110.
Lundberg, J.M. & Hökfelt, T. (1986) In *Progress in Brain Research* (eds Hökfelt, T., Fuxe, K. & Pernow, B.) vol. 68, Elsevier, Amsterdam, pp. 241–262.
Martin, W. & Gillespie, J.S. (1991) In *Novel Peripheral Neurotransmitters* (ed. Bell, C.) Pergamon, Oxford, pp. 65–79.
Martin, W., Smith, J.A., Lewis, M.J. & Henderson, A.H. (1988) *Br. J. Pharmacol.* **93**, 579–586.
Martin, W., Gillespie, J.S. & Gibson, I.F. (1993) *Br. J. Pharmacol.* **108**, 242–247.
Mayer, B., John, M. & Böhme, E. (1990) *FEBS Lett.* **277**, 215–219.
Meyhoff, H.H., Rosenkilde, P. & Bodker, A. (1992) *Br. J. Urol.* **69**, 88–90.
Murad, F. (1986) *J. Clin. Invest.* **78**, 1–5.
Murad, F., Arnold, W.P., Mittal, C.K. & Braughler, J.M. (1979) *Adv. Cyclic Nucleotide Res.* **11**, 175–204.
Nikolsky, W. (1879) *Arch. Anat. Physiol. (Lpz)* **Jahrgnag 1879**, 209–221.
Ottesen, B., Wagner, G., Virag, R. & Fahrenkrug, J. (1984) *Br. Med. J.* **288**, 9–11.
Ottesen, B., Wagner, G. & Fahrenkrug, J. (1988) In *Handbook of Sexology, vol. 6. The Pharmacology and Endocrinology of Sexual Function.* (ed. Sitsen, J.M.A.) Elsevier, Amsterdam, pp. 66–97.
Palmer, R.M.J., Ferrige, A.G. & Moncada, S. (1987) *Nature (Lond.)* **327**, 524–526.
Pickard, R.S., Powell, P.H. & Zar, M.A. (1991) *Br. J. Pharmacol.* **104**, 755–759.
Pickard, R.S., Powell, P.H. & Zar, M.A. (1993) *Br. J. Pharmacol.* **108**, 497–500.
Piotrowski, G. (1894) *Pflügers Arch.* **55**, 240–302.
Porst, H. (1993) *J. Urol.* **149**, 1280–1283.

Rajfer, J., Aronson, W.J., Bush, P.A., Dorey, F.J. & Ignarro, L.J. (1992) *N. Engl. J. Med.* **326**, 90–94.
Rapoport, R.M. & Murad, F. (1983) *J. Cyclic Nucleotide Res.* **9**, 281–296.
Rees, D.D., Palmer, R.M.J., Hodson, H.F. & Moncada, S. (1989) *Br. J. Pharmacol.* **96**, 418–424.
Saenz de Tejada, I., Blanco, R., Goldstein, I., Azadzoi, K., de las Morenäs, A., Krane, R.J. & Cohen, R.A. (1988) *Am. J. Physiol.* **254**, H459–H467.
Samuelson, U., Sjöstrand, N.O. & Klinge, E. (1983) *Acta Physiol. Scand.* **119**, 335–345.
Sanders, K.M. & Ward, W.M. (1992) *Am. J. Physiol.* **262**, G379–G392.
Sheng, H., Schmidt, H.H.H.W., Nakane, M., Mitchell, J.A., Pollock, J.S. Förstermann, U. & Murad, F. (1992) *Br. J. Pharmacol.* **106**, 768–773.
Sjöstrand, N.O. & Klinge, E. (1979) *Acta Physiol. Scand.* **106**, 199–214.
Sjöstrand, N.O., Klinge, E. & Himberg, J.-J. (1981) *Acta Physiol. Scand.* **113**, 403–405.
Sjöstrand, N.O., Eldh, J., Alaranta, S. & Klinge, E. (1988) *Acta Physiol. Scand.* **132**, 19A.
Sjöstrand, N.O., Eldh, J., Samuelson, U.E., Alaranta, S. & Klinge, E. (1990) *Acta Physiol. Scand.* **140**, 297–298.
Sjöstrand, N.O., Beckett, S.D. & Klinge, E. (1993) *Acta Physiol. Scand.* **147**, 403–415.
Sneddon, P. & Graham, A. (1992) *J. Autonom. Pharmacol.* **12**, 445–456.
Stief, C., Diederichs, W., Benard, F., Bosch, R., Aboseif, S. & Lue, T.F. (1989a) *Urol. Int.* **44**, 357–363.
Stief, C., Benard, F., Bosch, R., Aboseif, S., Nunes, L., Lue, T.F. & Tanagho, E.A. (1989b) *J. Urol.* **141**, 1444–1448.
Stief, C.G., Holmquist, F., Djamilian, M., Krah, H., Andersson, K.-E. & Jonas, U. (1992) *J. Urol.* **148**, 1437–1440.
Traish, A.M., Carson, M.P., Kim, N., Goldstein, I. & Saenz de Tejada, I. (1990) *J. Urol.* **144**, 1036–1040.
Trigo-Rocha, F., Hsu, G.L., Donatucci, C.F. & Lue, T.F. (1993a) *J. Urol.* **149**, 872–877.
Trigo-Rocha, F., Aronson, W.J., Hohenfellner, M., Ignarro, L.J., Rajfer, J. & Lue, T.F. (1993b) *Am. J. Physiol.* **264**, H419–H422.
von Anrep, B. & Cybulski, N. (1884) *St. Petersb. Med. Wschr.* **20**, 215–221.
Vardi, Y. & Siroky, M.B. (1993) *J. Urol.* **149**, 910–914.
Vincent, S.R. (1986) In *Neurohistochemistry: Modern Methods and Applications* (eds Panula, P., Päivräinta, H. & Soninila, S.) Liss, New York, pp. 375–396.
Virag, R., Ottesen, B., Fahrenkrug, J., Levy, C. & Wagner, G. (1982) *Lancet* **ii**, 1166.
Vizzard, M.A., Erdman, S.L. & de Groat, W.C. (1993a) *Neurosci. Lett.* **152**, 77–80.
Vizzard, M.A., Ermand, S.L. & de Groat, W.C. (1993b) *J. Auton. Nerv. Syst.* **44**, 85–90.
von Kölliker, A. (1852) *Verh. phys.-med. Ges. Würzburg.* **2**, 118–133.
Wagner, G. & Sjöstrand, N.O. (1988) In *Handbook of Sexology, vol. 6. The Pharmacology and Endocrinology of Sexual Function* (ed. Sitsen, J.M.A.) Elsevier, Amsterdam, pp. 32–43.
Waldman, S.A. & Murad, F. (1987) *Pharmacol. Rev.* **39**, 163–196.
Wang, R., Domer, F.R., Sikka, S.C., Kadowitz, P.J. & Hellstrom, W.G.J. (1994) *J. Urol.* **151**, 234–237.
Willis, E., Ottesen, B., Wagner, G. & Fahrenkrug, J. (1981) *Acta Physiol. Scand.* **113**, 545–547.
Willis, E.A., Ottesen, B., Wagner, G., Sundler, F. & Fahrenkrug, J. (1983) *Life Sci.* **33**, 383–391.
Yui, Y., Ohkawa, S., Ohnishi, K., Hattori, R., Aoyama, T., Takanishi, M., Morishita, H., Terao, Y. & Kawai, Ca. (1989) *Biochem. Biophys. Res. Commun.* **164**, 544–549.

Index

Acetylcholine
  and cerebral artery relaxation, 210
  nitrergic transmission interaction, 268
  in penile smooth muscle, 294–297
α_2-Adrenoceptor agonists, nitrergic transmission modulation, 264
Adrenocorticotropin, in hypothalamus, 152–154
Adult respiratory distress syndrome (ARDS), 194
Ageing
  and EDRF-mediated arterial relaxation, 220
  and vasodilator nerve stimulation, 213
Alcohols, and nitrergic transmission, 242
Alzheimer's disease, 184
  brain neurons in, 91
Amiloride, cGMP ion channel inhibition, 67
Amnesia, and hippocampal damage, 126
AMP, *see* Cyclic AMP
AMPA receptors
  in cerebellum, 107–109
  in hippocampus, 115–116
Amygdala, NADPH diaphorase in, 88
Amyl nitrite, 238, 242
Amyotrophic lateral sclerosis (ALS), *see* Motor neuron disease
Angina pectoris, and amyl nitrite, 238
Angiotensin II, 156–157
Anococcygeus muscle
  nitrergic innervation, 245, 246, 247
  NO-mediated transmission, 228
Arginine
  and corticotropin-releasing hormone release, 153
  extracellular accumulation, and NOS, 105
  N^G derivatives, as NOS inhibitor, 35–37
  and nitrate generation, 2
  and nitrergic transmission, 261–262
  in NO formation, 22
  in NOS activity regulation, 31–32
  oxidation in NO synthesis, 27–28
  structure, 231
  see also *N*^G-Methyl-L-arginine
Argininosuccinate (AS), in cerebellum, 115
Argininosuccinate synthetase in hippocampus, 117
Arteries, *see* Blood vessels; Cerebral arterial tone
Aspartate, as EAA receptor ligand, in cerebellum, 110, 111
Assays for NO, 22–34, 236
Astrocytes
  function, 164
  NOS in, 106–107, 165–166
    in vitro induction, 168–169
Atherosclerosis, NO oxidants in, 201
ATP
  in caecum, 258
  as co-transmitter, 266
    in penile erection, 300–301
  in colon, 258
  and inhibitory junction potentials, 251
  in stomach, 256
Autoimmune disease, and NOS induction, 179–180
Automodulation of nitrergic transmission, 264
Axotomy
  and NOS expression, 15
  and NOS upregulation, 34

2-(2'-Benzothiazolyl-5-styryl-3 (4'-phthalhydrazidyl) tetrazolium chloride (BSPT), in NADPH diaphorase detection, 85–86
Biopterins, redox cycling, in NO synthesis, 29–31
  see also Dihydrobiopterin, in NO synthesis; Tetrahydrobiopterin
Blood vessels
  cGMP kinase in, 59
  nitrergic innervation, 249
  NOS in, 10
  penile
    anatomy, 282–284
    in erection, 284, 285
  see also Cardiovascular system, nitrergic innervation; Cerebral arterial tone
Borna disease, and NOS induction, 177, 179

Brain
  cGMP in, 76
  NOS in, 11
    functions, 10–16, 12–15
    localization, 9
    molecular cloning, 3–5
    purification, 23–24
  see also Central nervous system, NO neurons in; Spinal cord
  see also particular brain areas
Brainstem, NOS in, 91

Caecum, nitrergic transmission in, 258
Calcineurin, and NOS phosphorylation, 33
Calcium
  intracellular, and cyclic GMP, 183–184
  and nitrergic transmission, 262
  and NOS, 8
  and vasodilator nerve stimulation, 211–212
Calcium currents, neuronal, and cGMP, 77
Calmodulin
  NO interaction, gene region, 3–4
  NOS binding, 25
    and calcium, 8
Calmodulin inhibitors, NOS inhibition, 234–235
Calphostin C, and NOS induction, 171
Carbon monoxide
  in LTP, 130, 131
  in soluble guanylyl cyclase regulation, 48
Cardiovascular system, nitrergic innervation, 249
  see also Blood vessels; Heart
Catabolite activator protein (CAP), cGMP kinase homology, 55
Cell death, and NOS, 15
Central nervous system, NO neurons in, 84–102
  during development, 94
  immunohistochemical/hybridization studies, 84
  and injury, 94–95
  NADPH diaphorase histochemistry, 84–86
  NO receptor localization, 95
  and NO transmission, 96
  non-mammalian species, 93–94
  NOS localization, 86–93
    cerebellum, 92
    diencephalon, 89–90
    midbrain/pons/medulla, 90–92
    spinal cord, 92–93
    telencephalon, 86–89
Central nervous system, NO neurons in, (*continued*)
  see also Brain; Spinal cord
  see also particular brain area
Cerebellum
  cGMP kinase in, 76
    substrates, 76–77
  NOS in, 92
  signal transduction, 107–115
    citrulline recycling, 115
    EAA receptor populations, 107–110
    endogenous EAA receptor ligands, 110–112
    long-term depression, 115, 144–147
    neurotoxicity, 115
    noradrenergic afferents, 114–115
    NOS/cGMP compartments, 112–114
Cerebral ischaemia
  and NMDA receptor, 192
  NO synthesis inhibition in, 200
  sound analogy, 196
  superoxide dismutase and, 200–201
  superoxide in, 197–198
  NO interaction in, 199–200
Cerebral vascular tone, 207–225
  and endothelium-derived NO, 217–222
    and ageing, 220
    basal release, 220, 222
    EDRF, 217–218
    and histamine vascular response, 220
    and NOS immunoreactivity, 218
    physical stimuli, and blood flow, 220
    in vivo response, 222
  'nitroxidergic' nerves, 214–216
    nerve function, 216–217
  NOS-positive perivascular nerve histology, 213–214
  vasodilator nerve stimulation, 209–213
    and ageing, 213
    cyclic GMP increase, 211
    mechanical response, 209–210
    NO release, 210–211
    response to NO, 211–212
    species variation, 212
  vasodilator substances, 208–209
cGMP, *see* Cyclic GMP
Chemiluminescence assay for NO, 236
Cholinergic effects, in penile smooth muscle, 294–297
Chondrocytes, NOS in, 5
Chromatographic purification of NOS, 23–24

Citrulline
  formation during NO synthesis, 27–28
  localization, 233
  recycling
    in cerebellum, 115
    in hippocampus, 117
  structure, 231
3H-Citrulline assay for NOS, 23
Clonidine, and nitrergic transmission, 264
Colon, nitrergic transmission in, 258–259
Cone photoreceptor, type V phosphodiesterases in, 73
Cortex, NOS in, 86–88
Corticosterone, 153–154
Corticotropin-releasing hormone neurosecretory system, 152–154
Cyclic AMP
  and aldosterone secretion, 70
  cGMP kinase activation, 63–64
  cyclic nucleotide binding sites, 55, 57
  inhibition by cGMP, 71
  olfactory epithelium ion channel regulation, 68
Cyclic AMP kinase, 61
Cyclic GMP, 43–44
  in astrocytes, 168
  in cerebellum, 112–114
  from glial cells, and NO, 183
  in hippocampus, and NMDA receptors, 117
  and intracellular calcium, 183–184
  localization, 243
  and vasodilator nerve stimulation, 211
Cyclic GMP receptor proteins, 51–81
  ion channel regulation, 52, 64–69
    olfactory epithelium, 68–69
    renal, 67
    rod photoreceptor, 64–67
    spermatozoa/testis, 67–68
  kinases, 51–64
    activity regulation, 58
    biologic roles, 62–63
    cAMP cross-activation, 63–64
    distribution, 58–59
    and long-term depression, 145
    structure, 52–57
    substrates, 59–62
    types, 52
  in nervous system, 74–78
    long-term potentiation/depression, 75–76
    neuronal ion fluxes, 77–78
    in neurons, 76–77
Cyclic GMP receptor proteins (*continued*)
  phosphodiesterases, 52, 69–74
    classes, 69
    inhibitors, 244–245
    properties, 73–74
    type II, 70–71
Cyclo-oxygenase, activity, and NO, 184
D-Cycloserine, and long-term depression in hippocampus, 117
Cystamine, 244
Cysteine, and relaxant action of NO, 239–240
Cytochrome P450
  NOS gene homology, 3, 4
  NOS isozymes as, 26
Cytochrome P450 enzymes, as NOS cofactors, 6–7
Cytokines
  and corticotropin-releasing hormone neurosecretory system, 153–154
  NO regulation, 184
  and NOS induction in glial cells
    astrocytes, 169
    gliomas, 169–170
    mechanisms, 170–172, 173
    microglia, 167
    Muller cells, 170
  and transcriptional NOS induction, 174

DARPP-32, phosphorylation, and cAMP/cGMP, 77
Dentate granular cells, excitatory amino acid receptors, 115–116
Dentate gyrus, NOS/NADPH diaphorase in, 88
Diencephalon, NOS in, 89–90
Diffusion of NO, 196
  into blood vessels, 197
Dihydrobiopterin, in NO synthesis, 30–31
Dimethyl L-arginine (L-NDMA), 234
Diphenyl iodonium (DPI), NOS inhibition, 234
Down's syndrome mouse model, 201
Duodenum, nitrergic transmission in, 256
Dyes, NOS inhibition, 34

Ectromelia virus infection, and NOS induction, 177
Endothelial cells
  NOS cloning, 5–6
  NOS in, 2, 3
  NOS isolation from, 24–25

Endothelium-derived hyperpolarizing factor (EDHF), 220
Endothelium-derived NO (EDNO), 218, 220
Endothelium-derived relaxing factor (EDRF), 217–218
  cerebral artery relaxation, 218, 219
Enteric nervous system, 250
Ethacrynic acid, NOS inhibition, 235
7-Ethoxyresorufin, NOS inhibition, 235
Excitatory amino acids
  in hippocampus, 115–116
  in NO synthesis in hypothalamus, 158–159
  receptor populations, 107–110
Exocytosis in nitrergic transmission, 261–263
  and L-arginine, 261–262
  and calcium, 262–263
Experimental allergic encephalitis (EAE), and NOS mRNA, 179–180

Feedback, NO on NOS activity, 176
Fetal NOS expression, 94
Fick's laws of diffusion, 196
FK506, and NOS phosphorylation, 33
Flavins
  NOS, 25–26
    catalytic domain, 26–27
  as NOS cofactors, 6
  nucleotides, NOS, binding sites, 3, 4, 5–6
Free radical scavengers, and nitrergic stimulation, 242–243

G proteins, as cGMP kinase substrates, 59–60
G-substrate, 76
GABA, and nitrergic transmission, 267
Gall bladder, nitrergic transmission in, 257
Gastrointestinal tract, nitrergic transmission in, 10, 250–259
  caecum, 258
  colon, 258–259
  duodenum, 256
  electrophysiological observations, 251
  enteric nervous system, 250
  extrinsic innervation, 250
  gall bladder, 257
  ileum, 257–258
  jejunum, 257
  oesophagus, 254–256
  sphincters, 252–254
  stomach, 256
Genistein, and NOS induction, 172
Glial cells, 163–190
  astrocytes
    function, 164
    NOS in, 165–166
  microglial lineage, 164
  as NO targets, 183–185
  NOS activity in, 106–107
  NOS expression, and injury, 94
  NOS induction in vitro, 167–172
    astrocytes, 168–169
    gliomas, 169–170
    mechanisms, 170–172, 173
    microglia, 167–168
    Muller cells, 170
  NOS induction in vivo, 176–183
    autoimmune disease and, 179–180
    ischaemia and, 180–182
    opioid tolerance and, 182–183
    spinal hyperalgesia and, 182
    toxoplasmosis and, 179
    trauma and, 180
    viral infection and, 177–179
  regulation of inducible NOS, 172, 174–176
    post-translational, 175–176
    transcription/translation, 174–175
Gliomas
  NOS in vitro induction in, 169–170
  proliferation, and NOS activity, 184
Glutamate
  in cerebellum, 110, 111
  in hippocampus, 116
  in hypothalamus, and NO production, 157
Glutamate receptors
  and long-term depression, 144–145
  quisqualate desensitization, 146
Glutamate-activated ion channels, and long-term potentiation, 127–128
Glyceryl trinitrate (GTN), 238
Glycine
  in cerebellum, 111
  in hippocampus, 116
    and long-term depression, 117
GMP, *see* Cyclic GMP
Guanylyl cyclase
  and cGMP, 43–44
  inhibitors, 243–244
  localization, 243
  and long-term depression, 145
  membrane-bound isoforms, 44
  and NO in brain, 11

Guanylyl cyclase (*continued*)
see also Soluble guanylyl cyclase

Haber-Weiss reaction, 197–198
Haem
catalytic domain of NOS, 26–27
cytochrome P450/NOS binding, 7
inhibitors, NOS inhibition, 235
NO binding, 195
NOS as, 26
in soluble guanylyl cyclase regulation, 48–49
see also Iron, NO binding
Haemoglobin, NO reaction, 197
Half-life of NO, 193
Harmaline, and EAA release, 109–110
Heart
cGMP kinase in, 59
nitrergic innervation, 249
stimulation, and hypothalamic NMDA receptors, 157–158
see also Cardiovascular system, nitrergic innervation
Hepatocytes, NOS in, 5
Herpes simplex infection, and NOS induction, 177
Hippocampus
long-term depression in, 117, 143–144
long-term potentiation in
distributed potentiation, 134–142
and glutamate-activated ion channels, 127–128
and memory, 126–127
NO and, 130–133
quantal analysis, 128–129
NADPH diaphorase in, 88
signal transduction, 115–116
citrulline recycling, 117
EAA populations, 115–116
endogenous EAA receptor ligands, 116
long-term depression, 117
NO neurotoxicity, 118
noradrenergic afferents, 117
NOS/cGMP compartments, 116
Histamine, arterial response, EDNO mediation, 220, 221
Homocysteic acid, 112
Hydrogen peroxide, in NO formation, 28–29
Hydroquinone
and nitrergic stimulation, 242–243
and NO relaxation inhibition, 260
Hydroxocabalamin, and nitrergic transmission, 241
Hydroxylamine, 240
Hyperalgesia, and NOS induction in glial cells, 182
Hypothalamus
functions, 151
NO signalling in, 151–162
autonomic regulation, 156–157
corticotropin-releasing hormone neurosecretory system, 152–154
functional considerations, 158–159
luteinizing hormone-releasing hormone neurosecretory system, 155
somatostatin release, 156
vasopressin neurosecretory system, 155–156
NOS in, 89–90, 152
Hypoxia, NOS inhibition, 235

Ileum, nitrergic transmission in, 257–258
Impotence, 282
treatment, 301–302
Inhalation of NO, 194
Inhibitory junction potentials (IJPs), NO mediated
and cystamine, 244
in gastrointestinal tract, 251
in jejunum, 257
and superoxide dismutase, 242
Injury, neuronal NOS expression and, 94–95
see also Ischaemia; Trauma
Inositol 1,4,5,-trisphosphate, receptor protein, 60
Insects, NADPH diaphorase activity in, 93–94
Interferon-γ, and NOS induction in glial cells, *see under* Cytokines
Interleukins, and NOS induction in glial cells, *see under* Cytokines
Ion channels
cGMP-regulated, 43–44, 64–69
neuronal, 77
olfactory epithelium, 68–69
renal, 67
rod photoreceptor, 64–67
spermatozoa/testis, 67–68
glutamate-activated, and long-term potentiation, 127–128
IP3 receptor protein, cerebellar, phosphorylation, 77

Iron, NO binding, 183, 195
see also Haem
Ischaemia, and NOS induction in glial cells, 180–182
see also Cerebral ischaemia

Jejunum, nitrergic transmission in, 257
Junction potentials, *see* Inhibitory junction potentials (IJPs), NO mediated

Kainate receptors
in cerebellum, 107–108
in hippocampus, 116
Kidney
brain NOS in, 11
cGMP kinase in, 59
see also Renal cells, cGMP-gated channels
Kinases
see Protein kinases
see also under Cyclic GMP receptor proteins
'Knock-out' mice, for NOS study, 15

L-NAME, *see* N^G-Nitro-L-arginine methyl ester (L-NAME)
L-NMA, as NOS inhibitor, 35–36
L-NMMA, *see* N^G-Monomethyl-L-arginine (L-NMMA)
L-NNA, *see* N^G-Nitro-L-arginine (L-NNA)
Learning, and long-term potentiation, 126–127
Linsidomine, 240, 301
Lipopolysaccharide, and NOS induction in glial cells
astrocytes, 169
gliomas, 169–170
mechanisms, 170–172, 173
microglia, 167
Muller cells, 170
Lipoxygenase, activity, and NO, 184
Long-term depression (LTD)
in cerebellum, 115, 144–147
NO in, 145–147
cGMP in, 75
in hippocampus, 117, 143–144
Long-term potentiation (LTP), 126–142
cGMP in, 75
distributed potentiation, 134–142
functional significance, 142
neighbouring cell in, 139–142
NO in, 138–139
Long-term potentiation (LTP) (*continued*)
paired cell experimental procedure, 134–136
spatial restrictions, 136–138
and glutamate-activated ion channels, 127–128
in hippocampus, and memory, 126–127
NO and, 11, 130–133
quantal analysis, 128–129
retrograde signal, 129–130
Lung
cGMP kinase in, 59
nitrergic innervation, 248
Luteinizing hormone-releasing hormone neurosecretory system, 155
LY-83583, 244

Macrophages
NOS cloning, 5
NOS in, 2–3
NOS isolation from, 25
transcriptional NOS induction in, 174
Medial septum, NOS/NADPH diaphorase in, 88
Medulla, NOS neurons in, 91
Membranes, and NOS, 24
Memory, and hippocampus, and LTP, 126–127
Metal-NO complexes, 183
N-Methyl-D-aspartate receptors, *see* NMDA receptors
Methylene blue, 243–244
N-Methylhydroxylamine, 244
Microglia, NOS in vitro induction in, 167–168
see also Glial cells
Mitochondria
in glial cells, NO toxicity, 185
oxygen inhibition in, 195–196
Molecular cloning, NOS, 3–6
brain NOS, 3–5
inducible NOS, 5–6
N^G-Monomethyl-L-arginine (L-NMMA), 233–234
and cerebral arterial stimulation, 210
and corticotropin-releasing hormone release, 153
and luteinizing hormone-releasing hormone, 155
and nitrergic muscle stimulation, 247
structure, 231

Motor neuron disease, and superoxide dismutase, 202–204
mRNA, NOS
 astrocyte-inducible, 169
 and autoimmune disease, 179–180
 brain, 4
 and toxoplasmosis, 179
 and viral infection, 177, 179
Muller cells, NOS induction in, 170
Muscarinic relaxation, in penile smooth muscle, 294–297
Myristoylation, NOS gene sequence for, 6

NADPH, NOS, consensus binding sites, 3, 4, 5–6
NADPH diaphorase activity, 11, 24
 in brain
  cerebellum, 92
  cerebral perivascular nerves, 214
  diencephalon, 89–90
  hippocampal neurons, 130
  midbrain/pons/medulla, 90–92
  telencephalon neurons, 86–89
 in glial cells, 165
 induction in ischaemia, 181–182
 and neuronal NOS localization, 231–233
 in non-mammalian species, 93–94
 and NOS, 83–84
 in pyloric stenosis, 15
 in spinal cord, 92–93
 staining interpretation, 84–86
 and trauma, 180
Neurodegenerative diseases, 184
Neurogenesis, NO in, 12–15
Neurons
 cGMP kinase in, 76–77
 NO availability to, 191–192
Neurotoxicity of NO
 in cerebellum, 115
 in hippocampus, 118
 see also Toxicity
Nicardipine, and vasodilator nerve stimulation, 211
Nicotine
 cerebrovascular dilatation, 208, 209, 210
 and penile smooth muscle relaxation, 298–299
Nicotinic agonists of nitrergic transmission, 266–267
Nitration, superoxide dismutase catalysed, 201–202
Nitrergic transmission, 228–229
 agonists, 266–267
 in anococcygeus/retractor penis muscles, 245–247
 in cardiovascular system, 249
 co-transmitters, 265–266
  ATP, 266
  VIP, 265–266
 criteria for establishing, 230–245
  effector mechanisms, 243–245
  mimicry, 237–240
  modifying agents, 240–243
  NOS inhibition, 233–235
  synthesis of transmitter, 230–233
  transmitter release, 236–237
  transmitter storage, 235–236
 in gastrointestinal tract, 250–259
 mechanisms, 229, 260–267
  exocytosis in, 261–263
  hypotheses, 260
 modulation, 263–264
 in respiratory tract, 248
 transmitter, 260–261
  interactions, 267–268
  spontaneous release, 263
 in urinary tract, 247–248
 see also Signal transduction
Nitric oxide synthase, 1–19, 21–42
 activity assays, 22–23
 activity regulation, 31–34
  and L-arginine, 31–32
  gene expression, 33–34
  NO feedback, 33
  phosphorylation, 32–33
 L-arginine concentration in, 262
 in brain
  cerebellum, 92, 105, 112–113
  diencephalon, 89–90
  functions, 10–16
  hippocampus, 116
  localization, 9
  midbrain/pons/medulla, 90–92
  telencephalon, 86–89
 cofactors, 6–10, 25–26
  calcium, 8
  cytochrome P450 enzymes, 6–7
  flavins, 6
  phosphorylation, 8, 10
  tetrahydrobiopterin, 7–8
 in glial cells, 106–107, 165–166
  see also under Glial cells
 isoforms, 105

Nitric oxide synthase (*continued*)
  isolation, 2–3
    from cerebellum, 105
  molecular cloning, 3–6
    brain NOS, 5–6
    inducible NOS, 5–6
  mRNA, NOS
    astrocyte-inducible, 169
    and autoimmune disease, 179–180
    brain, 4
    and toxoplasmosis, 179
    and viral infection, 177, 179
  and NADPH diaphorase reaction, 83–84
    interpretation cautions, 84–86
  neuronal localization, 231–233
  neurotransmitters with, 96
  in nitrergic transmission, 230, 231
  NO synthesis mechanisms, 26–31
    L-arginine oxidation, 27–28
    NOS haeme/flavin components in, 26–27
    oxygen reduction, 28–29
    tetrahydrobiopterin in, 29–31
  in penile nerves, 293–294
  purification/characterization, 23–25, 105
    brain NOS, 23–24
    endothelial NOS, 24–25
    inducible NOS, 25
  regulation, 1
  in spinal cord, 92–93
  see also under Glial cells
Nitric oxide synthase inhibition/inhibitors, 34–38, 233–235
  L-arginine analogues, 35–37, 233–234
  calmodulin inhibitors, 234–235
  cofactors, 234
  dyes, 34
  ethacrynic acid, 235
  7-ethoxyresorufin, 235
  haem inhibition, 235
  hypoxia, 235
  and long-term potentiation, 130–133
  and nitrergic transmission, 236, 237
  7-nitro indazole, 37–38
  and smooth muscle tone, 263
  see also by particular inhibitor
Nitrites, nitrergic transmitter mimicry, 238
Nitro blue tetrazolium, NOS inhibition, 234
7-Nitro indazole, 37–38
N^G-Nitro-L-arginine (L-NNA), 36–37, 210, 233–234
  and nitrergic muscle stimulation, 247
  structure, 231
  and vasodilator nerve stimulation, 210, 211
N^G-Nitro-L-arginine methyl ester (L-NAME), 233
  and corticosterone release, 153, 154
  ischaemia protection, 181
  and photic signalling, 157
Nitroarginine, and cerebral ischaemia injury, 200
Nitroglycerin, topical, for impotence, 302
Nitroprusside, *see* Sodium nitroprusside
S-Nitrosocysteine, and long-term depression, 144
Nitrosothiols
  as nitrergic tranmsitters, 239–240, 261
  for NO delivery in vivo, 195
  oxidation by NO, 194
Nitrotyrosine, 201–202
'Nitroxidergic' nerves, cerebral, 214–215
  nerve function, 216–217
NMDA receptors
  in cerebellum, 107–110
  and cerebral ischaemia, 192
  in hippocampus, 115–116
  ligands, 110–111
  and long-term potentiation, 127
  in noradrenaline release, 114, 117
  in photic signalling, 157
Nociceptive processing, spinal cord, 182
Noradrenaline
  nitrergic transmission interaction, 267–268
  NMDA receptor mediated release
    in cerebellum, 114
    in hippocampus, 117
  NOS activation in astrocytes, 166
  and transcriptional NOS induction, 174
Noradrenergic afferents
  cerebellar, 114–115
  hippocampal, 117
Null mutants, for NOS study, 15

Oesophagus, nitrergic transmission in, 254–256
Olfactory epithelium, cGMP-gated channels, 68–69
Oligodendrocytes, NO toxicity, 185

Opioid receptor agonists, and nitrergic transmission, 264
Opioid tolerance, and NOS induction in glial cells, 182
Oxygen, molecular, reactivity, 194
Oxyhaemoglobin, and nitrergic transmission, 240–241
Pain, and NOS induction, 182
Paired-pulse facilitation, 125
  and long-term depression, 143
Parkinson's disease, 184
  brain neurons in, 91
Penile erection, mechanisms, 282–284
Penile erection, neural regulation, 281–306
  clinical implications, 301–302
  co-transmitters, 300–301
  early studies, 282, 286
  efferent nerves, 284–285
  muscle, innervation, 286–288
  NO as cholinergic mediator, 294–299
    muscarinic relaxation, 294–297
    nicotine relaxation, 298–299
  NO as transmitter, 10–11, 290–294
    candidate transmitter, 290–291
    NOS in penile nerves, 293–294
    pharmacological/biochemical evidence, 291–293
  process of elucidation, 299–300
  vasoactive intestinal polypeptide as (candidate) transmitter, 288–290
Penis muscle, NO-mediated transmission, 228
Pentylenetetrazol (PTZ), and EAA release, 110
Peristalsis, 259
Peroxynitrite, 185
  from superoxide, 198
    in mitochondria, 195–196
  in motor neuron disease, and superoxide dismutase, 202–204
  as oxidant, 198–199
  rate of formation, 199
  superoxide dismutase reaction, 201
Phosphodiesterases, cGMP-regulated, 69–74
  classes, 69
  properties, 73–74
  type II, 70–71
  type V, 61
Phospholamban, 60
Phosphorylation
  DARPP-32, and cAMP/cGMP, 77
  IP3 receptor protein, cerebellar, 77
Phosphorylation (*continued*)
  in NOS regulation, 8, 10, 32–33
  protein kinases, by NOS, 175–176
Photic signals, NMDA receptors in, 157
Photoreceptor cells, type V phosphodiesterases in, 73
  see also Retina; Rod photoreceptor; Visual system
PMA, and NOS induction, 171
Polyethylene glycol/superoxide dismutase, and cerebral ischaemia injury, 197
  from head trauma, 200
Potassium, extracellular, and cerebellum fibre stimulation, 146
Potassium channels, neuronal, and cGMP, 77
Progressive supranuclear palsy, brain neurons in, 91
Protein kinases
  and long-term depression, 145
  NOS phosphorylation, 175–176
  PKC
    as NOS cofactor, 8, 10
    and NOS induction in glial cells, 171–172
  see also under Cyclic GMP receptor proteins
Purkinje cells, cGMP in, and NO, 75–76
Pyloric stenosis, 15
Pyocyanin, 244
Pyramidal cells, EAA receptors, 115–116
Pyrogallol, and nitrergic stimulation, 242

Quantal analysis
  for long-term depression, 143
  for long-term potentiation, 128–129
Quinolinic acid, 112
Quisqualate
  and glutamate receptor desensitization, 146
  NOS activation in astrocytes, 166
Quisqualate receptors
  in cerebellum, 108–109
  in hippocampus, 116

Rabies infection, and NOS induction, 177
Redox cycling of biopterins, in NO synthesis, 29–31
Renal cells, cGMP-gated channels, 67
  see also Kidney
Respiratory tract, nitrergic innervation, 248

Retina, interneurons, NADPH diaphorase in, 90
 see also Photoreceptor cells; Rod photoreceptor; Visual system
Retractor penis muscle, 282, 283
 nitrergic innervation, 245–247
Rhizotomy, 95
Rhodnius prolixus, NOS activity in, 94
RNA, *see* mRNA, NOS
Rod photoreceptor
 cGMP-gated channel, 64–67
 type V phosphodiesterases in, 73
 see also Retina; Visual system

Schizophrenia, brain neurons in, 91
Sciatic nerve injury, and NADPH diaphorase induction, 95
D-Serine
 in cerebellum, metabolism, 111–112
 in hippocampus, 116
Serotonin, and nitrergic transmission, 267
Signal transduction, 103–123
 cerebellum, 107–115
 hippocampus, 115–118
 NO sites of action, 107, 108
 see also Nitrergic transmission
Skeletal muscle, brain NOS in, 11
Smooth muscle
 cGMP kinase in, 58–59
 vascular, cGMP regulation, 62–63
 see also Penile erection, neural regulation
Sodium azide, 240
Sodium nitroprusside, 238
 and long-term depression, 144
Soluble guanylyl cyclase, 44–50
 catalytic activity, 46–47
 central nervous system localization, 95
 isoforms, 47–48
 regulation, 48–49
  carbon monoxide, 48
  haem group, 48–49
 structure, 44–46
 see also Guanylyl cyclase
Somatostatin, from hypothalamus, and NO, 156
Spermatozoa, cGMP-gated channels, 67–68
Sphincters, gastrointestinal, nitrergic transmission in, 252–254
Spinal cord
 hyperalgesia, 182
 NOS localization, 92–93
Spinal cord (*continued*)
 see also Brain; Central nervous system, NO neurons in
Staurosporine, and NOS induction, 171
Stomach, nitrergic innervation in, 256
Stress, and corticotropin-releasing hormone neurosecretory system, 153–154
Striatum, NADPH diaphorase in, 89
Superoxide
 in cerebral ischaemia, 197–198
  NO interaction, 199–200
 in mitochondria, 195–196
 and nitrergic transmission, 242
Superoxide dismutase
 and cerebral ischaemia, 197, 200–201
 in motor neuron disease, and peroxynitrite, 202–204
 nitration catalysis, 201–202
 and nitrergic transmission, 242
Swallowing, 259
Synaptic activity, *see* Long-term depression; Long-term potentiation; Signal transduction
Synaptic plasticity, brain NOS in, 11
Synaptic transmission, 125–126
Synthesis of NO, 26–31
 L-arginine oxidation, 27–28
 NOS haem/flavin components in, 26–27
 oxygen reduction, 28–29

Telencephalon, NOS in, 86–89
Testis, cGMP-gated channels, 67–68
Tetraethylammonium chloride (TEA), 262–263
Tetrahydrobiopterin
 in NO synthesis, 29–31
  and oxygen reduction, 28
 in NOS activity regulation, 32
 as NOS cofactor, 7–8, 26
Thalamus, NADPH diaphorase in, 90
Theta burst stimulation, and long-term potentiation, and NOS inhibitors, 133
Thiols
 nitrosothiols, nitrergic transmitter mimicry, 239–240
 oxidation by NO, 194
Toxicity
 of NO, 194–195
  in motor neuron disease, 203–204
  and NO synthesis inhibition, 200
  oxidants, 201–202

Toxicity (*continued*)
and superoxide/NO interaction, 199–200
to glial cells, 185
of peroxynitrite, 199
of superoxide dismutase, 197–198
see also Neurotoxicity of NO
Toxoplasmosis, and NOS induction in glial cells, 179
Trachea, nitrergic innervation, 248
Transgenic mice
Down's syndrome model, 201
for NOS study, 15
Transition metals, NO and, 195–196
Transmission, *see* Nitrergic transmission
Trauma
head, PEG-SOD protection, 200
and NOS induction in glial cells, 180
see also Injury
Tumour necrosis factor-α, and NOS in glial cells
astrocytes, 169
gliomas, 169–170
mechanisms, 170–172, 173
microglia, 167
Muller cells, 170
Tyrosine kinase, and NOS induction, 172
Tyrosines, nitration, 201
Tyrphostin, and NOS induction in glial cells, 172

UK-14304, and nitrergic transmission, 264
Urinary tract, nitrergic innervation, 247–248
Urine
nitrate excretion, 2
nitrotyrosine in, 201–202
Vasoactive intestinal polypeptide (VIP)
in caecum, 258
as co-transmitter, 265–266
in penile erection, 288–290, 300–301
in colon, 258
and inhibitory junction potentials, 251
as NO-releasing transmitter, 260
in respiratory tract, 248
in stomach, 256
Vasodilator-specific phosphoprotein (VASP), 60
Vasopressin
cerebral artery responses, 222
from hypothalamus, and NO, 155–156
Vestibular ocular reflex (VOR), and long-term depression, 144
Vimentin, cGMP kinase targeting, 62
Viral infection, and NOS induction, 177–179
Visual system
cGMP in, 43
NADPH diaphorase in, 90
photic signalling, NMDA receptors in, 157
vestibular ocular reflex, 144
see also Photoreceptor cells; Retina, interneurons, NADPH diaphorase in; Rod photoreceptor

Zaprinast, 244